MW01165030

DATA AND COMPUTER COMMUNICATIONS

DATA AND COMPUTER COMMUNICATIONS

TERMS, DEFINITIONS AND ABBREVIATIONS

Gilbert Held

4-Degree Consulting
Macon, Georgia,
USA

JOHN WILEY & SONS
Chichester · New York · Brisbane · Toronto · Singapore

Wiley Editorial Offices

John Wiley & Sons Ltd, Baffins Lane, Chichester,
West Sussex, PO19 1DU, England

John Wiley & Sons, Inc., 605 Third Avenue,
New York, NY 10158-0012, USA

Jacaranda Wiley Ltd, G.P.O. Box 859, Brisbane,
Queensland 4001, Australia

John Wiley & Sons (Canada) Ltd, 22 Worcester Road,
Rexdale, Ontario M9W 1L1, Canada

John Wiley & Sons (SEA) Pte Ltd, 37 Jalan Pemimpin #05-04,
Block B, Union Industrial Building, Singapore 2057

Library of Congress Cataloging-in-Publication Data:

Held, Gilbert, 1943–
Data and computer communications : terms, definitions, and abbreviations / Gilbert Held.
p. cm.
ISBN 0 471 92066 5
1. Telecommunication—Dictionairies. 2. Data transmission systems—Dictionairies. I. Title.
TK5102.H45 1989 89–14796
384′.03—dc20 CIP

British Library Cataloguing in Publication Data:

Held, Gilbert, *1943–*
Data and computer communications : terms, definitions and abbreviations.
1. Computer-telecommunication systems
I. Title
621.38′0413

ISBN 0 471 92066 5

Printed in Great Britain by Courier International Ltd, Tiptree, Essex

PREFACE

The telecommunications field is perhaps the most rapidly evolving area of all disciplines. In just a few years the use of microprocessors, fiber optics and satellites has dramatically changed the structure of the telecommunications industry, the product offerings of vendors and the facilities provided by common carriers. Accompanying this rapid evolution are a wealth of terms, definitions and abbreviations which are probably beyond the capability of any one individual to remember. Thus, my goal in developing this dictionary was to provide the reader with a comprehensive up-to-date reference that could be used to refresh one's memory or to determine the specific meaning of a term or what an abbreviation might represent.

In developing this dictionary I attempted to make its use as easy as possible by including both terms and abbreviations together instead of separating the two. As you thumb through this dictionary you will note that all entries are arranged in a structured, alphabetical order.

Although we have attempted to develop a most comprehensive dictionary of telecommunications terms, definitions and abbreviations, it is most probable that we missed some vendor specific entries. In addition, new telecommunications terms and abbreviations are constantly being added to reflect the introduction of new products and technologies. Due to this, any information readers may care to provide is welcomed and can be sent to me directly or through my publisher.

GILBERT HELD
Macon, Georgia

ACKNOWLEDGEMENTS

The author is most appreciative of the cooperation and assistance of many organizations that resulted in the comprehensiveness of this book. Specifically, I would like to acknowledge the following organizations that granted permission to extract glossary information from previously published manuals, books and trade literature.

Atlantic Research Corporation
AT&T Network Systems
Auerback Publishers, Inc.
Codex Corporation
Digilog, Inc.
Digital Equipment Corporation
Dynatech Communications
GENERAL DATACOMM INDUSTRIES, Inc.
International Business Machines Corporation
Micom Systems, Inc.
Navtel
Siemens Information Systems, Inc.
Telefile, Inc.
Telenet
Tymnet

In addition to the previously mentioned organizations two individuals deserve special mention for their efforts. I would like to thank Ms. Patricia Peel for doublechecking the efforts of a young typist as well as for compiling several glossaries into the original electronic draft. Concerning the young typist, I would like to thank my son, Jonathan, for spending a portion of his summer vacation using his computer to convert several shoeboxes of 3 × 5 index cards into the book you are now reading.

A

AB Signaling A technique for carrying "on-hook/off-hook" telephony signaling information over T1 spans that use the D4 framing (12-frame superframe) format. In AB Signaling, the circuit provider (telephone company or PTT) "robs" the least significant bit of each channel octet in the 6th and 12th frames of each superframe to convey signaling information. Bits robbed from the 6th frame are called "A" bits and correspond to signals on the E-wire of an analog telephone. Bits robbed from the 12th frame are called "B" bits, and correspond to signals on the M-wire of an analog telephone.

abbreviated addressing In packet switched networks, addressing in which a simple mnemonic code is used in lieu of the complete addressing information; the cross reference to complete address is stored in the packet assembler/disassembler (PAD).

abbreviated dialing The feature of some PABX systems and other switches whereby frequently-called numbers can be dialed by means of a brief dial code which the switch translates into a conventional telephone number with appropriate access and area codes.

ABCD Signaling A technique for carrying "on-hook/off-hook" telephony signaling information over T1 spans that use the Extended Superframe (24-frame superframe) format. In ABCD Signaling, the circuit provider (telephone company or PTT) "robs" the least significant bit of each channel octet in the 6th, 12th, 18th, and 24th frames of each superframe to convey signaling information. Bits robbed from the 6th and 18th frames are called "A" bits and correspond to signals on the E-wire of an analog telephone. Bits robbed from the 12th and 24th frames are called "B" bits, and correspond to signals on the M-wire of an analog telephone. ABCD Signaling is used only as an interim technique during conversions from D4 framing to Extended Superframe format.

ABM Asynchronous Balanced Mode.

ABORT A predefined software controlled or operator initiated action which results in the immediate cessation of an activity.

ABR Automatic Baud Rate.

absolute address A reference to a storage location that has a fixed displacement from absolute memory location zero.

absolute calling Coding in which instructions are written in machine language (i.e. coding using absolute operators and addresses); coding that does not require processing before it can be understood by the computer.

ABT Abort Timer or Answer Back Tone.

ACB 1. In IBM's VTAM, Access Method Control Block. 2. In IBM's NCP, Adapter Control Block.

ACB address space In IBM's VTAM, the address space in which the ACB is opened.

ACB name In IBM's VTAM: 1. The name of an ACB macro instruction. 2. A name specified in the ACBNAME parameter of an APPL statement. Contrast with network name. *Note*: This name allows an ACF/VTAM application program that is used in more than one domain to specify the same application program identification (pointed to by the APPLID parameter of the program's ACB statement) in each copy. ACF/VTAM knows the program by both its ACB name and its network name (the name of the APPL statement). Program users within the domain can request a session using the ACB name or the network name; program users in other domains must use the network name (which must be unique in the network).

ACB-based macro instruction In IBM's VTAM, a macro instruction whose parameters are specified by the user in an access method control block.

Accent The AT&T telephone trademark.

accept In an ACF/VTAM application program, to accept a CINIT request from an SSCP to establish a session with a logical unit; the application program acts as the primary end of the session. *Note*: The accept process causes a BIND request to be sent from the primary end of the session to the logical unit that will act as the secondary end of the session, requesting that the session be established and passing session parameters. For example, the session-initiation request that originally caused the SSCP to send the CINIT request may have resulted from a logon by the terminal operator, from a macro instruction issued by an ACF/VTAM application program, or from an ACF/VTAM operator command.

acceptor of data (acceptor) A term used to describe any device capable of accepting data in a controlled manner; it is used in British Standard Interface Specifications to refer to devices which take data from a source.

access In a local area network, the ability to work with files. Usually used in connection with files stored on network disks. Various access rights may be assigned to users.

access charges Charges levied by local telephone operating companies for providing their customers with access to long distance (interexchange) carrier services.

access code 1. A series of digits which must be dialed to link your telephone line to a specific type of telephone service. 2. A series of letters and/or numbers which must be entered into a computer terminal or personal computer to provide access to specific databases or mainframe computer software. 3. The five-digit code you must dial to use a long distance service other than your primary long distance company. 4. With some PBX equipment, the code you must dial to access special services such as WATS lines or private lines.

access control The management of rights to use resources.

access group All stations which have identical rights to make use of computer, network, or data PABX resources.

access line The connection between a subscriber's facility and a public network—either a PDN, public switched network, or public telephone network. Also called local line or local loop.

access method 1. In IBM environments, a host program managing the movement of data between the main storage and an input/output device of a computer system; BTAM, TCAM, VTAM are common data communications access methods. 2. In LAN technology, a means to allow stations to gain access to—to make use of—the network's transmission medium; classified as shared access (which is further divided into explicit access or contended access) or discrete access method.

Access Method Control Block (ACB) In IBM's VTAM, a control block that links an application program to VSAM or ACF/VTAM.

Access Process Parameter (APP) In packet switching, a list of parameters governing the operation of a specified port. (This list is communicated to the remote Packet Assembler/Disassembler (PAD) or Data Terminal Equipment (DTE) by the Data Qualifier (DQ) packet (X.29 protocol).)

Access Request The event that notifies the system of a user's desire to initiate a data communications session. It begins the access function and starts the counting of access time. Two specific examples of Access Request events are the off-hook event in the public switched telephone network, and the completion of a Connect request by a terminal operator in the ARPANET.

access rights Privileges which are granted or not granted in order to control how users may work with files. For example, you must have the proper rights before you may read a file, delete a file, or modify a file.

Access Tandem (AT) An AT&T ESS switch used to provide carrier access to end offices (and possibly collocated stations).

access time The time required to retrieve information from a computer or computer component.

accounting exit routine In IBM's ACF/VTAM, an optional, installation exit routine that collects statistics about session initiation and termination.

Acculink An AT&T trademark for a series of multiplexers.

Accunet Digital service from AT&T Communications, including Accunet T1.5, terrestrial wideband at 1.544 Mbps, Accunet Reserved T1.5, satellite-based channels at 1.544 Mbps primarily for video teleconferencing applications; Accunet Packet Services, packet-switching services; Accunet Dataphone digital service (DDS), private-line digital circuits at 2400, 4800, 9600, and 56 Kbps; Accunet switched service providing 56 Kbps dial digital transmission.

Accunet Packet Service A data communications

service offered by AT&T Communications that is based on packet switching technology. It permits customers to efficiently transmit bursts of data through a switched network.

Accunet Switched 56 Service A data communication service offered by AT&T Communications that provides full-duplex, digital data transmission at 56 Kilobits per second (Kbps) via terrestrial digital facilities, which can be accessed through dedicated lines.

Accunet T1.5 Service A data communication service offered by AT&T Communications that provides full-time, full-duplex, dedicated, point-to-point, transmission of digital information at 1.544 Mbps. The service supports applications that require transmissions of voice, data, video, or any signal that can be digitally encoded—in any combination—entirely over dedicated terrestrial channels.

accuracy A general performance criterion expressing the correctness with which a specific communication function is accomplished.

ACD Automatic Call Distributor.

ac/dc ringing A widely-used technique for causing a telephone to ring, in which an ac voltage is used to power the bell and the dc voltage is used to power a relay which cuts off the ring when the receiver is taken off-hook.

ACF Advanced Communications Function.

ACF/NCP In IBM's VTAM, Advanced Communications Function for the Network Control Program.

ACF/SSP In IBM's VTAM, Advanced Communications Function for the System Support Programs. Synonym for SSP.

ACF/TAP In IBM's VTAM, Advanced Communications Function for the Trace Analysis Program. Synonym for TAP.

ACF/TCAM In IBM's VTAM, Advanced Communications Function for the Telecommunications Access Method.

ACF/VTAM In IBM's VTAM, Advanced Communications Function for the Virtual Telecommunications Access Method.

ACF/VTAM application program In IBM's VTAM, a program that has opened an ACB to identify itself to ACF/VTAM and can now issue ACF/VTAM macro instructions.

ACF/VTAM definition In IBM's VTAM, the process of defining the user application network to ACF/VTAM and modifying IBM defined characteristics to suit the needs of the user.

ACF/VTAM definition library In IBM's VTAM, the operating system files or data sets that contain the definition statements and start options filed during ACF/VTAM definition.

ACF/VTAM operator A person or program authorized to issue ACF/VTAM operator commands.

ACF/VTAM operator command A command used to monitor or control an ACF/VTAM domain.

ACF/VTAME In IBM's VTAM, Advanced Communications Function for the Virtual Telecommunications Access Method Entry.

acknowledgment (ACK) A control character used (with NAK) in BSC communications protocol to indicate that the previous transmission block was correctly received and that the receiver is ready to accept the next block. Also used as a ready reply in other communications protocols, such as Hewlett-Packard's ENQ/ACK protocol (see following diagram) and the ETX/ACK method of flow control.

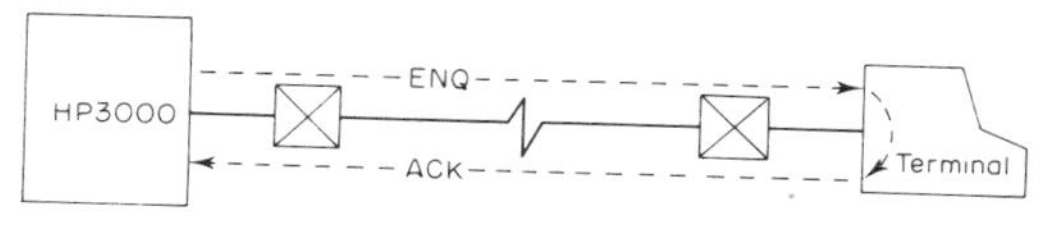

ENQ/ACK protocol diagram

acknowledgment (ACK)

ACM Association of Computing Machinery.

ACOnet Akademisches Computer Netz.

Acorn A trademark of AT&T for a network control system.

acoustic coupler A device that converts electrical signals into audio signals, enabling data to be transmitted over the public telephone network via a conventional telephone handset; it also converts the audio signals back into electrical signals at the receiving end. A kind of modem.

ACP Advanced Communications Package.

acquire 1. In IBM's VTAM, the operation in which an authorized ACF/VTAM application program initiates and establishes a session with another logical unit; the application program acts as the primary end of the session. *Note*: The acquire process causes an Initiate request to be sent to the SSCP which causes the SSCP to return a CINIT request to the application program (the PLU); this in turn causes the PLU to send a BIND request to the SLU. Contrast with accept. 2. In relation to ACF/VTAM resource control, to take over resources (communication controllers or other physical units) that were formerly controlled by a data communication access method in another domain, or to assume control of resources that were controlled by this domain but released.

ACR Abandon Call and Retry.

AC Signaling The use of alternating current signals or tones to accomplish transmission of information and/or control signals.

ACS Advanced Communications Services.

activate An operator action resulting in the resumption of a previously suspended task.

activate/passive device In current loop applications, a device capable of supplying the current for the loop (active) and a device that must draw its current from connected equipment (passive).

active In IBM's VTAM, pertaining to a major or minor node for which a VARY NET, ACT command has been issued. Also, a major or minor node in a list of major nodes to be activated when ACF/VTAM is started. Contrast with inactive. *Note*: For a major node, this makes the node and its minor nodes known to ACF/VTAM. For a minor node, this generally results in the execution of an SNA protocol to make the minor node usable by the network. For an LU minor node, this indicates that the ACF/VTAM operator has given permission for the LU to participate in an LU–LU session.

ACU Automatic Calling Unit.

A/D Analog/Digital.

adapter A device that (1) enables different sizes or types of plugs to mate with one another or to fit into an information outlet; (2) provides for the rearrangement of leads; (3) allows large cables with numerous wires to fan out into smaller groups of wires; or (4) makes interconnections between cables.

Adapter Control Block (ACB) In IBM's NCP, a control block that contains line control information and the states of I/O operations for BSC lines, start–stop lines, or SDLC links.

Adaptive Differential Pulse Code Modulation (ADPCM) An encoding technique (CCITT) that allows an analog voice conversation to be carried within a 32 KB digital channel; 3 or 4 bits are used to describe the difference between two adjacent samples at 8000 times a second.

adaptive equalizer An equalizer that adjusts to meet varying line conditions; most operate automatically.

adaptive routing Message routing which is automatically adjusted to compensate for changes in network traffic patterns and channel availability.

ADCCP Advanced Data Communications Control Procedure.

ADCU Association of Data Communications Users.

additional facilities In packet-switched networks, standard network facilities which are selected for a given network but which may or may not be selected for other networks. Contrast with essential facilities.

address 1. (*noun*) A unique designation for the location of data or the identity of an intelligent device. Multiple devices on a single communications line must have unique addresses to allow each to respond to its own messages (see polling). 2. (*verb*) To add or include the coded representation of the desired receiving device (as in to "address a message").

address prefix In Digital Equipment Corporation Network Architecture (DECnet), any leading portion of an NSAP address.

address resolution In Digital Equipment Corporation Network Achitecture (DECnet), the Session Control function which maps from a Naming Service object name to the identifiers of protocols and corresponding addresses which are mutually supported by the local system and the remote system(s) on which the named object resides.

address selection In Digital Equipment Corporation Network Architecture (DECnet), the Session Control function which provides transparent selection of protocols and addresses for Transport Connection establishment based upon the destination object name.

address space The complete range of addresses that is available to a programmer.

addressing authority In Digital Equipment Corporation Network Architecture (DECnet), the authority responsible for the unique assignment of Network layer addresses within an addressing domain.

addressing domain In Digital Equipment Corporation Network Architecture (DECnet), a level in the hierarchy of Network layer addresses. Every NSAP address is part of an addressing domain that is administered directly by one and only one addressing authority. If that addressing domain is part of a hierarchically higher addressing domain (which must wholly contain it), the authority for the lower domain is authorized by the authority for the higher domain to assign NSAP addresses from the lower domain.

adjacent Network devices or programs that are directly connected by a data link.

adjacent NCPs In IBM's VTAM, network control programs (NCPs) that are connected by subarea links with no intervening NCPs.

adjacent networks Two SNA networks joined by a common gateway NCP.

adjacent nodes In IBM's VTAM, two nodes that are connected by one or more data links with no intervening nodes.

adjacent SSCP table A list of SSCPs that can be used to determine the next SSCP on the session-initiation path to a same-network destination SSCP or to a destination network for an LU–LU session. The table is filed in the VTAM definition library.

adjacent subareas In IBM's VTAM, two subareas connected by one or more links with no intervening subareas.

ADMD Administration Management Domain.

Administration Management Domain (ADMD) In packet switching, a management domain managed by an Administration (X.400 specific).

administration subsystem That part of a premises distribution system that includes the distribution hardware components where you can add or rearrange circuits. The components include cross-connects, interconnects, information outlets, and the associated patch cords and plugs. Also called administration points.

administrative domain In Digital Equipment Corporation Network Architecture (DECnet), a collection of End Systems, Intermediate Systems, and Subnetworks operated by a single organization or administrative authority. It may be subdivided into a number of routing domains.

Administrative Module (AM) The AM is part of the AT&T 5ESS switch which performs the part of call processing, administration, and maintenance which cannot be economically distributed to switching modules. The AM consists of the processor, disk storage, and tape backup units. The AM processor performs the centralized processing functions, high-speed tape, and controls the flow of data between the other dedicated processors distributed throughout the remaining units. The processor functions are fully duplicated (except for the port switch) in order to ensure continued processing capability.

ADP Automatic Data Processing.

ADPCM Adaptive Differential Pulse Code Modulation.

ADU Automatic Dialing Unit.

Advanced Communications Function (ACF) IBM communications software for mainframes and frontend communications controllers, which implements various aspects of IBM's Systems Network Architecture.

Advanced Communications Function for the Network Control Program (ACF/NCP) In IBM's VTAM, a program product that provides communication controller support for single-domain and multiple-domain data communication.

Advanced Communications Function for the Telecommunications Access Method (ACF/TCAM) In IBM's VTAM, a program product that provides single-domain data communication capability and, optionally, multiple-domain capability.

Advanced Communications Function for the Virtual Telecommunications Access Method (ACF/VTAM) In IBM's VTAM, a program product that provides single-domain data communication capability and, optionally, multiple-domain capability.

Advanced Communications Function for the Virtual Telecommunications Access Method Entry (ACF/VTAME) In IBM's VTAM, a program product that provides single-domain and multiple- domain data communication capability for IBM 4300 systems that may include communication adapters.

Advanced Communications Package (ACP) An AT&T 3B-based set of Centrex features.

Advanced Communications Service (ACS) A shared data communications network service formerly proposed by the Bell System.

Advanced Data Communications Control Procedures (ADCCP) The USA Federal Standard communications protocol. A standard which specifies a bit-oriented protocol similar to SDLC and HDLC.

Advanced Program to Program Communication (APPC) IBM SNA facility for communicating between programs rather than between a human operator and a program. APPC uses SNA Logical Unit Type 6.2.

Advanced Program-to-Program Communications/Personal Computer (APPC/PC) An IBM product that runs on PCs on the Token-Ring Network; an implementation of the LU6.2 protocol.

Advanced Research Projects Agency (ARPA) Agency that developed the first major packet-switched network, ARPANET.

Advanced Speech Processor (ASP) A device proprietary to General DataComm, Inc. to compress the 64K, PCM derived bandwidth into 16K thereby using 1/4 of the TDM bandwidth to carry a voice circuit while retaining the voice quality.

AFI Authority and Format Indicator.

AFIPS American Federation of Information Processing Societies.

AGC Automatic Gain Control.

agent In Digital Equipment Corporation Network Architecture (DECnet), that part of an entity which provides the interface to network management.

aggregate The total bandwidth (expressed in bits per second) of a multiplexed bit stream, or the collection of all channels within that bit stream. The aggregate bandwidth of a North American T1 stream is 1.544 Mbps.

aggregate input rate The sum of all data rates of the terminals or computer ports connected to a multiplexer or concentrator; burst aggregate input rate refers to the instantaneous maximum.

aggregate user A collection of entities outside a defined subsystem, comprising one or more end users and the data communication system elements that connect those users with the subsystem.

AIOD Automatic Identification of Outward Dialing.

Air Call A partially owned subsidiary of Bell South Enterprises, the holding company for all unregulated Bell South companies which market cellular, paging and telephone answering services in the United Kingdom.

airline mileage The distance between two points in the U.S. as determined by a standard set of vertical and horizontal (V-H) coordinates for the major cities; the basis for distance-sensitive circuit service rates.

AIS 1. Alarm Indication Signal. 2. Automatic Intercept System.

alarm An asynchronous event that implies abnormal operation.

Alarm Indication Signal (AIS) An all-ones pattern which is transmitted out when the incoming signal has failed.

A-LAW An algorithm used in Europe for the digitization of voice signals by PCM encoding of PAM samples.

ALB Analog Loopback.

alert In IBM's VTAM NPDA, a high priority event that warrants immediate attention. The NPDA data base record is generated for certain event types that are defined by user-constructed filters.

ALGOL Algorithmic Language.

algorithm A logical or mathematical model that incorporates a specific set of rules that tell how information is to be manipulated to give a desired result.

Algorithmic Language (ALGOL) A computer language used to precisely present complex mathematical procedures and algorithms.

alias name In IBM's VTAM, a name defined in a host used to represent a logical unit name, logon mode table name, or class of service name in another network. This name is defined to a name translation program when the alias name does not match the real name. The name translation program is used to associate the real and alias names.

alias name translation facility In IBM's VTAM, a function of the Network Communications Control Facility (NCCF) program product for converting logical unit names, logon mode table names, and class of service names used in one network into equivalent names to be used in another network.

alias network address In IBM's VTAM, an address used by a gateway NCP and a gateway SSCP in one network to represent an LU or SSCP in another network.

ALIT Automatic Line Insulation Test.

all ones In T1 transmission, 1024 or more consecutive ones. When a device loses synchronization, it will send all ones to keep the network up and, at the same time, indicate that there is a problem in transmission.

Alliance A teleconferencing service offered by AT&T. Operator setup is obtained by dialing 1-800-544-6363 or you can dial 0-700-456-1000 to do it yourself.

allocate To assign a resource for use in performing a specific task.

Aloha An experimental packet-switched network implemented on radio by the University of Hawaii in the mid-1970s.

alphabet A table of correspondence between an agreed set of characters and the signals which represent them. The best known, standard alphabet is International Alphabet No. 5 (IA5) or CCITT or ISO 7-bit code.

alphageometric Pertaining to a scheme for displaying letters, numbers, and various graphic elements by means of combining small geometric building block patters in Videotex and similar information display systems.

alphamosaic Pertaining to a scheme for displaying letters, numbers, and various graphic elements by means of assembling arrays of individual "tiles" of identical shapes in Videotex and similar information display systems.

alphanumeric Describing a character set that contains letters, numerals (digits), and other characters such as punctuation marks.

alternate/no answer option A PBX and switch feature in which calls are automatically transferred to another line within the system after a preset number of rings.

alternate buffer A section of memory in a communications device set aside for the transmission or reception of data. The alternate buffer is used in conjunction with a primary buffer. For example, when the alternate buffer is empty (transmission) or full (reception), data transmission continues using the primary buffer, while the associated computer or terminal device transfers data to or from the alternate buffer in anticipation of its use when the primary buffer is empty or full.

Alternate Mark Inversion (AMI) A physical technique for bipolar transmission of digital signal. In AMI, the logical value of zero is represented by bit spaces with neutral polarity, and the logical value of 1 is represented alternately by pulses of positive and negative polarity.

alternate mode A mode of using a virtual terminal by which each of two interacting systems or users has access to its data structure in turn. The associated protocols include facilities to allow the orderly transfer of control from one user to the other. This is in contrast to free running mode.

alternate path In IBM's VTAM: 1. Another channel an operation can use after a failure. 2. In CCP, one of two paths that can be defined for information flowing to and from physical units attached to the network by means of an IBM 3710 Network Controller.

alternate path retry In IBM's VTAM, a facility that allows an I/O operation that has failed to be retried on another channel assigned to the device performing the I/O operation. It also provides the capability to establish other paths to an online or offline device.

alternate route A secondary communications path used to reach a destination if the primary path is occupied or otherwise unavailable.

alternating current (ac) Electrical current which is used for analog signaling.

Alternative Operator Services (AOS) A service for O-plus calls provided by non-communications carriers for large institutions to include hotels, motels, hospitals, universities, and other establishments with non-Bell pay telephones.

ALU Arithmetic Logic Unit.

Alvyn Aluminum PVC sheath used on cable for building risers or other similar areas where a flame-resistant sheath is required to meet the National Electrical Code standards.

AM 1. Administrative Module. 2. Amplitude Modulation.

AMA Automatic Message Accounting.

AMACC Automatic Message Accounting Collection Center.

AMATPS Automatic Message Accounting Teleprocessing System.

ambient noise Communications interference which is present in a communications line at all times. Also known as background, Gaussian, or white noise, as distinct from impulse noise.

American Cellular Communications A partially owned subsidiary of Bell South Enterprises, the holding company for all unregulated Bell South companies which provides cellular mobile telephone service primarily outside the southeastern United States.

American National Standards Institute (ANSI) Voluntary organization that represents the USA in the ISO; defined USASCII (now ASCII). Founded in 1918. Membership includes manufacturers, common carriers, and other standards organizations such as the IEEE. ANSI also produces Federal Information Processing Standards (FIPS) for the DoD.

American Standard Code for Information Interchange (ASCII) A 7-bit-plus-parity character set or code established by ANSI to achieve compatibility between data services; sometimes called USASCII, the USA Standard Code for Information Interchange; normally used for asynchronous transmission. Equivalent to the ISO 7-bit code.

American Telephone and Telegraph (AT&T) The USA's major common carrier for long distance telephone lines.

American Wire Gauge (AWG) A numbering system used to express the size (diameter) of electrical wires. The AWG number is inversely proportional to the diameter of the wire. Heavy industrial wiring, as an example, may be AWG#0 or 2, whereas telephone systems use AWG#22, 24, or 26.

Ameritech One of seven regional Bell operating companies (BOCs), covering the mid-Western United States, based in Chicago, IL.

AMI Alternate Mark Inversion.

amplifier Electronic component used to boost (amplify) signals. Performance (called gain) measured in deciBels.

amplitude The maximum departure of a waveform from its average value.

amplitude distortion An unwanted change in signal amplitude, usually caused by non-linear elements in the communications path.

Amplitude Modulation (AM) One of three basic ways (see also FM and phase modulation) to add information to a sine wave signal: the magnitude of the sine wave, or carrier, is modified in accordance with the information to be transmitted.

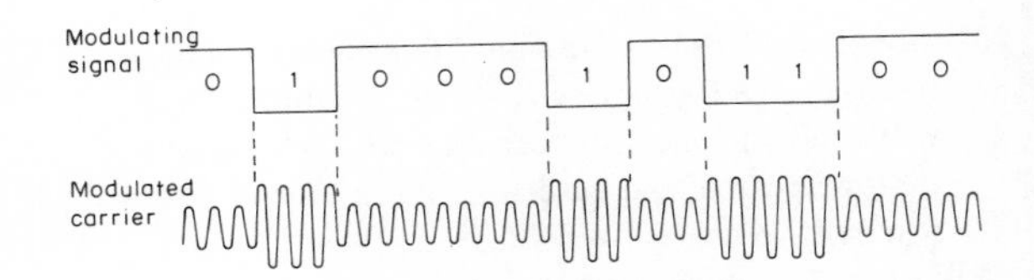

Amplitude Modulation (AM)

amplitude (peak) The maximum departure of the value of signal from its reference point. It can be referred to as the strength of the signal.

amplitude variation (ripple) Unwanted variation of signal voltage at different frequencies on a communications line.

ANAC Automatic Announcement Circuit.

analog Continuously variable as opposed to discretely variable. Physical quantities such as temperatures are continuously variable and so are described as analog; analog signals vary in accordance with the physical quantities they represent. The public telephone network was designed to transmit voice in analog form.

analog data Data in the form of continuously variable physical quantities.

analog–digital converter A device that converts a signal that is a function of a continuous waveform into a representative number sequence.

analog extension Use of analog transmission facilities (lines and modems) to connect a station not on a digital network (DDS).

analog loopback A diagnostic test that forms the loop at the modem's telephone line interface.

analog loopback testing In data communications, analog loopback testing is a technique whereby a local modem's transmitter is connected to the same modem's receiver input. (For full duplex modems, the transmitter and receiver must be switched into the same channel since they are normally in opposite channels.) In the local modem, the test pattern

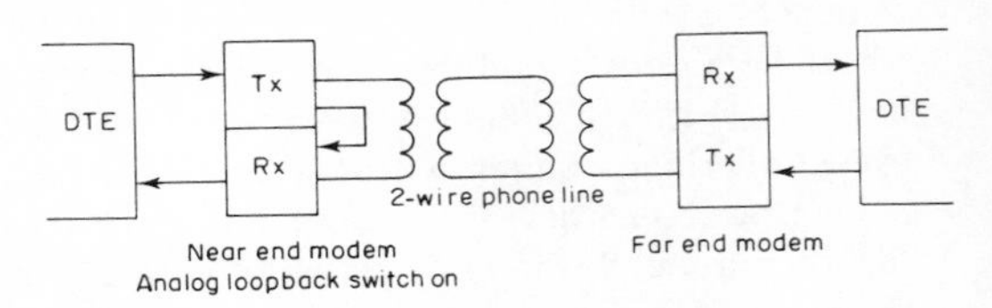

analogue loopback testing

is passed through most of the modem circuits and then transmitted back to the test device. The received signal in the test device is compared with the original transmitted signal and any errors induced by the modem are detected.

analog signaling An analog signal is one that varies in a continuous manner, such as voice or music.

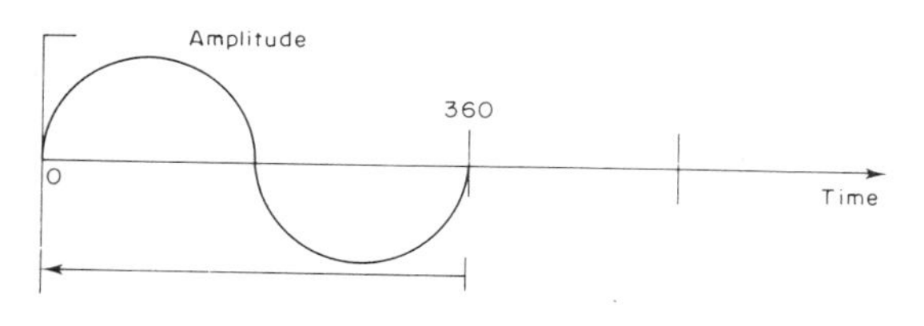

analog signaling

analog switch Any of a variety of switching devices which operate without converting the analog signal into a digital signal. Most newer switches operate digitally, converting the analog signal into a digital signal, switching the signal and reconverting to analog for further transmission.

analog to digital (A/D) conversion Conversion of an analog signal to a digital signal.

analog transmission Transmission of a continuously variable signal as opposed to a discretely variable signal. Physical quantities such as temperature are continuously variable and so are described as "analog."

ancillary equipment In IBM's VTAM, equipment not under direct control of the processing unit.

ANI Automatic Number Identification.

ANIK A series of satellites used for communications throughout Canada as well as for cross-border services to the US.

anisochronous signal A signal which is not related to any clock, and in which transitions could occur at any time.

AN/PSC-3 A portable man-pack tactical satellite communications system used by the U.S. military.

ANSI American National Standards Institute.

ANSI Standards

X3.15 Bit sequencing of ASCII in serial-by-bit data transmission
X3.16 Character structure and character parity sense for serial-by-bit data communications in ASCII
X3.36 Synchronous high-speed data signaling rates between data terminal equipment and data circuit-terminating equipment
X3.41 Code extension techniques for use with 7-bit coded character set of ASCII
X3.44 Determination of the performance of data communications systems
X3.79 Determination of performance of data communications systems that use bit-oriented control procedures
X3.92 Data encryption algorithm.

answer only A data terminal or data set which can accept, but not originate, calls on the switched telephone network.

answerback A response from a data transmission device in response to a request from another transmitting data processing device that is ready to accept or has accepted data.

answering machine Equipment which automatically answers a telephone after a preset number of rings. The caller hears the message previously prerecorded and is able to leave a recorded message.

answering tone A signal sent by the called modem (the "answer" modem) to the calling modem (the "originate" modem) on public telephone networks that indicate the called modem's readiness to accept data.

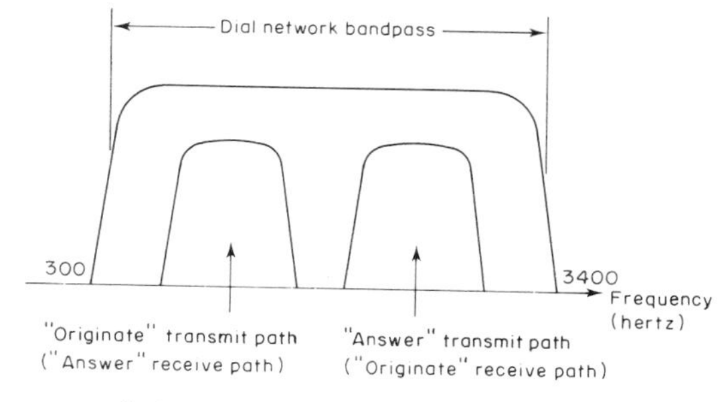

answering tone

antenna A circuit element designed for radiating or receiving electromagnetic waves, such as radio waves.

anti-streaming A feature in a modem that enables it to ignore a RTS signal from a DTE if it is held on for longer than a specified amount of time.

AN/VSC-7 A vehicle-mounted tactical satellite communications system used by the U.S. military.

any-mode In IBM's VTAM: 1. The form of a receive request that obtains input from any one (unspecified) session. 2. The form of an accept request that completes the establishment of a session by accepting any one (unspecified) queued CINIT request.

AOS Alternative Operative Services.

APD Avalanche Photodiode.

API Application Program Interface.

APP Access Process Parameter.

apparatus boxes Metallic structures which provide mechanical protection for equipment on a customer's premise.

apparatus closet Traditionally, in commercial buildings, the wiring closet, or other enclosed space, on each floor where backbone cables are connected to the switching and power devices associated with key telephone systems. This term is largely outdated, since the original distinction between "apparatus" and "satellite" closets no longer holds true: "apparatus" closets contained key switching and power devices and "satellite" closets did not. Today, backbone and satellite closets are constructed to hold modern power and multiplexing equipment, which is smaller, lighter, and more sophisticated than earlier types of switching and power equipment.

apparent power The power in a circuit when the load is reactive, where reactance due to inductance and capacitance in the load causes a shift in the phase angle between current and voltage. The reactance changes with each change of frequency, causing the power in the load to vary with frequency. This causes the measured power output level of a telephone line to vary as the signal frequency is varied.

APPC Advanced Program-to-Program Communications.

APPC/PC Advanced Program-to-Program Communications/Personal Computer.

Appleshare Apple Computer's file server software that lets a Macintosh use a file server (usually another Mac with a hard disk drive) to store files for a network of other Macs.

Appletalk Apple Computer's local area network.

application A definable set of tasks to be accomplished as part of the work of an enterprise. An application may be accomplished through manual or computerized procedures including communications control, or both.

application layer Highest (seventh) layer in OSI model, containing all user or application programs.

application program 1. In general, a program that is designed to perform a specific user function. 2. In data communications, a program (which frequently resides in data communications equipment) used to connect and communicate with terminals that performs a set of specified activities for terminal users.

application program exit routine In IBM's VTAM, a user-written exit routine that performs functions for a particular application program and is run as part of the application program. Examples are RPL exit routines, EXLST exit routines, and the TESTCB exit routine.

application program identification The symbolic name by which an application program is identified to IBM's ACF/VTAM. *Note*: It is specified in the APPLID parameter of the ACB macro instruction. It corresponds to the ACBNAME parameter in the APPL statement or, if ACBNAME is defaulted, to the name of the APPL statement.

Application Program Interface (API) A set of software calls and routines that can be referenced by an application program to access predefined network services.

application program major node In IBM's VTAM, a member or book of the ACF/VTAM definition library that contains one or more APPL statements, each representing an application program.

APR In IBM's VTAM, Alternate Path Retry.

APS Automatic Protection Switching.

ARB Arbitrator.

arbitrator (ARB) A circuit board that resides in the chassis of a Telenet Processor 4000 (TP4). Its specific function is to allocate main memory addresses among the microprocessor boards.

architecture The manner in which a system (such as a network or a computer) or program is structured. See also closed architecture, distributed architecture, and open architecture.

archive To back up, or store, data.

arcing A luminous passage of current through ionized gas or air.

ARCnet Attached Resources Computing Network.

ARCTS Automatic Reporting Circuit Test Set.

area In Digital Equipment Corporation Network Architecture (DECnet), the group of systems which constitute a single Level 1 routing subdomain.

area address In Digital Equipment Corporation Network Architecture (DECnet), the concatenation of the IDP and LOC-AREA fields of an NSAP address. A system may have more than one area address, but the systems making up an area must have at least one area address in common with each of their neighbors.

area code A 3-digit number that identifies a particular geographic location. The area code is used in many types of telecommunications for access to hardware devices physically or logically situated at a particular geographic location.

argument User selected items of data that are passed to a subroutine, program, or procedure.

Ariel A trademark and interactive video system service of AT&T.

ARISTOTE Association de Réseaux Informatiques en Système Totalement Ouvert et Très Elaboré.

ARPA Advanced Research Projects Agency.

ARPANET A large scale wideband data network interconnecting a large population of terminals and computers at U.S. and European universities and commercial institutions. Sponsored by the Advanced Research Projects Agency; a pioneering use of packet-switching network technology.

ARQ Automatic Request for Retransmission.

array connector A connector for use with ribbon fiber cable that joins 12 fibers simultaneously. A fan-out array design can be used to connect ribbon fiber cables to non-ribbon cables.

arrester 1. Device which diverts high voltage to ground and away from equipment. 2. The voltage limiting portion of a protector. 3. A lightning arrester.

ARS Automatic Route Selection.

artificial intelligence (AI) The quality attributed to programs and computers that gives the illusion of human intellectual activity.

ARTL Analog Responding Test Line.

ASCII American Standard Code for Information Interchange (7 level).

HIGH / LOW $B_3B_2B_1B_0$		0000	0001	0010	0011	0100	0101	0110	0111	1000	1001	1010	1011	1100	1101	1110	1111
$B_7B_6B_5B_4$		0	1	2	3	4	5	6	7	8	9	A	B	C	D	E	F
0000	0	NUL	SOH	STX	ETX	EOT	ENQ	ACK	BEL	BS	HT	LF	VT	FF	CR	SO	SI
0001	1	DLE	DC1	DC2	DC3	DC4	NAK	SYN	ETB	CAN	EM	SUB	ESC	FS	GS	RS	US
0010	2	SP	!	"	#	$	%	&	'	(	)	*	+	,	—	.	/
0011	3	0	1	2	3	4	5	6	7	8	9	:	;	<	=	>	?
0100	4	@	A	B	C	D	E	F	G	H	I	J	K	L	M	N	O
0101	5	P	Q	R	S	T	U	V	W	X	Y	Z	[	\	]	^	—
0110	6	\	a	b	c	d	e	f	g	h	i	j	k	l	m	n	o
0111	7	p	q	r	s	t	u	v	w	x	y	z	{	|	}	~	DEL

BINARY — HEX — ASCII

ASCII character set

ASCII terminal A terminal that uses ASCII; usually synonymous with asynchronous terminal and with dumb terminal.

ASCNET Australian Computer Science Network.

ASD Auto-Speed Detect.

ASMDR Advanced Station Message Detail Reporting.

ASP 1. Aluminum, Steel, and Polyethylene, the preferred sheath for waterproof (filled) cable. 2. Auto-Speed Port. 3. Advanced Speech Processor.

ASR Automatic Send/Receive.

assembler A program that translates an assembly programming language into the code of zeros and ones used by computers.

assembly level code A low-level language for computer programming that uses mnemonic commands to represent machine language operation codes. This language allows the programmer to write programs for a specific machine with greater ease than using binary-number machine language instructions.

associated address space In IBM's VTAM, the address space in which RPL-based requests are issued that specify an ACB opened in another address space.

associated signaling Signaling system in which signaling bits are included in the data channels, i.e. bit 8 of frames 6 and 12 in North America (D4 format).

Astrotec AT&T trademark for a laser transmitter system.

asymmetrical transmission The process by which a modem obtains full duplex transmission by transmitting data in two directions at different speeds. The modem monitors the quantity of data being transmitted in order to assign the higher speed channel to the major data transmission flow direction.

ASYNC Short for asynchronous or for asynchronous transmission.

asynchronous character A binary character used in asynchronous transmission which contains equal-length bits, including a start bit and one or more stop bits which define the beginning and end of the character.

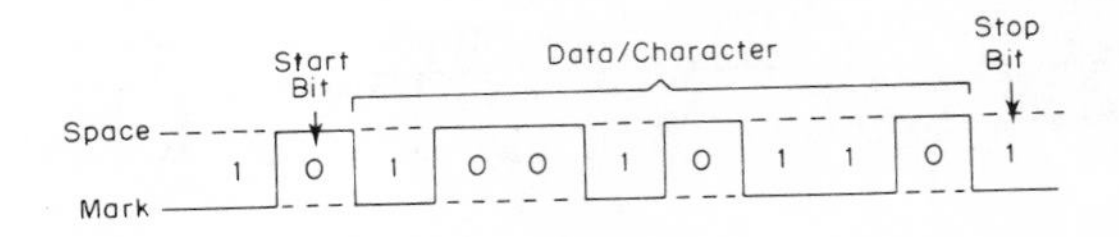

asynchronous character

asynchronous (character framed) Having a variable or random time interval between successive characters, operations, or events. Transmission in which each character, word, or small block, is individually synchronized (timed), usually by the use of start and stop bits. This is in contrast to synchronous operation. Also called Start/Stop (S/S).

asynchronous communication A method of communication where the time synchronization of the transmission of data between the sending and receiving stations is set by start and stop bits and the baud rate.

asynchronous data Data transmitted character by character where each character has a start bit and one or more stop bits. The transition from stop to start triggers the receiver's internal bit rate clock. The receiver then accepts and discards the start bit, accepts the number of bits appropriate to the code set, and is conditioned by the stop bit for the next character. Start-stop synchronization enables characters to be sent at random, since each character carries the necessary synchronizing information with itself.

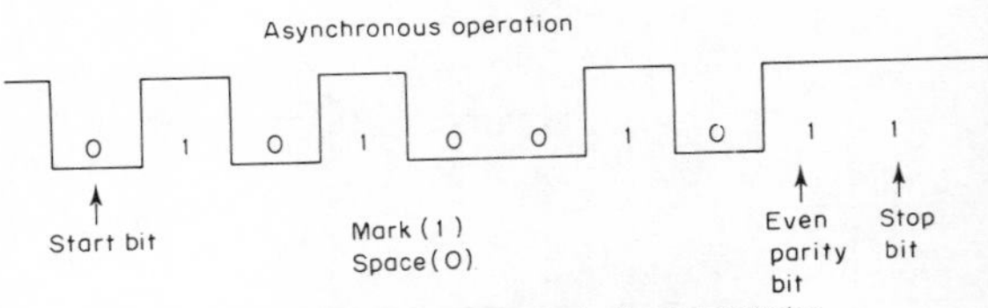

asynchronous data

asynchronous data channel A communications channel capable of transmitting data but not timing information. Sometimes called an anisochronous data channel.

asynchronous exit routine In IBM's VTAM, an RPL exit routine or an EXLST exit routine other than LERAD or SYNAD.

asynchronous link A communications link using asynchronous transmission, in which each transmitted character is framed with start and stop bits. The time interval between characters may be of unequal length.

asynchronous modem A modem that uses asynchronous transmission and, therefore, does not require timing synchronization with its attached DTE or remote modem; also used to describe a modem which converts asynchronous inputs from the DTE to synchronous signals for modem-to-modem transmission.

asynchronous operation In IBM's VTAM, an operation, such as a request for session establishment or data transfer, in which the application program is allowed to continue execution while VTAM performs the operation. VTAM informs the program after the operation is completed.

asynchronous request In IBM's VTAM, a request for an asynchronous operation.

asynchronous routine A set of computer instructions that a computer program calls into service on an as-needed basis.

asynchronous terminal A terminal that uses asynchronous transmission; usually synonymous with ASCII terminal and with dumb terminal.

Asynchronous Time-Division Multiplexer (ATDM) A TDM that multiplexes asynchronous signals by oversampling; also, infrequently used to mean concentrator.

asynchronous transmission Method of sending data in which the interval between characters may be of unequal length; since asynchronous characters are used, no additional synchronizing or timing information need be sent. Also called start–stop transmission.

AT Access Tandem.

AT command set A set of commands used by personal computers to direct dialing operations in a modem. Although the intelligent modem patent is held by BIZcom, it was Hayes Microcomputer Products which popularized the use of intelligent modems. Thus, sometimes also called the "Hayes command set."

AT commands The "AT" commands are subdivided into three major groups: configuration, immediate action, and diagnostic commands.

The "AT" prefix begins every command line with the exception of the +++ (escape) and the A/ (repeat) commands. "AT," often referred as attention code, delivers to the modem information about the data rate and the parity setting of the local DTE.

Multiple commands can be placed on a single line. Space characters are allowed between commands to improve readability. A command line must be terminated with the ASCII carriage return character. A line feed character following the carriage return character is optional. The backspace or delete key can be used to delete any character entered from the keyboard, except the "AT" prefix. Upon execution of the command the result code is returned by the modem.

Example:	Terminal:	AT DT 998 1234
	Modem:	OK

This example demonstrates the dial command. A string of ASCII characters (the "AT" command) followed by the digits (a number to dial) is entered from the keyboard. Upon successful execution of the dial command, an OK message is returned to the user.

RESULT CODES

Short	*Long Form*	*Description*
0	OK	Command line executed without errors
1	CONNECT	Connected at 300 bps
2	RING	Local telephone line ringing
3	NO CARRIER	Carrier lost, or never received
4	ERROR	Error in command line, invalid command line, command line exceeds command buffer, invalid character format
5	CONNECT 1200	Connected at 1200 bps data rate
6	NO DIALTONE	No dial tone received within time-out period
7	BUSY	Called line busy
8	NO ANSWER	Called line not answered within time-out period
9	CONNECT 600	Connection established at 600
10	CONNECT 2400	Connection established at 2400

DIAL MODIFIERS

P	Pulse dial
R	Originate call in Answer mode
T	Tone dial
S	Dial stored number
W	Wait for dial tone
,	Pause
;	Return to command state
!	Flash
@	Wait for silence

"AT" COMMANDS SUMMARY

AT	Attention code preceded command lines except +++ (escape) and A/ (repeat)
A	Go off hook in answer mode
A/	Repeat previous command line
B	CCITT V.22 operation at 1200 bps
B1	Bell 212A operation at 1200 bps (default)
D	Dial a number (0–9 ABCD*#)
E	Turn Echo off
E1	Turn Echo on (default)
H	Go on hook (hang up) (default)
H1	Go off hook and switch the auxiliary relay
I	Request product code
I1	Compute and return checksum (firmware ROM)
I2	Compute and return checksum with OK or ERROR message
L,L1	Low speaker volume
L2	Medium speaker volume (default)
L3	High speaker volume
M	Speaker is off
M1	Speaker is off while carrier is present (default)
M2	Speaker is always on
M3	Speaker is disabled while dialing or receiving carrier
O	Return to On-line mode
O1	Return to On-line mode and initiate retrain sequence (in 2400 bps only)
Q	Return result codes (default)
Q1	Do not return result codes
Sn=x	Write x in S-register n
Sn?	Read S-register n
V	Enable short form result codes
V1	Enable full word result codes (default)
X	CONNECT result code enabled (300 bps operation)
X1	All CONNECT result codes enabled; dial blind; busy signal is not recognized
X2	All CONNECT result codes enabled; wait for dial tone before dialing; busy signal is not recognized
X3	All CONNECT result codes enabled; dial blind; busy signal is recognized
X4	All CONNECT result codes enabled; wait for dial tone before dialing; busy signal is recognized (default)
Y	Enable long space disconnect (default)
Y1	Disable long space disconnect
Z	Fetch configuration profile contained in external nonvolatile memory
+++	The default escape code
&C	DCD always ON (default)
&C1	DCD tracks the state of the carrier signal
&D	DTR ignored (default)
&D1	Assume command state on ON-to-OFF transition of DTR
&D2	Go o-hook, disable auto-answer, assume command state on ON-to-OFF transition of DTR
&D3	Assume initialization state on ON-to-OFF transition of DTR
&F	Fetch factory configuration profile from internal ROM
&G	No guard tone (default)
&G1	550 Hz guard tone
&G2	1800 Hz guard tone
&M	Asynchronous mode (default)
&M1	Synchronous mode 1 (sync/async modes are supported)
&M2	Synchronous mode 2 (dial stored number mode)
&M3	Synchronous mode 3 (manual dial with DTR off)
&P	Make/Break pulse ratio = 39/61 (US/Canada) (default)
&P1	Make/Break pulse ratio = 33/67 (CCITT)
&R	CTS OFF-to-ON transition follows RTS OFF-to-

	ON transition (default)
&R1	CTS always ON; RTS ignored
&S	DSR always ON (default)
&S1	DSR operates accordingly to CCITT V.22 bis/V.22 recommendation
&T	Terminate any test currently in process
&T1	Initiate local analog loopback
&T3	Initiate local digital loopback
&T4	Grant request from remote modem for remote digital loopback
&T5	Deny request from remote modem for remote digital loopback
&T6	Initiate remote digital loopback
&T7	Initiate remote digital loopback with self test
&T8	Initiate local analog loopback with self test
&W	Write configuration to nonvolatile memory
&X	Transmit clock source is modem (default)
&X1	Transmit clock source is DTE
&X2	Transmit clock source is derived from received carrier signal
&Z	Store telephone number

ATC Automatic Technical Control.

ATDM Asynchronous Time-Division Multiplexer.

ATM Automated Teller Machine.

AT&T American Telephone and Telegraph.

AT&T ALL PRO America A calling plan which integrates AT&T PRO America I and the AT&T PRO State plan in a customer's own state. AT&T ALL PRO offers a discount on direct-dialed AT&T Long Distance calls to locations within your state and across the country, 24 hours a day, seven days a week.

AT&T Alliance Teleconferencing Service An AT&T service which allows customers to dial directly conference calls with as few as 3 or as many as 59 locations.

AT&T Call Me card A card which permits toll-free calling to your business number, and no other. Suitable for low-volume calling, the card holder controls how it is publicized and distributed.

AT&T card A free calling card which enables the user to charge all AT&T Long Distance calls from virtually any U.S. location to his/her office number. It also works from most foreign countries for calls back to the U.S.

AT&T Communications The regulated subsidiary of post-divestiture AT&T responsible for long distance operations; the successor to AT&T Long Lines.

AT&T Information Systems The unregulated subsidiary of post-divestiture AT&T responsible for equipment sales and leasing.

AT&T International 800 Service An AT&T service which permits customers in more than 20 countries to call a U.S. location toll free. The 800 Service provider specifies the service by the country in which calls originate.

AT&T Mail An electronic mail service marketed by AT&T. Dial 1-800-MAIL-123 to register electronically or 1-800-367-7225, extension 600 to register by voice.

AT&T Mail Access I A software package marketed by AT&T for operation on personal computers which is designed to enhance the use of that vendor's electronic mail service.

AT&T Mail Talk A voice response system service marketed by AT&T which permits subscribers of that firm's electronic mail system to read, delete, and save messages via a touch tone telephone. The following table lists AT&T Mail Talk Commands.

AT&T MAIL TALK COMMANDS

Key	*Command*
1	Read first message
2	Read previous message
3	Delete/Restore current message
4	Help
5	Repeat current message
6	Save current message
7	Stop/Start
8	Change rate of speed
9	Read last message
0	Exit system
#	Read next message/terminate login
*	Stop current message, await next command

AT&T Merlin Communications System A family of flexible business communications systems. The smallest MERLIN system starts at 4 telephone lines and 10 phones; the largest can handle up to 32 lines and 72 phones.

AT&T Network Services AT&T Technologies' subdivision that manufactures and supplies equipment primarily to telephone organizations.

AT&T PRO America Family A series of calling plans for out-of-state direct-dialed AT&T Long Distance calls. AT&T PRO America includes calls to 39 countries. Businesses which spend more than $120 a month on a combination of AT&T Long Distance interstate calls and AT&T International Long Distance calls to eligible countries generally qualify for PRO America.

AT&T PRO State A calling plan for AT&T Long Distance calls within a state. Only calls outside the subscriber's local calling area are included.

AT&T Pub 54106 "Requirements for Interfacing Digital Terminal Equipment to Services Employing the Extended Superframe Format": a basic reference source on ESF in the US.

AT&T Pub 62411 "ACCUNET T1.5 Service Description and Interface Specifications": a basic reference source on AT&T T1 implementations.

AT&T Reach Out America Bonus Plan A special plan marketed by AT&T for people who do most of their calling in the evenings and weekends. Under this plan, AT&T customers pay a flat hourly rate for any calls made after 10 PM on weekdays or during weekend hours.

AT&T 800 Readyline A toll-free service for moderate call volume. It looks like regular AT&T 800 Service to callers, but there are no special lines or equipment to install, making it easy to adjust for changes in geographic range or seasonal calling patterns.

AT&T 800 Service Also known as WATS, this toll-free service is for high-customer call volume. AT&T 800 Service can be designed for in-state callers only or in widening bands for regional or national coverage and can begin with a single line.

AT&T Spirit Communications System A basic, easy-to-use phone system with many built-in features. Spirit supports from 3 to 24 lines and from 8 to 48 telephones.

AT&T System 25 A state-of-the-art digital PBX designed to bring sophisticated telephone system benefits to small and growing businesses with 20 to 150 telephone stations.

AT&T Technologies AT&T equipment manufacturing and supply division; includes AT&T Information Systems and AT&T Network Services (formerly Western Electric).

AT&T WATS Service Low cost, high-quality long distance service for outgoing calls. AT&T WATS requires special access lines, but can provide significant savings for companies with heavy outgoing call volume. Coverage can be tailored from single-state to nationwide in a series of increasing bands.

Attached Resources Computing Network (ARCnet) A networking architecture marketed by Datapoint Corporation and other vendors which uses a token-passing bus architecture.

attack dialing The process of immediately repeating a dialed call in which a busy signal is encountered, to minimize the time required to eventually connect the call.

attended An operation undertaken by, or under the control of, a human operator.

attention In the context of the virtual terminal, the attention or interrupt facility allows a user operating in alternate mode to regain access to the virtual terminal data structure when it is under the control of a remote computer. The receipt of an attention signal is acknowledged by a mark.

attention code A request for service signal.

attenuation Deterioration of signals as they pass through a transmission medium; generally, attenuation increases (signal level decreases) with both frequency and cable length. Measured in terms of levels or deciBels.

attenuation distortion The relative loss at discrete frequencies across the bandwidth when compared to a 1004 Hz reference signal. For example, if a circuit had 6 dB more loss at 2804 Hz than it did at 1004 Hz, it is said to have an attenuation distortion of 6 dB. Attenuation distortion is also referred to as "frequency response" or "slope."

ATTIS (AT&T Information Systems) Division of AT&T Technologies that supplies and manufactures customer premises equipment.

attribute Column in a data table having unique names.

audio frequencies The frequencies that can be heard by the human ear (usually between 30 and 20 000 hertz) when transmitted as sound waves.

audio response unit An output device which provides synthesized or prerecorded speech as the response to inquiries. The response may be transmitted over communications facilities to a remote location from which the inquiry originated.

audiographics The technology which allows sound

and visual images to be transmitted simultaneously. Generally refers to single frame or slow frame visual images as opposed to continuous frame images such as television. Audiographic transmission is often used to teach or train people located remotely from an educational institution or business training center, saving travel and housing expense.

audiometer A calibrated audio oscillator that can be used to vary the amplitude and frequency of sound. The audiometer is normally used for conducting hearing tests.

Audiotex A service which permits a database host to pass data to a voice-mail computer, where it is interpreted and delivered over the telephone as a natural voice spoken message.

audit trail A printed record of system commands and events identified by date and time of occurrence. The audit trail consists of operator-initiated commands and system-generated events.

Auseanet Austroasian Network.

Australia Overseas Telecommunications Commission The communications carrier which provides all of Australia's international satellite communications services.

authentication The process of identifying a user to a network or computer.

authentication of messages Addition of a check field to a block of data so that any change to the data will be detected. A secret key enters into the calculation and is known to the intended receiver of the data. In a different form of authentication, the whole block is transformed and this can be a public key cryptosystem.

Authority and Format Indicator (AFI) In Digital Equipment Corporation Networks Architecture (DECnet), the part of a NSAP address which indicates the addressing authority responsible for the assignment of the IDP and its format. It also indicates the format (binary or decimal) of the DSP.

authorization exit routine In IBM's VTAM, an optional installation exit routine that approves or disapproves requests for session initiation.

authorized path In IBM's VTAM for MVS, a facility that enables an application program to specify that a data transfer or related operation be carried out in a privileged and more efficient manner.

authorized user A person or organization authorized to use a network or system.

autoanswer A machine feature which allows a station to automatically respond to call over the public switched network.

autobaud See automatic baud rate detection.

Autobaud rate detection A devices's ability to adjust itself to the incoming baud rate from the customer's equipment.

Autoconnect A feature in which the connecting processor automatically places the call to a pre-defined address, generally to a host, thus eliminating the need for the terminal user to enter a call command manually each time the terminal is powered on. Autoconnect is implemented on a terminal that accesses one particular host and is usually effected as soon as the terminal is turned on.

Autodelivery A type of electronic mail delivery where a message is sent to a user's catalog and a copy is also automatically generated and sent to a station (hardcopy device).

Auto-Dial A type of modem used on the public telephone network that automatically originates calls (dials the desired number).

Autodin The Automatic Digital Network, a data communications network developed and operated by contractors for the U.S. Department of Defense.

Automated Coin Toll Service (ACTS) The feature that automatically computes charges on coin toll calls, announces charges to the customer, counts coin deposits, and sets up coin calls—all without need of an operator.

automatic activation In IBM's VTAM, the activation of links and link stations in adjacent subarea nodes as a result of channel device name or RNAME specifications related to an activation command naming a subarea node.

automatic answering modem A modem that will answer a phone call in an unattended mode and prepare for receipt of data.

Automatic Answering Unit (AAU) A hardware device which performs the autoanswer function.

Automatic Baud Rate Detection (ABR) A process by which a receiving device determines the speed, code level, and stop bits of incoming data by examining the first character—usually a preselected sign-on character. ABR allows the receiving device to accept data from a variety of transmitting devices operating at different data rates without needing to configure the receiver for each specific data rate in advance. Also called autobaud.

Automatic Call Director (ACD) Device used for telemarketing applications to direct incoming calls to the next free agent position. This function is also a feature of many PBX products.

automatic call distribution A PABX feature that allows incoming calls to be automatically connected to an agent's telephone without the agent taking any action.

Automatic Call Distributor (ACD) A hardware device which performs the autoanswer function and switches the calls to the first available station or communications port.

Automatic Calling Unit (ACU) A hardware device which permits a business machine to automatically originate calls over the public switched network.

automatic deactivation In IBM's VTAM, the deactivation of links and link stations in adjacent subarea nodes as a result of a deactivation request naming a subarea node. *Note*: Automatic deactivation occurs only for automatically activated links and link stations that have not also been directly or indirectly activated.

automatic dialer 1. A device which automatically dials preset telephone numbers, usually those dialed frequently. 2. A device which automatically dials a series of random numbers; most often used in computerized telephone solicitations.

Automatic Dialing Unit (ADU) A hardware device which is capable of automatically generating dialing pulses, or signals.

automatic equalization Equalization of a transmission channel which is adjusted automatically while sending special signals.

Automatic Identification of Outward Dialing (AIOD) A PBX service feature that identifies the calling extension, permitting cost allocation.

automatic logon In IBM's VTAM, a process by which VTAM creates a session-initiation request (logon) for a session between a secondary logical unit (other than a secondary application program) and a designated primary logical unit whenever the secondary logical unit is not in session with, or queued for a session with, another primary logical unit. *Note*: Specifications for the automatic logon can be made when the secondary logical unit is defined or can be made using the VARY NET,LOGON command.

Automatic Message Accounting (AMA) The automatic collection, recording, and processing of call-related information for billing by a telecommunications provider.

Automatic Message Accounting Teleprocessing System (AMATPS) A billing system feature where the AT&T 5ESS switch forwards billing information over a data link to a centralized AMA data collection system. The AMA data collection system interfaces with an RAO.

automatic modem selection A PABX feature that allows a subscriber to be connected to a modem from a modem pool so an internal or external data call can be completed.

automatic number identification Means provided at a central office switch in which the switch is able to determine the identity of a calling party and forward that information to the called party.

Automatic Protection Switching (APS) Switch equipment designed to reroute communications automatically to a secondary circuit should the primary fail.

automatic repeat request Same as automatic request for retransmission.

Automatic Request for Repetition (ARQ) Same as automatic request for retransmission.

Automatic Request for Retransmission (ARQ) An error control method in which the receiving device informs the transmitting device which transmission blocks were received successfully; the transmitting device retransmits any blocks not successfully received.

Automatic Route Selection (ARS) The capability of a PABX and other network devices to select the optimum route for a call (in terms of cost, reliability, quality, etc.) without operator intervention.

Automatic Send/Receive teleprinter (ASR) Teleprinter equipped with paper tape, magnetic tape, or a solid state buffer that allows it to transmit and receive data unattended.

automatic switching system A communications system in which the switching operations are performed by electrically controlled devices without human intervention.

Automatic Technical Control (ATC) A computer system used to maintain operational control of a data communications network.

Automatic Teller Machine (ATM) The remote

terminal used by banks to allow customers to perform banking transactions.

Autoplex A trademark of AT&T cellular telecommunications system service.

Auto-Speed Port (ASP) A network access port which automatically detects the speed of the terminal and matches operation at that speed. Also called Auto-Speed Detect (ASD).

Autovon The Automatic Voice Network, a voice network developed and operated by contractors for the U.S. Department of Defense.

auxiliary equipment Equipment, such as a printer, not under direct control of the processing unit.

auxiliary network address In IBM's ACF/VTAM, any network address, except the main network address, assigned to a logical unit capable of having parallel sessions.

availability A measure of equipment, system, or network performance—usually expressed in percent; the ratio of operating time to the sum of operating time plus down time. Based on MTBF and MTTR. Availability = MTBF/(MTBF + MTTR).

available In IBM's VTAM, pertaining to a logical unit that is active, connected, enabled, and not at its session limit.

Avalanche Photodiode (APD) A diode which increases its electrical conductivity by a multiplication effect when struck by light. Used in receivers for lightwave transmission.

AVD Alternate Voice Data.

average busy-hour traffic count The average number of calls received during the busy hour.

AWG American Wire Gauge.

AXE Ericsson's digital switching system.

B

B channel "bearer channel" In ISDN, the 64-Kbps channel of a Basic Rate Interface (BRI) (where there are two) or a Primary Rate Interface (PRI) (where there are 23) that is circuit-switched and can carry either voice or data transmission.

B7 stuffing A simple technique for maintaining ones density on digital circuits. With B7 stuffing, each string of 8 consecutive zero bits is replaced by a string of 7 zeros and a 1.

Babbling tributary In a local area network, a station that continuously transmits meaningless messages.

Backboard A construction of wooden or metal panels used for mounting miscellaneous apparatus. May be equipped with premounted connecting blocks, mounting brackets, and/or predrilled holes.

Backbone In packet-switched networks, the major transmission path for a PDN.

Backbone closet The closet where backbone cable is terminated and cross connected to either horizontal distribution cable or other backbone cable. The backbone closet houses cross-connect facilities, and may contain auxiliary power supplies for terminal equipment located at the user work location. Sometimes called a riser closet.

Backbone network A transmission facility designed to interconnect generally lower-speed distribution networks or clusters of dispersed user devices.

Backbone subsystem The part of a premises distribution system that includes a main cable route and facilities for supporting the cable from an equipment room (often in the building basement) to the upper floors, or along the same floor, where it is terminated on a cross-connect in a backbone closet, at the network interface, or distribution components of the campus subsystem. The subsystem can also extend out on a floor to connect a backbone and satellite closet or other satellite location.

back-level host In SNA network interconnection, a host processor containing TCAM, VTAME, or a release of VTAM prior to the current release.

back-off In a local area network, the process of delaying an attempt to transmit. Usually results when an earlier attempt to transmit encountered some difficulty, such as a collision on a CSMA/CD network.

backplane bus A collection of electrical wiring and connectors that interconnect modules inserted into a computer system or other electronic device.

back-to-back gateways Two gateways separated by one intervening network that contains no gateway SSCP function involved with either of the two gateway NCPs.

backup 1. (*noun*) The hardware and software resources available to recover after a degradation or failure of one or more system components. 2. (*verb*) To copy files onto a second storage device so that they may be retrieved if the data on the original source is accidentally destroyed.

backward channel Also known as a reverse channel, this type of channel is used to send data in the direction opposite to the primary (forward) channel. The backward channel is often used for low-speed data as may be found in keyboarded or supervisory controls.

backwards learning A method of routing in which nodes learn the topology of the network by observing the packets passing through, noting their source and the distance they have travelled.

balanced circuit A circuit terminated by a network whose impedance balances the impedance of the line so the return losses are negligible.

balanced mode An arrangement in which two stations are able to transmit to each other without requiring that either be fixed as a "master" or "secondary" station.

balanced-to-ground A two-wire circuit, in which the impedance-to-ground on one wire equals the impedance-to-ground on the other wire, a favorable condition for data transmission (compare with unbalanced-to-ground).

balancing network Electronic circuitry used to

match two-wire facilities to four-wire toll facilities. This balancing is necessary to maximize power transfer and minimize echo. Sometimes called hybrid.

balun (balanced/unbalanced) Impedance matching devices used to connect balanced twisted pair cable to unbalanced coaxial cable. They are required on each end of a twisted pair cable when converting a coax to twisted pair medium.

band The frequencies lying between two defined limits.

band-limited A signal or channel of restricted frequency range.

bandpass The portion of a band, expressed in frequency differences (bandwidth), in which the signal loss (attenuation) of any frequency when compared with the strength of a reference frequency is less than the value specified in the measurement.

band splitter A multiplexer (commonly an FDM or TDM) designed to divide the composite bandwidth into several independent, narrower bandwidth channels, each suitable for data transmission at a fraction of the total composite data rate.

bandsplit The technique of subdividing a channel (band) into subchannels.

bandwidth The range of signal frequencies which can be carried by a communications channel subject to specified conditions of signal loss or distortion. Measured in hertz (cycles per second). Bandwidth is an analog term which provides a measure of a circuit's information capacity. There are three general ranges of frequencies for transmission: narrow band, 2 to 300 Hz; voice band, 300 to 3000 Hz; wide band, over 3000 Hz

Bandwidth Management Service (BMS) An AT&T Accunet T1.5 service option which enables customers to use an on-premises terminal or personal computer to reconfigure channels within T1 trunks.

banner 1. A predefined identifier which is displayed on a terminal screen under certain conditions. 2. A one-page information sheet, printed as the first page of a printout. A banner identifies the printout's creator.

bantam connector A miniature tip-and-ring jack connector used in digital telephony. Equivalent to WECO 310, but physically smaller.

baseband Direct transmission method used for short distances (less than 10 miles). Uses a bandwidth whose lowest frequency is zero (dc level), that is, transmits of raw (carrier-less) binary data. The transmission medium carries only one signal at a time.

baseband (digital) A digital stream that carries just one information channel.

baseband modem A modem which does not apply a complex modulation scheme to the data before transmission, but which applies the digital input (or a simple transformation of it) to the transmission channel. This technique can only be used when a very wide bandwidth is available and only operates over short distances where signal amplification is not necessary. Sometimes called a limited distance or short-haul modem.

baseband signal The original signal from which a transmission waveform is produced by modulation. In telephony it is the speech waveform. In data transmission many forms are used, and the baseband signal is usually made of successive signal elements. Also called basic signal.

baseband signaling Transmission of a signal at its original frequencies, without modulation.

baseband system A system that transmits at only one frequency and speed.

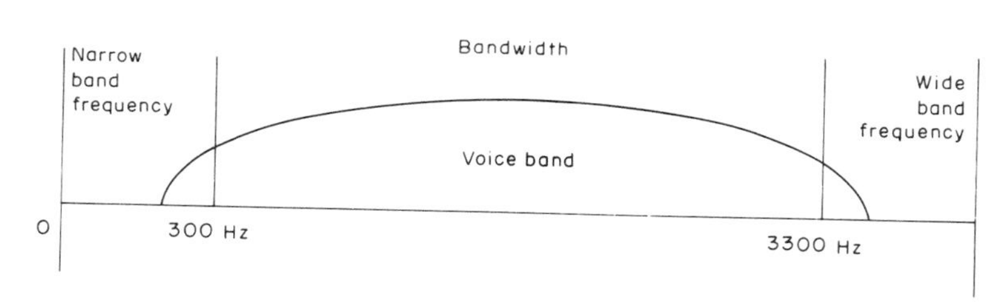

bandwidth

base group Twelve communications VF paths. A unit of frequency division multiplexing systems bandwidth allocation.

BASIC Beginners All-purpose Symbolic Instruction Code. Basic Information Unit (BIU). In SNA, the unit of data and control information that is passed between half-sessions. It consists of a request/ response header (RH) followed by a request/ response unit (RU).

basic mode In ACF/VTAM, Release 1 and in VTAM, a mode of data transfer in which the application program can communicate with non-SNA terminals without using SNA protocols.

Basic Rate Interface (BRI) The end-user ISDN service containing two Bearer channels (B-channels) at 64 Kbps each, and 1 Delta channel (D-channel) at 16 Kbps. Also called 2B+D Service. Transmitted at 144 Kbps.

basic services Refers to transport level services provided by Bell operating companies (BOCs).

Basic Signal See baseband signal.

Basic Telecommunications Access Method (BTAM) IBM teleprocessing access method. Used to control the transfer of data between main storage and local or remote terminals. BTAM provides the applications program with macro instructions for using the capabilities of the devices supported. BTAM supports binary synchronous (BSC) as well as start/stop communication.

Basic Transmission Unit (BTU) In SNA, the unit of data and control information passed between path control components. A BTU can consist of one or more path information units (PIUs).

batch accounting A facility implemented in the Telenet Processor (TP) accounting system that reduces overhead, enabling accounting packets to be transferred in groups (batches) rather than individually.

batch processing A data processing technique in which input data is accumulated and prepared off-line and processed in batches.

batched communications Transmission of large blocks of information without interactive responses from the receiver.

baud Unit of signaling speed. The speed in baud is the number of discrete conditions or signal events per second. If each signal event represents only one bit, the baud rate is the same as bps; if each signal event represents more than one bit (such as in a dibit), the baud rate is smaller than bps.

Baudot code A code (named after Emile Baudot, a pioneer in printing telegraphy) for asynchronous transmission of data in which 5 bits represent a single character. Use of Letters Shift and Figures Shift enables 64 alphanumeric characters to be represented. Used mainly in teleprinter systems which add one start bit and 1.5 stop bits.

BCC Block Check Character.

BCD Binary Coded Decimal.

B-channel See bearer channel.

BCM Bit Compression Multiplexer.

BCH An error detecting and correcting technique used at the communications receiver to correct errors as opposed to retry attempts as in ARQ. Named for Messrs Bose, Chaudhuri, and Hocquencgham who developed it.

BCS Block Check Sequence.

BDLC Burroughs Data Link Control.

beam Microwave radio systems which use ultra/ super high frequencies (UHF, SHF) to carry communications, where the signal is a narrow beam rather than a broadcast signal.

beam splitter A device used to divide an optical beam into two or more separate beams. Often a partially reflecting mirror.

bearer channel (B-channel) A 64 Kbps information channel, time-division-multiplexed with other bearer channels and a Delta channel to form an Integrated Services Digital Network interface.

beeper A slang term for a radio pager.

begin bracket In SNA, the value (binary 1) of the begin-bracket indicator in the request header (RH) of the first request in the first chain of a bracket. The value denotes the start of a bracket.

Bell 103 An AT&T, 0–300 bps modem providing asynchronous transmission with originate/answer capability. Also often used to describe any Bell 103-compatible modem.

Bell 113 An AT&T, 0–300 bps modem providing asynchronous transmission with originate or answer capability (but not both). Also often used to describe any Bell 113-compatible modem.

Bell 201 An AT&T, 2400 bps modem providing synchronous transmission. Bell 201 B was designed for leased line applications (the original Bell 201 B was designed for public telephone network applications). Bell 201 C was designed for public telephone network applications. Also often used to describe any Bell 201-compatible modem.

Bell 202 An AT&T, 1800 bps modem providing asynchronous transmission that requires 4-wire circuit or full-duplex operation. Also an AT&T 1200 bps modem providing asynchronous transmission over 2-wire, full-duplex, leased line, or public telephone network applications. Often used to describe any Bell 202-compatible modem.

Bell 208 An AT&T, 4800 bps modem providing synchronous transmission. Bell 208 A was designed for leased line applications; Bell 208 B was designed for public telephone network applications. Also often used to describe any Bell 208-compatible modem.

Bell 209 An AT&T, 9600 bps modem providing synchronous transmission over 4-wire leased lines. Also often used to describe any Bell 209-compatible modem.

Bell 212, Bell 212 A An AT&T, 1200 bps full-duplex modem providing asynchronous transmission or synchronous transmission for use of public telephone network. Also often used to describe any Bell 212-compatible modem.

Bell 310 A large tip-and-ring jack connector used in digital telephony. Also called WECO 310.

Bell 2296 A A V.32 modem capable of operating at 9.6 Kbps over the switched telephone network.

Bell 43401 A Bell publication which defines requirements for transmission over telco-supplied circuits that have dc continuity (that are metallic).

BEL 1. A control character that is used when there is a need to call for attention. It may control alarm or attention devices. 2. Equal to 10 decibels.

Bell Atlantic One of the seven regional Bell operating companies, formed as a result of divestiture, whose service area covers the middle-Atlantic states.

Bell Operating Company (BOC) One of the 22 local telephone companies divested from AT&T and grouped under seven regional Bell holding companies.

Bellcore (Bell Communications Research) The standards organization of the Bell operating companies. Because of its market position within the industry, most manufacturers attempt to emulate Bell standards to produce Bell-compatible equipment.

BellSouth Corporation One of the seven regional holding companies (ROC) formed by the divestiture of AT&T. Consists of Southern Bell, BellSouth, and South Central Bell.

BellSouth Enterprises, Inc. The holding company for all unregulated BellSouth companies.

BellSouth International A subsidiary of BellSouth Enterprises, the holding company for all unregulated BellSouth companies, which manages BellSouth activities outside the U.S.

BellSouth Mobility Inc. A subsidiary of BellSouth Enterprises, the holding company for all unregulated BellSouth companies, which provides cellular mobile telephone and paging service in the southeastern United States.

BER Bit Error Rate.

BERNET I, II Berlin Network.

BERT Bit-Error-Rate-Test (set).

Best-1 A network design tool of BGS Systems of Waltham, MA.

beta test site The testing of a product prototype and/or early release at client locations prior to general public marketing.

BEX Broadband exchange.

bias A type of telegraph that occurs when the significant intervals of the modulation do not all have their exact theoretical durations.

bias distortion Communications signal distortion with respect to unequal mark and space durations in a digital transmission. The formula to calculate bias distortion is:

Bias Distortion = [(mark pulse width − space pulse width)/total pulse width] * 100%

If mark > space, the result is positive. If space > mark, the result is negative

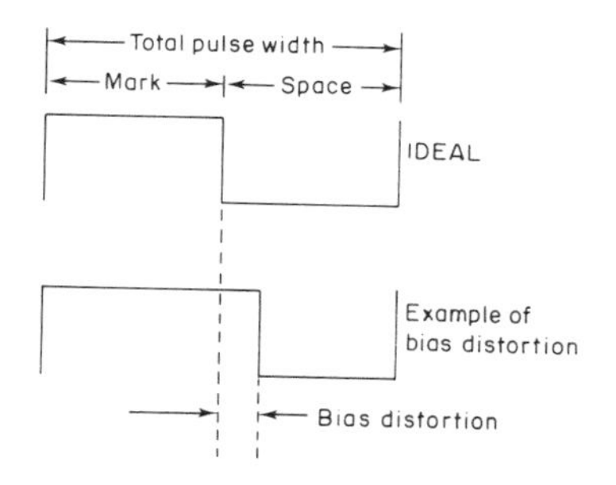

bias distortion

bidder In SNA, the LU–LU half-session defined at session activation as having to request and receive permission from the other LU–LU half-session to begin a bracket.

Billed Telephone Number (BTN) The primary telephone number used for billing, regardless of the number of telephone lines associated with that number.

billing increment The segment of telephone line connection time measured for billing purposes. For example, some services are measured and billed in 1 minute increments; some are measured and billed in 6- or 10-second increments.

binary Digital system with two states, 1 and 0.

binary code Computer language made up of ones and zeros arranged to represent words of computer instruction.

Binary Coded Decimal (BCD) A digital system that uses binary codes to represent decimal digits, where the numbers zero through nine have a unique 4-bit representation.

binary signal characteristics In a digital communication signal, either one state or the other appears on the line but never both at the same time. Further, the signals may be either normal or inverted depending on the circuit specifications. A comparison of circuit conditions necessary for binary signals in normal or inverted form is given in the chart below.

	NORMAL		INVERTED	
Binary digits	1	0	1	0
Applied voltage	on	off	off	on
Code element	mark	space	space	mark
Circuit contacts	closed	open	open	closed
Current identification	current	no current	no current	current

Binary Synchronous Communications (BSC; Bisync) A set of rules for the synchronous transmission of binary-coded data. All data in BSC is transmitted as a serial stream of binary digits. Synchronous communication means that the active receiving station on a communications channel operates in step with the transmitting station through the recognition of SYN bits.

One of the more common modes of employment of BSC is in a multipoint environment where the polling or selection of tributary stations is done. Polling is an "invitation to send" transmitted from the control station to a specific tributary station. Selection is a "request to receive" notification from the control station to one of the tributary stations instructing it to receive the following message(s).

SYN SYN is used to establish and maintain synchronization and as a time fill in the absence of any data or other control character.

SOH: Start of Heading precedes a block of heading characters. A heading consists of auxiliary information, such as routing and priority, necessary for the system to process the text portion of the message.

STX: Start of Text precedes a block of text characters. Text is that portion of a message treated as an entity to be transmitted through to the ultimate destination without change. STX also terminates a heading.

ETX: End of Text character terminates a block of characters started with STX or SOH and transmitted as an entity. The BCC (Block Check Character) is sent immediately following ETX. ETX requires a reply indicating the receiving station's status.

ETB: End of Transmission Block indicates the end of a block of characters started with SOH or STX. The BCC is sent immediately after ETB. ETB requires a reply indicating the receiving station's status.

SYN	SYN	SOH	Header	STX	TEXT	ETX or ETB	BCC

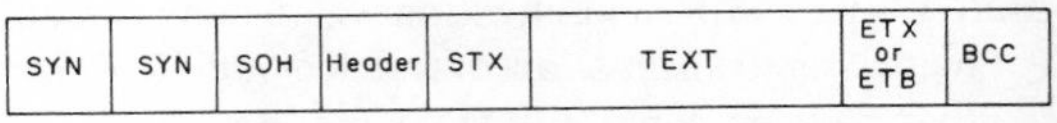

IBM Bisync (BSC) message format

Binary Synchronous Communications

binary synchronous transmission Data transmission in which synchronization of characters is controlled by timing signals generated at the sending and receiving stations.

Binary 8-Zero Suppression (B8ZS) A technique, used widely in North America, for maintaining ones density on digital circuits. In B8ZS, any string of 8 consecutive "zero" bits is replaced by a special bit pattern that contains two bipolar violations in defined positions. Receiving hardware recognizes this pattern by the bipolar violations and substitutes a string of 8 "zero" bits to restore the original bit stream.

bind In SNA, a request to activate a session between two logical units (LUs).

binding posts A small screw type terminal used to make electrical connections to wires. Usually it is part of a terminal or connecting block.

bipolar coding A method of transmitting a binary stream in which binary 0 is sent as no pulse and binary 1 is sent as a pulse which alternates in sign for each 1 that is sent. The signal is therefore ternary.

bipolar modulation A coding technique for binary signals consisting of three states (zero, positive and negative signal levels). Binary 0 in the input is represented by a 0 in the output, and binary 1 is represented by either of the other two levels. Each successive 1 that occurs is coded with the opposite polarity.

bipolar non-return to zero signaling In bipolar non-return to zero signaling, alternating polarity pulses are used to represent marks while a zero pulse is used to represent a space.

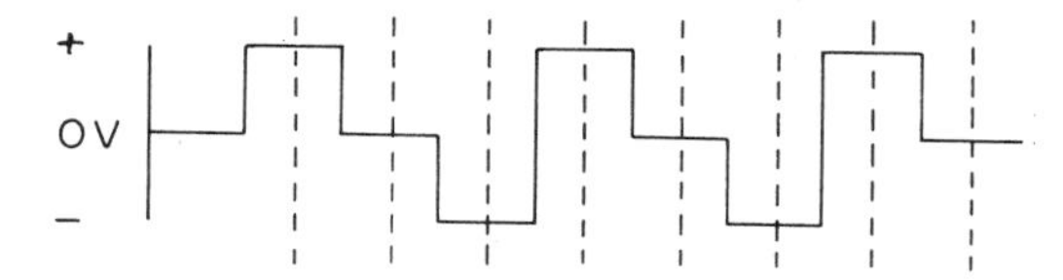

bipolar non-return to zero signaling

bipolar return to zero signaling In bipolar return to zero signaling, alternate polarity pulses are used to represent marks while a zero pulse is used to represent a space. Here the bipolar signal returns to zero after each mark. As this signal ensures that there is no dc voltage buildup on the line, repeaters can be placed relatively far apart in comparison to other signaling techniques and this signaling technique is employed in modified form on digital networks. Since two or more spaces cause no change from a normal to zero voltage, sampling is required to determine the value of bits.

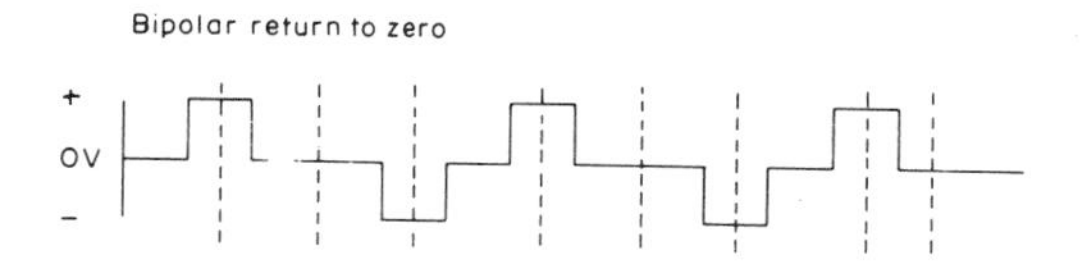

bipolar return to zero signaling

bipolar transistor A transistor whose operation depends on the flow of both negatively charged electrons and positively charged holes.

bipolar transmission Method of sending binary data in which negative and positive states alternate. Used in digital transmission facilities such as DDS and T1. Bipolar is a three-level signal format consisting of high, zero, and low states where the high and low states are of opposite voltage potential (i.e. + and − voltages). Sometimes known as polar transmission.

Bipolar Violation (BPV) The occurrence of consecutive "one" bits with pulses of the same polarity. A bipolar violation is usually regarded as a transmission error, but in some formats, such as B8ZS, specific bipolar violations are used for low-level signaling.

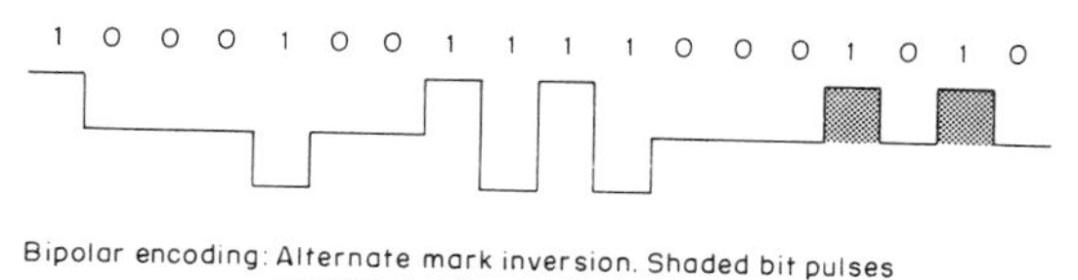

Bipolar Violation (BPV)

bis Appended to a CCITT network interface standard, it identifies a second version of the standard.

bisynchronous control characters

BISYNCHRONOUS CHARACTER	HEX VALUE	CHARACTER DESCRIPTION
SYN	32	Synchronous idle
PAD	55	Start of frame pad
PAD	FF	End of frame pad
DLE	10	Data line escape
ENQ	2D	Enquiry
SOH	01	Start of heading
STX	02	Start of text
ITB	1F	End of intermediate block
ETB	26	End of transmission (block)
ETX	03	End of text

bit The smallest unit of the information that the computer recognizes. A bit (short for binary digit) is represented by the presence or absence of an electronic pulse, 0 or 1. The term was originated at AT&T Bell Laboratories by John Tukey.

bit bucket A slang term used to describe the discarding of binary information by relating it to the practice of throwing paper or other information into a wastebucket.

bit duration 1. The time it takes one encoded bit to pass a point on a transmission medium. 2. In serial communications, a relative unit of time measurement used for comparing delay times due to propagation.

BITCOM A menu-driven PC communications program from BIT Software, Inc., of Milpitas, CA.

bit interleaving A multiplexing technique in which data from the low-speed channels is integrated into the high-speed bit stream one bit at a time (i.e. one bit from channel 1, one bit from channel 2, ..., one bit from channel *n*). Usually used for digital transmission at rates above T1.

bit masking Senses specific binary conditions and ignores others. "Don't care" characters are placed in bit positions of *not* interest, and zeros and ones in positions to be sensed.

bit stream A binary signal without regard to groupings by character. The stream is also used to identify synchronous transmission.

bit stripping Removing the start/stop bits on each ASYNC character and transmitting the data using synchronous techniques. Common with statistical multiplexers.

bit stuffing A process in bit-oriented protocols, where a string of "one" bits is broken by an inserted "zero", to prevent user data containing a series of 6 "one" bits. Also called zero insertion.

Bit-Error-Rate (BER) The ratio of bits received in error to the total bits received. If one error occurs in every 100 000 bits, the bit error rate is said to be 10^{-5}.

Bit-Error-Rate Test (BERT) A data communications measurement in which the total number of received errors are divided by the total number of received data bits. The smaller the resultant number, the higher is the quality of the communications path.

BERT DEFINITIONS

Bits received The total number of bits received while the test set is in sync with the received data.

Bit errors Any received data bit in error while the test set is in sync with the incoming data.

Blocks received The total number of blocks received while the device is in sync with the received data.

Block errors A block error is any block containing one or more bit errors while in sync. The block size is user-definable.

Error free sec Number of seconds during which the receiver is in sync with the data and no bit errors are encountered.

Asynchronous errored sec The test interval is divided into uniform 1-second intervals which are independent of the incoming data. If an error occurs during an interval while in sync, the second is counted as an errored second.

Synchronous errored sec If an error occurs during an interval in which the test set is in sync, with the incoming data, a 1-second interval clock is started at the beginning of the detected transmission error. With this method, the errored second is considered to be synchronized with the incoming data errors.

Time outs Applies only to half duplex circuits. When running a half duplex test, a user defined period of time is entered to prevent the tester from locking up if a reply is not completely received (i.e. the line must be turned around or SYNC may never be re-established).

BCC errors The total number of Block Check Character errors resulting from the receiver checking for transmission errors.

Frame errors Applies to asynchronous protocols. The total number of start and or stop bits received in error.

Parity errors The total number of parity errors resulting during the transmission and reception of data.

In sync with data In BERT applications this will occur when the incoming bit pattern matches the test pattern.

No data condition Applies only to the beginning of the BERT and indicates that the receiver is unable to get into sync with the incoming data.

Sync loss sec The number of seconds that elapse during a sync loss. Any loss of sync will be counted as at least 1 sync loss second.

Elapsed sec Refers to the amount of time that the receiver is enabled. It is comprised of the addition of sync loss secs + ERRORED SECS + ERROR-FREE SECS. For FDX testing this is equivalent to the actual duration of the test after sync has been established. For HDX the receiver is enabled at most 50% of the time, so the TOTAL ELAPSED TIME is less than 50% of the actual duration of the test.

% error sec Refers to the percentage of errored seconds. It is calculated by the following division: (ERRORED SEC/TOTAL TEST DURATION) * 100.

BER The Bit Error Rate is the ratio of the total number of bit errors divided by the total number of bits received. The user can decide whether or not to include the errors that occur during a sync loss.

BLER The Block Error Rate is the ratio of the total number of block errors divided by the total number of blocks received. The block length can be defined by the user.

Sync loss Synchronization loss occurs when the test set, which was previously in sync with the received data, detects eight consecutive characters in error.

BITNET Because It's Time Network.

bit-oriented Pertaining to communications protocols (such as SDLC) in which control information may be coded in fields as small as a single bit in length.

Bit-Oriented Protocol (BOP) The most widely known BOP is IBM's Synchronous Data Link Control (SDLC). In SDLC, information is sent in frames. Within each frame are fields which have specific functions.

The flag field is a unique combination of bits (01111110) which lets the receiving device know that a field in SDLC format is about to follow.

The address field specifies which secondary station is to get the information.

The control field is used by the primary station (which maintains control of the data link at all times) to tell the addressed (or secondary) station what it is to do: to poll devices; transfer data; or retransmit faulty data, among other things. The addressed station can use the control field to respond to the primary station: to tell it which frames it has received or which it has sent. The control field is also used by both the primary and secondary stations to keep track of the sequence number of the frames transmitted and received at any one time. SDLC frames which contain information fields and supervisory fields have sequence numbers.

The information field may be any length and may be comprised of any code structure.

The frame check sequence (FCS) is sixteen bits long and contains a CRC. All data transmitted between the start and stop flags is checked by the FCS.

The flag field also marks the end of a standard SDLC frame.

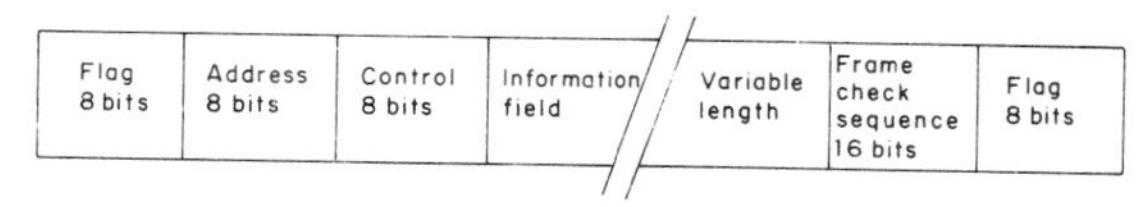

Bit-Oriented Protocol (BOP)

Bit-Oriented Signaling (BOS) A common signaling protocol used in AT&T's Digital Multiplexed Interface (DMI). In Bit-Oriented Signaling, values of bits in the signaling channel represent physical signals on the E and M wires of an analog telephone. BOS is used only as an interim provision on systems migrating between D4 framing (which uses AB signaling) and DMI (which uses Message-Oriented Signaling according to CCITT Common Channel Signaling System No. 7).

bit-rate The speed (or rate) at which the individual bits are transmitted across a digital transmission system or circuit, usually expressed in bits per second.

bit-robbing A technique for carrying "on-hook/off-hook" telephony signaling information over T1 spans. With bit-robbing, the circuit provider (telephone company or PTT) "robs" the least significant bit of each channel octet in the 6th and 12th frames of each superframe (6th, 12th, 18th, and 24th frames in Extended Superframe Format) to convey signaling information. Bits robbed from the 6th (and 18th) frame are called "A" bits, and correspond to signals on the E-wire of the analog telephone. Bits robbed from the 12th (and 24th) frame are called "B" bits, and correspond to signals on the M-wire of an analog telephone. Bit-robbing reduces the channel bandwidth available for digital data from 64 Kbps to 56 Kbps.

bits/s (bps) Bits per second.

bits per second (bps) A measure of the rate of information transfer. Often combined with metric prefixes as in Kbps for thousands of bits per seconds (K for kilo-) and Mbps for millions of bits per second (M for mega-).

BIU Basic Information Unit.

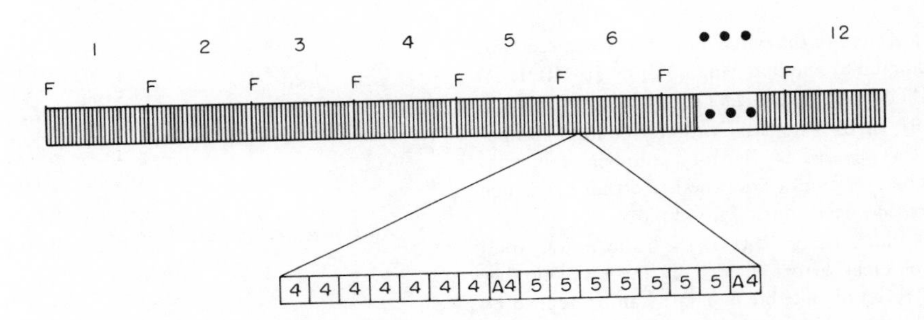

Bit-robbing: A 12-frame superframe-Two octets from Frame 6 have been enlarged to illustrate robbed-bit signalling-the least significant bit of each octet in Frames 6 and 12 is "robbed" to provide signalling: robbed bits in Frame 6 are "A" bits, those in Frame 12 are "B" bits.

bit-robbing

BIU segment In SNA, the portion of a basic information unit (BIU) that is contained within a path information unit (PIU). It consists of either a request/response header (RH) followed by all or a portion of a request/response unit (RU), or only a portion of an RU.ent user data containing a series of 6 "one" bits. Also called zero insertion.

black box A device that performs certain functions on the input(s) before outputting the result(s).

Black Box Corporation The leading data communications and computer device mail order company, publishes and distributes the *Black Box Catalog*.

blank A condition of "no information" in a data recording medium or storage locations. This can be represented by all spaces or all zeros.

BLAST 1. Blocked Asynchronous Transmission. 2. A communications software program designed for use on personal computers and mainframes. Marketed by Communications Research Group.

BLDG/PWR Building/Power.

BLERT Block-Error-Rate Test.

block A group of characters transmitted as a unit, over which a coding procedure is usually applied for synchronization and/or error control purposes.

block chaining Linking blocks of message data in main or secondary storage through pointer arrangements to provide queueing and flexible memory use.

block check That part of the error control procedure used for determing that a data block is structured according to a given set of rules.

Block Check Character (BCC) A character added to the end of a transmission block for the purpose of error detection—such as a CRC or LRC.

BCC	ETX or ETB	Text	STX	Header	SOH	SYN	SYN

Direction of data flow ⟶

IBM Bisync message format

Block Check Character (BCC)

Block Check Sequence (BCS) At least two characters added to the end of a data block for error checking. The BCS is often a cyclic redundancy check (CRC) but can be a checksum result.

block encryption Encryption of a block of data or text as a unit. Successive blocks are transformed independently but dependency may then be introduced by block chaining.

Block Error Rate Test (BLERT) A data communications testing measurement in which the total number of received block errors are divided by the total number of received blocks. The smaller the resultant number, the higher is the quality of the communications path.

block mode A method of data transmission where data is accumulated in a buffer within the terminal. This accumulation block of data is transmitted from the buffer to the network or the terminal screen.

block multiplexer channel In IBM systems, a multiplexer channel that interleaves bytes of data. Also called byte-interleaved channel.

block overhead The bits remaining in a cell after discarding the block payload.

block payload The user information bits within a cell.

blockage A condition which exists when all telephone circuits are busy so that no other telephone call can be carried. Blockage rates—that is, the frequency with which blockage occurs—is one of several standards used to measure performance of local telephone companies and long distance companies.

Blocked Asynchronous Transmission (BLAST) A means of sending asynchronous information synchronously.

blocking 1. The process of grouping data into transmission blocks. 2. In LAN technology, the inability of a PABX to service connection requests, usually because its switching matrix can only handle a limited number of connections simultaneously. Blocking occurs if a call request from a user cannot be handled because there is an insufficient number of paths through the switching matrix. Blocking thus prevents free stations from communicating.

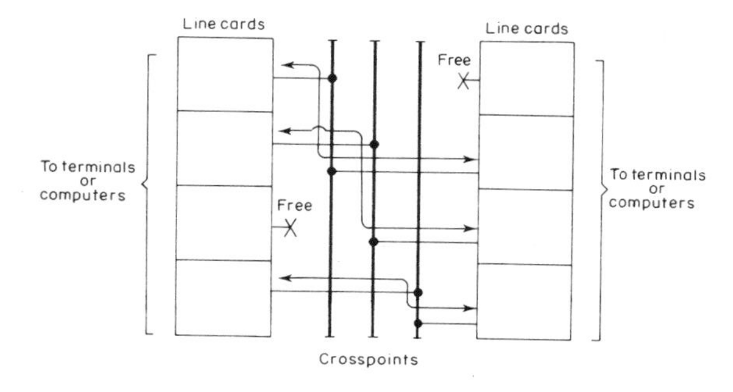

blocking

blocking of PIUs In SNA, an optional function of path control that combines multiple path information units (PIUs) into a single basic transmission unit (BTU). *Note*: When blocking is not done, a BTU consists of one PIU.

block-multiplexer channel In IBM systems, a multiplexer that interleaves bytes of data. Also called byte-interleaved channel.

blue alarm An alarm generated by the receiving device on a DS1 circuit when a Loss of Signal (LOS) condition lasts for more than 150 milliseconds. The receiving device signals a blue alarm by transmitting an unbroken stream of logical "one" bits until the original transmitting device clears the alarm (by restoring normal transmission).

BMS Bandwidth Management Service.

BNA Burroughs Network Architecture.

BNC A bayonet-locking connector for miniature coax. BNC is said to be short for bayonet-Neill-Concelman.

BNN In IBM's SNA, Boundary Network Node.

BOB Breakout Box.

BOC Bell Operating Company.

bond A low-resistance electrical connection between two cable sheaths, two pieces of apparatus, or two ground connections or between similar parts of two electrical circuits.

boot The process of loading a computer's operating system into the random memory of the computer.

bootstrap A technique or device designed to bring itself into a desired state by means of its own action; for example, a program whose first few instructions are sufficient to load an entire program into the computer from an input device.

bootstrap routine A routine contained in a single record which is read into memory by a bootstrap loader located in ROM. The bootstrap routine then reads the operating system into memory.

BOP Bit-Oriented Protocol.

BOS Bit-Oriented Signaling.

boundary function In IBM's SNA: 1. A capability of a subarea node to provide protocol support for adjacent peripheral nodes, such as: (a) transforming network addresses to local addresses, and vice versa; (b) performing session sequence numbering for low-function peripheral nodes; and (c) providing session-level pacing support. 2. The component that provides these capabilities.

Boundary Network Node (BNN) In IBM's TCAM, the programming component that performs FID2 (format identification type 2) conversion, channel data link control, pacing and channel/device error recovery procedures for a locally attached station. These functions are similar to those performed by a network control program for an NCP-attached station.

boundary node (IBM/SNA) A subarea node that

can provide certain protocol support for adjacent subarea nodes.

BPS Bits per second.

BPSK Binary Phase Shift Keying.

BPV Bipolar Violation.

BPV errors Bipolar Pulse Violations which are not part of a valid B8ZS or other code substitution technique.

BPV rate A measurement of accuracy in bipolar transmission, derived by dividing the number of bipolar violations by the duration of the test in seconds.

BPV-free second A second of time in which no Bipolar Violations occur.

BPV-free seconds rate A measure of the quality of a bipolar signal over time, obtained by dividing the number of BPV-free seconds by the duration of the test in seconds.

bracket In IBM's SNA, one or more chains of request units (RUs) and their responses that are exchanged between the two LU–LU half-sessions and that represent transaction between them. A bracket must be completed before another bracket can be started. Examples of brackets are data base inquiries/replies, update transactions, and remote job entry output sequences to work stations.

bracket protocol In IBM's SNA, a data flow control protocol in which exchanges between the two LU–LU half-sessions are achieved through the use of brackets, with one LU designated at session activation as the first speaker and the other as the bidder. The bracket protocol involves bracket initiation and termination rules.

branch A transfer of control from one line of code to one other than the next logical instruction. Branches may be conditional, based on some defined circumstance, or unconditional.

BRCS Business and Residence Customer Service.

breadth of bulletin board access code Defines the organizational bounds within which a Telemail user may access bulletin boards.

breadth of inquiry code Defines the organizational bounds within which information can be made available about a Telemail user.

breadth of posting code Defines the organizational bounds within which a Telemail user may send messages.

breadth of receipt code Defines the organizational bounds from which a Telemail user may receive messages.

break A space (or spacing) condition that exists longer than one character time (typical length is 110 milliseconds). Often used by a receiving terminal to interrupt (break) the sending device's transmission, to request disconnection, or to terminate computer output.

Breakout Box (BOB) (EIA monitor) Digital test equipment that monitors the status of signals on the pins of an RS-232C connector and allows signals to be broken, patched, or cross-connected.

BRI Basic Rate Interface.

bridge The interconnection between two networks using the same communications method, the same kind of transmission medium, and the same addressing structure; also the equipment used in such an interconnection. Bridges function at the data link layer of the OSI model.

bridge clip A clip that electrically interconnects two adjacent terminals for the purpose of providing a multiplying or testing point.

bridge tap Made when a technician bridges across the cable pair to bring it into a customer location. If the service is disconnected, the bridge tap may be left in place. Excessive bridge taps on a cable may be the cause of significant attenuation distortion.

broadband 1. In general, pertaining to communications channel having a bandwidth greater than a voice-grade channel and potentially capable of much higher transmission rates. Also called wideband. 2. In LAN technology, pertaining to a system in which multiple channels access a medium (usually coaxial cable) that has a large bandwidth (50 Mbps is typical) using radio-frequency modems.

Broadband Exchange (BEX) Public switched communication system of Western Union, featuring various bandwidth, full duplex connections.

broadband system A system that can transmit data at more than one frequency, making it possible for many transmissions to take place concurrently.

broadcast 1. Transmission of a message intended for general reception rather than for a specific station. 2. In LAN technology, a transmission method used in bus topology networks that sends all messages to all stations even though the messages are addressed to specific stations.

broadcast medium A transmission system in which all messages are heard by all stations.

broadcast message A message addressed to all users of a computer network.

broadcast subnetwork A multiaccess subnetwork that supports the capability of addressing a group of attached systems with a single message.

Brooklyn Bridge A file transfer utility program developed by White Crane Systems, Inc., of Norcross, GA, for transferring data between personal computers.

BSC Binary Synchronous Communications.

BTAM Basic Telecommunications Access Method.

BTN Billed Telephone Number.

BTU Basic Transmission Unit.

buffer A temporary storage device used to compensate for a difference in either the rate of data flow or the time of occurrence of events in transmissions from one device to another.

buffer group In IBM's VTAM, a group of buffers associated with one or more contiguous, related entries in a buffer list. The buffers may be located in discontiguous areas of storage and may be combined into one or more request units.

buffer list In IBM's VTAM, a contiguous set of control blocks (buffer list entries) that allow an application program to send function management data (FMD) from a number of discontiguous buffers with a single SEND macro instruction.

buffer list entry In IBM's VTAM, a control block within a buffer list that points to a buffer containing function management (FM) data to be sent.

buffered repeater A device which both amplifies and regenerates digital signals to enable them to travel farther along a cable. In a local area network, this type of repeater also controls the flow of messages to prevent collisions.

buffering Process of temporarily storing data in a register or in RAM, allowing transmission devices to accommodate differences in data rates and perform error checking and retransmission of data that has been received in error.

bug An error in a program or a malfunction in a piece of equipment.

building entrance area The area inside a building where cables enter the building and are connected to backbone cables and where electrical protection is provided. The network interface may be located here, as well as the protectors and other distribution components for the campus subsystem.

bulk redundancy A method of coding an anisochronous channel on a synchronous stream of bits in which a 1 state is represented by a string of 1s while it lasts, and an 0 state by a string of 0s. It is a redundant method because the strings must be long ones to reduce telegraph distortion when the anisochronous signal is reconstructed.

Burroughs Network Architecture (BNA) The network architecture developed by Burroughs Corporation for use with its mainframe computers and distributed processing products.

burst A group of events occurring together in time.

burst error A group of bits in which two consecutive erroneous bits are always separated by less than a given number of correct bits. In other words, the erroneous bits are more closely spaced than is statistically expected.

burst isochronous Pertaining to a signal consists of bursts of digits synchronized to a clock, interspersed by "silent" periods when no bits are presented. To indicate the bursts and silence a special clock may be provided which operates only when bits are present. This is called a "stuttering clock".

burst mode The transmission of bulk data in large, continuous blocks, in which the channel is exclusively used for the extent of the transmission.

bursty Not uniformly distributed in time.

bus, buss 1. In general, a data path shared by many devices such as the input/output bus in a computer. 2. In LAN technology, a linear network topology.

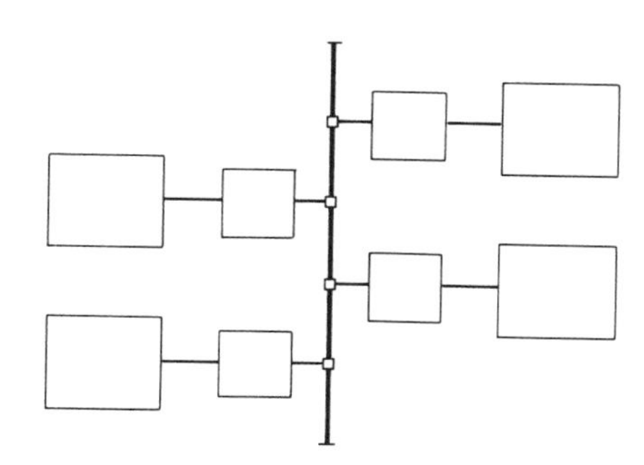

bus, buss

bus hog A method of operating direct memory access (DMA) devices such that the device obtains control of the backplane bus and remains in control until it has completed as many data transfers as it requires at a point in time.

bus network A network in which LAN nodes are all connected to a single length of cable.

bus segment A section of a bus that is electrically continuous, with no intervening components, such as repeaters.

Business and Residence Custom Services (BRCS) An approach that the 5ESS switch employs to provision revenue-generating services. BRCS features can be assigned to virtually any line type (or trunks) in almost any combination, removing the unnecessary restrictions found in other approaches.

busy A signal used to indicate the state of the modem. A modem is in the busy state during call connection or call processing.

busy hour The 60-minute period of a day in which the largest number of transactions occur. Also peak hour or peak period.

BX.25 AT&T's rules for establishing the sequence of events and the specific types and forms of signals required to transfer data between computers. BX.25 includes the international rules known as X.25, and more.

bypass 1. Refers in general to any private networking scheme used to access long distance transmission facilities without being routed through the local exchange carrier. 2. A T1 multiplexing technique similar to drop-and-insert, used when some channel in a DS1 stream must be demultiplexed at an intermediate node. With bypass, only those channels destined for the intermediate node are demultiplexed—ongoing traffic remains in the T1 signal, and new traffic bound from the intermediate node to the final destination takes the place of the dropped traffic. With drop-and-insert, all channels are demultiplexed, those destined for the intermediate node are dropped, traffic from the intermediate node is added, and a new T1 stream is created to carry all channels to the final destination.

bypass channels A term given to channels that are routed through a node without being demultiplexed.

bypass relay A relay in a ring network which permits message traffic to travel between two nodes that are not normally adjacent.

byte A collection of bits operated upon as a unit; most are 8 bits long; and most character sets use one byte per character. The capacity of storage devices is frequently given in bytes or in Kbytes (K meaning 1024).

byte multiplexer channel A mainframe input/output channel that provides multiplexing, of data in bytes.

byte multiplexing A byte (or character) for 1 channel is sent as a unit, and bytes from different channels follow in successive time slots.

Byte-Oriented Protocol (BOP) A protocol technique using defined characters (bytes) from a code set for communications control.

byte stuffing Insertion into a byte stream of some 'dummy' bytes so that the mean data rate is less than the rate of the channel. The qualifying bit, if used, can distinguish the dummy bytes, which then appear as a species of control signal.

B8ZS Binary 8 Zero Suppression.

C

C band A portion of the electromagnetic spectrum used for satellite and microwave transmission. Frequencies of approximately 3.9 to 6.2 GHz.

C conditioning C conditioning applies attenuation equalization and delay equalization to the line so that line quality falls within certain limits.

C-1 conditioning AT&T conditioning for two-point or multipoint channels. The attenuation distortion and envelope delay distortion parameters for C-1 conditioning are:

ATTENUATION DISTORTION

Frequency range (Hz)	*Variation (dB)*
1004–2404	−1 to +3
304–2704	−2 to +6
2704–3004	−3 to +12

ENVELOPE DELAY DISTORTION

Frequency range (Hz)	*Variation (ms)*
1004–2404	1000
804–2604	1750

C-2 conditioning AT&T conditioning for two-point or multipoint channels. The attenuation distortion and envelope delay distortion parameters for C-2 conditioning are:

ATTENUATION DISTORTION

Frequency range (Hz)	*Variation (dB)*
504–2804	−1 to +3
304–3004	−2 to +6

ENVELOPE DELAY DISTORTION

Frequency range (Hz)	*Variation (ms)*
1004–2604	500
604–2604	1500
504–2804	3000

C-3 conditioning AT&T conditioning for switched network access lines and switched network trunks. The attenuation distortion and envelope delay distortion parameters for switched network interoffice access lines are:

ATTENUATION DISTORTION

Frequency range (Hz)	*Variation (dB)*
504–2804	−0.5 to +1.5
304–3004	−0.8 to +3

ENVELOPE DELAY DISTORTION

Frequency range (Hz)	*Variation (ms)*
1004–2604	110
604–2604	300
504–2804	650

The attenuation distortion and envelope delay distortion parameters for switched network trunks are:

ATTENUATION DISTORTION

Frequency range (Hz)	*Variation (dB)*
504–2804	−0.5 to +1
304–3004	−0.8 to +2

ENVELOPE DELAY DISTORTION

Frequency range (Hz)	*Variation (ms)*
1004–2604	80
604–2604	260
504–2804	500

C-4 conditioning AT&T conditioning for two-point or multipoint channels having four or fewer customer's premises. The attenuation distortion and envelope delay distortion parameters for C-4 conditioning are:

ATTENUATION DISTORTION

Frequency range (Hz)	*Variation (dB)*
504–3004	−2 to +3
304–3204	−2 to +6

ENVELOPE DELAY DISTORTION

Frequency range (Hz)	*Variation (ms)*
1004–2604	300
804–2804	500
604–3004	1500
504–3004	3000

C-5 conditioning AT&T conditioning available for two-point channels where both customer premises are located in the mainland. The attenuation distortion and envelope delay distortion parameters for C-5 conditioning are:

ATTENUATION DISTORTION

Frequency range (Hz)	*Variation (dB)*
504–2804	−0.5 to +1.5
304–3004	−1 to +3

ENVELOPE DELAY DISTORTION

Frequency range (Hz)	*Variation (ms)*
1004–2604	100
604–2604	300
504–2804	600

C-7 conditioning AT&T conditioning available for two-point channels where both customer's premises are located in the mainland. The attenuation distortion and envelope delay distortion parameters for C-7 conditioning are:

ATTENUATION DISTORTION

Frequency range (Hz)	*Variation (dB)*
404–2804	−1 to +4.5

ENVELOPE DELAY DISTORTION

Frequency range (Hz)	*Variation (ms)*
1004–2604	550

C-8 conditioning AT&T conditioning available for two-point channels where both customer's premises are located in the mainland. The attenuation distortion and envelope delay distortion parameters for C-7 conditioning are:

ATTENUATION DISTORTION

Frequency range (Hz)	*Variation (dB)*
404–2804	−1 to +3.0

ENVELOPE DELAY DISTORTION

Frequency range (Hz)	*Variation (ms)*
1004–2604	125

C connector A bayonet-locking connector for coax. C is named after Carl Concelman.

cabinet A physical enclosure for rack-mount equipment. Standard cabinets have 1 and 3/4-inch vertical spacing between mounting holes and 19-inch wide horizontal spacing between mounting rails.

cable A group of conductive elements (wires) or other media (fiber optics), packaged as a single line to interconnect communications systems.

cable loss The amount of radio frequency (RF) signal attenuated by coaxial cable transmission. The cable attenuation is a function of frequency, media type, and cable distance.

cable system, cabling system In LAN technology the medium used to interconnect stations. Often called the premises network.

cable-based LAN A shared medium LAN that uses a cable for its transmission medium.

cache memory A high-speed computer memory which contains the next most likely instruction or sequence of instructions to be executed.

CAD Computer Aided Design.

CAE Computer Aided Engineering.

CAI Computer Aided Instruction.

CALC Customer Access Line Charge.

call A request for connection or the connection resulting from such a request.

call accounting, call accounting record In packet switched networks, the process of accumulating data on individual calls or of reporting such data. Usually includes start and end times, NTN or NUI, and number of data segments and packets transmitted for each individual call.

call add-on A PABX or Central Office function that permits adding another party to a conversation already under way.

call capacity The ability of a telephone system to provide a specific grade of service.

call clearing The orderly termination of a user's call on a network.

call detail recording The capability of a telephone branch to change to log calls for accounting purposes.

call establishment The routing of a user's call to its destination and the set-up of data transfer.

call forwarding A PABX or Central Office feature that permits a telephone subscriber to reroute incoming calls intended for the subscriber to a different telephone number, either all the time or only when the subscriber's number is busy or does not answer.

call forwarding when busy Similar to call forwarding, this programming mode routes calls to another designated extension when the primary extension is busy.

call gapping A control which regulates the maximum rate at which calls are released towards a destination code. Call gapping control performs the function of limiting access attempts during mass-call situations to specific destination codes.

call hold A PABX or Central Office feature that permits a telephone subscriber to place an existing call on "hold", e.g., in a waiting state, while handling an incoming call or otherwise attending to some other matter.

call pickup A PABX or Central Office feature that permits a subscriber to answer the telephone of another subscriber on his/her own instrument, by "picking up" the incoming call. Normally, several phone extensions can be programmed as a group to allow subscribers to answer any calls coming in on each telephone in the group.

call reference A unique value used locally to identify a Modem Connect or X.21 call.

call request A data packet that requests the establishment of a circuit between two network resources.

call request packet In packet switched networks, the packet sent by the originating DTE showing requested NTN or NUI, network facilities, and call user data.

call second A basic unit of measurement in telephone operations (actually, 100 call-seconds equals one CCS, the equivalent of one call held for 100 seconds).

call setup time The time required to set up communications between equipment on a switched network or line.

call sharing A form of switched line sharing in which many clients have access to the same call on that line.

call termination See call clearing.

call trace A PABX feature which can be used to identify a trunk or trunk group number on incoming calls and the called number and station on calls going out of the system.

call transfer A PABX or Central Office feature that permits a telephone subscriber or operator to redirect a call to another number.

call user data In packet switched networks, user information transmitted in a call request packet to the destination DTE.

call waiting A signal generated by a PABX or Central Office which notifies a subscriber currently engaged in an existing call that another call has arrived.

callback The capability for a caller to program a phone to retry another busy extension as soon as it is free.

callback modem A modem which functions as a limited security device, preprogrammed with identification codes, passwords, and telephone numbers. When a caller accesses the callback number and enters his or her identification code and password correctly, the modem will then hang up the connection and use the stored telephone number to dial the caller. Thus, security is effected by the telephone numbers callers are assigned to use. Also called a security modem.

call-by-call service selection An AT&T ISDN feature which allows T1 users to change the services to which the 23 B channels are dedicated, such as Megacom Service, Megacom 800 Service or Accunet Switched Digital Service.

Callcenter A trademark of Aspect Telecommunications of San Jose, CA, as well as a stand-alone, automatic call distributor from that company that can service up to 7000 calls per hour.

called, calling, or called/calling channel In LAN technology and packet switched networks, a called channel is a channel that can receive but not originate calls; a calling channel can originate but not receive calls; and finally, a called/calling channel can both originate and receive calls.

caller identification feature Customers and users, as specified by the customer, who access a network through public dial-in access ports may be identified to a packet network through the use of the caller identification feature. Each such user, upon verifying his identity by means of valid identification code and password, may establish virtual connections through the network with all associated public port and traffic charges billed to the responsible customer.

calling rate The number of calls per telephone. Determined by dividing the count of busy-hour calls by the number of telephones.

CAM Computer Aided Manufacturing.

CAMA Centralized Automatic Message Accounting.

Cambridge Monitor System (CMS) An IBM operating system for mainframe computers named after the University of Cambridge where a lot of the original development work was conducted.

Cambridge Ring In LAN technology, an empty slot ring LAN. It has not yet achieved a great deal of popularity outside its country of origin, England—where several near-Cambridge Ring systems are being marketed.

camp-on, camp-on-busy A PABX or cable-based LAN facility that allows users to wait on line (in queue) if the requested resource is busy and that connects the users in queue—on a first-come, first-served basis—when the requested resource becomes available.

campus In local area networks, a group of buildings or similar installations served by the same network.

campus subsystem The part of a premises distribution system which connects buildings together. The cable, interbuilding distribution facilities, protectors, and connectors that enable communication among multiple buildings on a premises.

CAN Channel Character.

cancel (CAN) A character indicating that the data preceding it is in error and should be ignored.

cancel closedown In IBM's VTAM, a closedown in which VTAM is abnormally terminated either because of an unexpected situation or as the result of an operator command.

capacitor An electronic circuit component that temporarily stores electric charges. The capacitor's ability to block direct current flow and pass alternating current flow makes the capacitor useful for power supply filters.

capacity An abbreviation of "channel capacity" which is the rate at which data are carried on the channel, measured in bit/s.

carbon block Surge limiting device that will be grounded by arcing across the air gap when the voltage of a conductor exceeds a predetermined level. If the current flow across the gap is large or persists for an appreciable time, the protector mechanism will operate and the protector will become permanently grounded.

Carbon Copy Plus A software program from Meridian Technology which enables personal computers and microcomputer local networks to dial in to remote networks to access and share resources as well as to communicate with other non-networked PCs and mainframes.

card module A printed-circuit board that plugs into an equipment chassis.

CAROT Centralized Automatic Reporting on Trunks.

carriage return (CR) An ASCII or EBCDIC control character used to position the print mechanism at the left margin on a printer—or the cursor at the left margin on a display terminal.

carrier A continuous signal which is modulated with a second, information-carrying signal.

carrier, communications common A company which furnishes communications services to the general public, and which is regulated by appropriate local, state, or federal agencies.

carrier band A band of continuous frequencies that can be modulated with a signal.

carrier detect (CD) An RS-232 control signal (on pin 8) which indicates that the local modem is receiving a signal from the remote modem. Also called Received Line Signal Detector (RLSD) and Data Carrier Detect (DCD).

carrier failure alarm A red or yellow alarm on a T1 circuit, or the end-to-end combination of a red and a yellow alarm.

carrier frequency The frequency of the carrier wave which is modulated to transmit signals.

carrier heterodyne Interference between conventional amplitude modulated (AM) signals caused when the carriers of two adjacent AM signals heterodyne at radio frequency, resulting in an audio beat after modulation falling within the bandpass of the radio receiver output.

carrier modulation A signal at some fixed amplitude and frequency which is combined with an information-bearing signal in the modulation process to produce an output signal suitable for transmission.

carrier selection As a result of the 1982 Federal court Modified Final Judgement, most Local Telephone Companies must offer residence and business customers the opportunity to select which long distance company they wish to use on a primary basis. The company selected is the one which will be used when dialing 1 + area code + telephone number. This opportunity is being phased in over time, as local telephone companies modify their Central Office equipment to accommodate access to more than one long distance company.

Carrier Sense Multiple Access (CSMA) In LAN technology, a contended access method in which stations listen before transmission, send a packet, and then free the line for other stations. With CSMA, although stations do not transmit until the medium is clear, collisions still occur. Two alternative versions (CSMA/CA and CSMA/CD) attempt to reduce both the number of collisions and the severity of their impact.

Carrier Sense Multiple Access with Collision Avoidance (CSMA/CA) In LAN technology, CSMA that combines slotted TDM (to avoid having collisions occur a second time) with CSMA/CD. CSMA/CA works best if the time slot is short compared to packet length and if the number of stations is small.

Carrier Sense Multiple Access with Collision Detection (CSMA/CD) A local area network technique, where all devices attached to a LAN listen for transmissions before attempting to transmit and, if two or more begin transmitting at the same time, each backs off for a random period of time (determined by a complex algorithm) before attempting to retransmit.

carrier signaling Any signaling technique used in multichannel carrier transmission.

carrier system A system used to maximize the utilization of cable pairs used in the telephone network. Rather than having each call dedicated in individual cables throughout the nation, through the use of electronics, carrier systems allow as many as 24 individual voice/data customers to utilize the same cable pair.

1. Analog carrier systems are an older technology and allows either 12 or 24 customers to use two cable pairs. The channelization of these systems is accomplished in the frequency domain.

2. Digital carrier systems are the latest technology and allow 24 to 96 customers on two cable pairs. The channelization of these systems is accomplished in the time domain. The following figure describes the process of the digital carrier system.

carrier telegraphy, carrier current telegraphy A method of transmission in which the signals from a telegraph transmitter modulate an alternating current.

carrier wave Wave upon which a signal is superimposed.

Carterphone decision A 1968 FCC decision which held that telephone company tariffs containing blanket prohibition against the attachment of customer provided terminal equipment to the telecommunications network were unreasonable, discriminatory, and unlawful. The FCC declared the telephone companies could set up reasonable standards for interconnection to insure the technical integrity of the network. Following Carterphone, the telephone companies filed tariffs for protective connecting arrangements to facilitate the interconnection of customer provided terminal equipment.

cascaded network A network divided into a number of linearly connected, contiguous segments.

cassette tape A slow-speed, low-capacity method of storing and retrieving data which uses a technology similar to audio cassettes.

cataloged data set A data set that is represented in an index, or hierarchy of indexes, in the system catalog; the indexes provide the means for locating the data set.

cataloged procedure A set of job control statements that has been placed in a library and that can be retrieved by name.

catanet A collection of networks that are interconnected through gateways.

Cathode Ray Tube (CRT) The video imaging (picture) tube used in a television or video display data terminal.

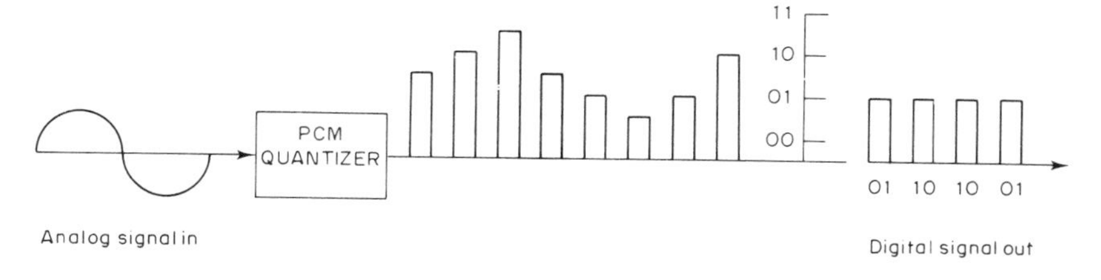

carrier system, digital

CATV Community Antenna Television.

CBEMA Computer and Business Equipment Manufacturers Association.

CBT Computer-Based Terminal.

CBX Computer Branch Exchange.

CCA Circuit Card Assembly.

CCB Central Control Box.

CCC Clear Channel Capability.

CCIA Computer and Communications Industry Association.

CCIR International Consultative Committee for Radio.

CCIS Common Channel Interoffice Signaling.

CCITT Consultative Committee for International Telegraph and Telephone.

CCITT #7 Signaling The newest standard for signaling within telecommunications networks being developed by CCITT. It will eventually replace the CCIS6 network in the U.S.

CCP In IBM's VTAM, Configuration Control Program.

CCR Customer Controlled Reconfiguration.

CCS Hundred-call seconds.

CCSA Common Control Switching Arrangement.

CCSS Common Channel Signaling System.

CCT Coupler Cut Through.

CCTV Closed Circuit Television.

CCU 1. Cluster Control Unit. 2. Communications Control Unit.

CD 1. Carrier Detect. 2. Collision Detection.

CDB Circuit Descriptor Block.

CDCCP Control Data Communications Control Unit Procedure (Control Data Corp.).

CDFP Centrex Facility Data Pooling.

CDI Control and Data Interface.

CDMA Call Division Multiple Access.

CDR Call Detail Recording.

CDRM In IBM's VTAM, Cross-Domain Resource Manager.

CDRSC In IBM's VTAM, Cross-Domain Resource.

CEB In IBM's VTAM, Conditional End Bracket.

CEKS Centrex Electronic Key Set.

cell A subdivision of a mobile telephone service area, containing a low-powered radio communications system connected to the local telephone service. A block of fixed length.

cellular mobile radio A system providing exchange telephone service to a station located in a mobile vehicle, using radio circuits to a base radio station which covers a specific geographical area. As the vehicle moves from one area to another different base radio stations handle the call.

cellular radio communications A mobile radio system that consists of hexagonal geographic areas with groups of frequencies allocated to each cell. Seven cells make up a block, and no adjacent cell uses the same set of frequencies. The frequency allocation pattern, however, is the same in each successive block.

cellular telephone technology A relatively new technology that vastly improves mobile telephone communications. Cellular towers receive and transmit telephone signals, and switch them to or from the nearest Central Office for further transmission. Cellular networks are being constructed in most major metropolitan areas; customers in most markets have or will have a choice of two cellular telephone companies.

Central Office The building where common carriers terminate customer circuits and where the switching equipment that interconnects those circuits is located. Sometimes also known as the central exchange—or just simply as exchange.

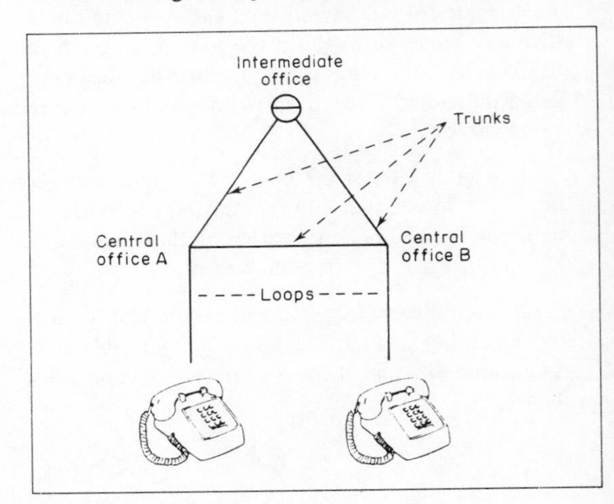

Central Office

Central Office Exchange Service (CENTREX) A service provided by telephone companies in which a portion of its central office is used to switch calls to and from individual stations at different user sites.

Functionally similar to an on-site PBX, typically providing direct inward dialing, direct distance dialing, and attendant switching.

Central Office Local Area Network (COLAN) A Centrex-like service marketed by several Bell operating companies (BOCs) in which LAN capabilities are provided to customers within a common local telephone serving area through the use of the BOC's Central Office switch.

Central Office switching equipment Electromechanical, or electronic, equipment that routes a call to its destination.

Central Office Terminal (COT) The terminating equipment for a digital line closest to the central office.

Central Processing Unit (CPU) The unit that executes programmed instructions, performs the logical and arithmetic functions on data and controls input/output functions. Often used as a synonym for a computer.

Central Processing Unit Master Switching Subsystem (CPMS) The series of software codes, as developed at Telenet, that provides the operating environment of a particular packet switching network (PSN). Software residing in the master central processing unit (CPU) that performs all packet level processing system control and resource management functions. CPMS40 denotes the static-environment subsystem. Dynamic series include CPMS60 and above. CPMS is composed of three software components: PROTO, SWITCH, and the Telenet Processor Operating System (TPOS).

centralized Pertaining to processing with one central processor, which may support remote terminals and/or remote job entry stations.

centralized adaptive routing A method of routing in which a network routing center dictates routing decisions, based on information supplied to it by each node.

Centralized Automatic Message Accounting (CAMA) A feature on an AT&T 5ESS switch which permits the switch office to collect and store toll information and message unit billing on calls originated by local offices served by the CAMA office.

Centralized Automatic Reporting On Trunks (CAROT) An AT&T computerized system that automatically accesses and tests trunks for a maximum of 14 offices simultaneously.

centralized management A form of network management where management is performed from a single point in the network.

centralized processing A data processing configuration where the processing for various locations is centralized in a single computer site (compare with distributed processing).

Centralized Trunk Test Unit (CTTU) An AT&T operational support system providing centralized trunk maintenance through a data link on a switch.

CENTREX Central Office Exchange Service.

Centronics Printer manufacturer that set the *de facto* interconnection standards for parallel printers, using a 36-pin, byte-wide connector.

CEPT Conference of European Postal and Telecommunications Administrations.

CERT Character Error Rate Testing.

chad The material removed when forming a hole or notch in a storage medium such as punched tape or punched cards.

chadless tape Perforated tape with the chad partially attached, to facilitate interpretive printing on the tape.

chain A series of processing locations in which information must pass through each location on a store and forward basis to get to a destination.

chaining I/O commands The linking together (in a chain) of the commands which initiate input/output operations. When one command is finished the next one in the chain begins operation.

Chameleon 32 A protocol analyzer with a T1 analysis option, produced by Tekelec, Inc., of Calabasas, CA.

change-direction protocol In SNA, a data flow control protocol in which the sending logical unit (LU) stops sending normal-flow requests, signals this fact to the receiving LU using the change-direction indicator (in the request header of the last request of the last chain), and prepares to receive requests.

channel 1. (CCITT standard): a means of one-way transmission. Compare with circuit. 2. (Tariff and common usage): as used in tariffs, a path for electrical transmission between two or more points without common carrier-provided terminal equipment, such as a local connection to DTE. Also called circuit, line, data link, path, or facility. 3. In an IBM host system, a high-speed data link connecting the CPU and its peripheral devices. 4. (T1 usage): a North American T1 circuit carries 24 channels; a European T1 circuit carries 31 channels; an ISDN Basic Rate circuit carries 3 channels. The designation "channel" does not indicate a specific bandwidth or bit rate (e.g. the ISDN Basic Rate has two B-channels at 64 Kbps and one D-channel at 16 Kbps.

channel, analog A channel on which the information transmitted can take any value between the limits defined by the channel. Typically, a channel that carries alternating current (AC) but not direct current (DC).

channel, digital A channel that carries pulsed, direct-current signals.

channel, four-wire A communications channel designed for full duplex operation. Four wires are provided at each termination—two for sending data and two for receiving data.

channel, grade A channel's relative bandwidth (i.e., narrowband, voice-grade, wideband).

channel, primary The higher speed of two channels used for transmitting the data. Also called forward or main.

channel, reverse The slower speed of two channels used for slow-speed data such as error-detection.

channel, two-wire A communications circuit designed for simplex or half-duplex operation. Two wires are provided at each termination point, and information is transmitted in only one direction at a time.

channel, voice-grade A channel suitable for transmission of speech, digital, or analog data, or facsimile, generally with a frequency range of about 300 to 400 Hertz.

channel adapter In IBM's VTAM, a communication controller hardware unit used to attach the controller to a System/360 or a System/370 channel.

channel associated signaling Call control signaling transmitted within the bandwidth of the call it controls. Also called in-band signaling. In T1 transmission, channel associated signaling is performed by bit-robbing.

channel-attached Pertaining to devices attached directly to the input/output channels of a mainframe computer—devices attached by cables, not by communications (locally attached IBM).

channel-attached communication controller In IBM's VTAM, an IBM communication controller that is attached to a host processor by means of a data channel.

channel-attached cross-domain NCP In IBM's VTAM, an NCP that is channel attached to a data host, but resides in the domain of another host. It has been contacted over the channel by the host, but it has not been activated. (That is, no SSCP-to-PU session exists for it.)

channel attachment major node In VTAM: 1. For MVS and VSE operating systems, a major node whose minor node is an NCP that is channel-attached to a data host. 2. A major node that may include minor nodes that are the line groups and lines that represent a channel attachment to an adjacent (channel-attached) host. 3. For the VSE operating system, a major node that may include minor nodes that are resources (host processors, NCPs, line groups, lines, SNA physical units and logical units, cluster controllers, and terminals) attached through a communication adapter.

channel bank A device for multiplexing and demultiplexing T1 circuits at a telephone company's Central Office. The channel bank also transmits and detects signaling and framing information.

channel bypass A T1 multiplexing technique similar to drop-and-insert, used when some channels in a DS1 stream must be demultiplexed at an intermediate node. Also called "passthrough." With bypass, only those channels destined for the intermediate node are demultiplexed—ongoing traffic remains in the T1 signal, and new traffic bound from the intermediate node to the final destination takes the place of the dropped traffic. With drop-and-insert, all channels are demultiplexed, those

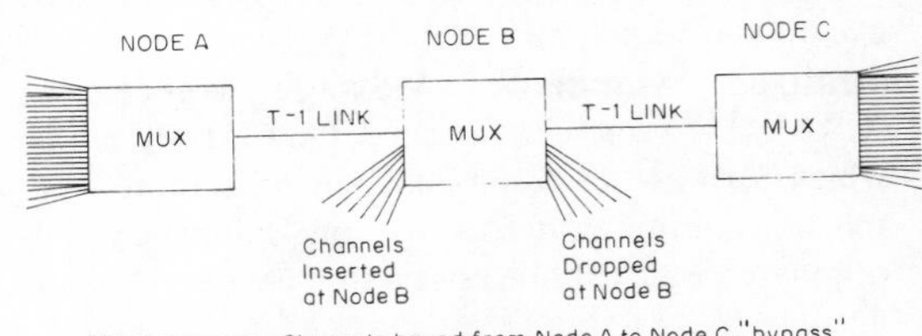

Channel bypass - Channels bound from Node A to Node C "bypass" the demultiplexing process at Node B

channel bypass

destined for the intermediate node are dropped, traffic from the intermediate node is added, and a new T1 stream is created to carry all channels to the final destination.

channel capacity The highest possible information-transmission rate through a channel at a specified error rate. Also the total individual channels in a system.

channel coding with redundant multilevel signals The original name used to describe Trellis Coded Modulation which produced a pattern of mapping that occurs within a finite state machine.

channel demultiplexing The extraction of one or more channels from a multiplexed bit stream.

channel group A grouping of 12 standard telephone circuits in a single carrier system, with each circuit frequency-multiplexed on the channel group, occupying adjacent bands in the frequency spectrum.

channel interface See channel and interface.

channel loopback A diagnostic test that forms the loop at the multiplexer's channel interface (refer to loopback).

Channel Service Unit (CSU) A device on a digital circuit that manages such physical characteristics of the signal as ones density, clocking, and bipolar signal format. On a T1 circuit, the CSU is installed between the customer's multiplexer and the circuit provider's network; according to the current reading of the rules governing customer/network boundaries, the CSU is Customer Premises Equipment. The CSU may be integrated into a DSU (digital service unit).

channel terminal That portion of the multiplexing equipment required to derive a subscriber channel from the carrier facility.

character A sequence of adjacent bits representing a unit of language.

1. Control character—a character used to convey status of functions, remote diagnostic information, or non print character information.

2. Data character—a character containing alphanumeric information mainly used in a communications link between terminals and/or computers.

3. Sync character—a character used to establish and maintain synchronization between computers, multiplexers, and/or synchronous terminals.

character code One of several standard sets of binary representations for the alphabet, numbers, common symbols, and control functions. Common character codes include ASCII, BAUDOT, BCD, and EBCDIC.

Character Error Rate Testing (CERT) Testing a data line with test characters to determine error performance.

character interleave A TDM technique in which a channel slot contains a full character of data.

character parity The addition of a redundant overhead bit to a character to provide error detection capability.

character set A collection of characters, such as ASCII or EBCDIC, used to represent data in a system. Usually includes special symbols and control functions. Often synonymous with code.

character synchronization A process used in a receiver to determine which bits, sent over a data link, should be grouped together into characters.

character-coded In IBM's VTAM, pertaining to commands (such as LOGON or LOGOFF) entered by an end user and sent by a logical unit in character form. The character-coded command must be in the syntax defined in the user's unformatted system services definition table. Synonym for unformatted.

characteristic distortion Distortion caused by transients which, as a result of modulation, are present in the transmission channel. The magnitude of distortion depends on the channel's transmission qualities.

characteristic impedance The impedance termination of an approximately electrically uniform transmission line which minimizes reflections from the end of the line.

character-oriented Pertaining to communications protocols, such as BSC, in which control information is coded in character-length fields. Contrast with bit-oriented.

Characters Per Second (CPS) A measure of data rate.

character-time The basic unit for coding time delays for flyback buffering delays. One character-time equals the reciprocal of the channel data rate in bits per second times the number of bits in the character.

chassis The main structure of a computer cabinet that supports the electronic components and holds the circuit boards.

Cheapernet Colloquial name for a thinner Ethernet

coaxial cable-based LAN.

check bit An additional bit, such as a parity bit, that is added to a message. This check bit will be used in later error-detection procedures.

check character A character transmitted for message validity purposes. If an error occurs, the check character will cause a check or compare procedure to fail in the receiver and an error will be reported.

check notice A note displayed to a user at sign-on asking the user to check a monitored bulletin board or other type of subsystem.

checkpoint 1. A place in a routine where a check, or a recording of data for restart purposes, is performed. 2. A point at which information about the status of a job and the system can be recorded so that the job step can be restarted later.

checkpointing Preserving processing information during a program's operation that allows such processing to be restarted and duplicated.

checksum A block check character or sequence that is computed using binary addition to determine the validity of data that has been received. The receiving device computed the checksum based on the data stream and compares it to the received checksum to determine validity.

child directory entry In Digital Equipment Corporation Network Architecture (DECnet), an entry in the Naming Service which points to a child directory of some directory in the namespace.

chip A flat, fingernail-size piece of material, usually silicon, in which tiny amounts of other elements with desired electronic properties are deposited and etched to form integrated circuits. These tiny chips are the key to microelectronic revolution in computers. Also called microchips. usually silicon, in which tiny amounts of other elements with desired electronic properties are deposited and etched to form integrated circuits. These tiny chips are the key to microelectronic revolution in computers. Also called microchips.

CI Control Interface.

CICS Customer Information Control System.

CICS/VS Customer Information Control System/Virtual Storage.

CID In IBM's SNA, Communication IDentifier.

CIM Computer Integrated Manufacturing.

CINIT In IBM's SNA, a network services request sent from an SSCP to an LU requesting that LU to establish a session with another LU and to act as the primary end of the session.

cipher A form of cryptography depending only on the sequence of characters or bits. On the other hand, in a cryptographic code, meaningful words or phrases are the units that are encoded.

ciphertext In IBM's SNA, synonym for enciphered data.

circuit 1. In data communications, a means of two-way communications between two points consisting of transmit and receive channels. 2. In electronic design, one or more components that act together to perform one or more functions.

circuit, four-wire A communications path in which four wires (two for each direction of transmission) are connected to the station equipment.

circuit, multipoint A circuit connecting three or more points.

circuit, two wire A transmission in which two wires are used to implement a half-duplex or simplex channel.

circuit arrangements 1. Two-wire circuits (2W) use a single cable pair for both transmit and receive.

2. Four-wire circuits (4W) use two individual cable pairs, one for transmit and one for receive.

Two- and four-wire circuits are used primarily for switched voice and data communications. In this configuration the circuit would start out as two-wire, change to a four-wire segment, and then change back to a two-wire section. Telephone message network circuits are configured in this way. A hybrid coil is at the two-wire/four-wire junction and is used to make the configuration and balance the circuit so that no energy is reflected back to the transmitter.

circuit board Plastic base on which microchips and other components are mounted to make up circuitry—for example, for computers and switching systems.

Circuit Descriptor Block (CDB) A temporary list that summarizes an individual virtual call within the Telenet Processor (TP). This list includes the Logical Channel Numbers (LCNs), the Circuit Line Block (CLB) address, the X.25 Line Block (XLB) address, the number of the next frame expected, and timing information.

Circuit Line Block (CLB) A table describing a non-X.25 port and line on the Telenet processor (TP).

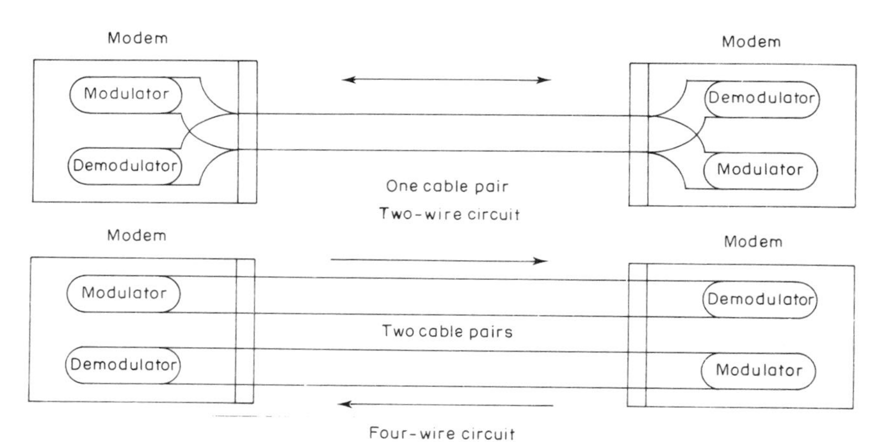

circuit arrangements

This table includes the X.25 address of the port, the Line Processing Unit (LPU) line number, and the status of the line. The CDB address is entered in the Circuit Line Block (CLB) for the duration of the call.

Circuit Quality Monitoring System (CQMS) A method for monitoring critical parameters of data communications lines and circuits.

circuit segment A single point-to-point circuit, usually one part of a multipoint circuit or network. Also called a link.

Circuit Switched Digital Capability (CSDC) A technique for making end-to-end digital connections. It lets customers place calls normally, then use the same connection to transmit high-speed data.

circuit switched network A type of communications network that establishes a physical connection—a circuit—between sender and receiver for the duration of a call. Telephone systems are circuit switched networks.

circuit switching A technique in which physical circuits, as opposed to virtual circuits, are transferred (switched) to complete connections. Sometimes called line switching.

circular buffer A form of queue in which items are placed in successive locations in a store, and are taken from these locations in the same sequence. Two pointers keep track of the head and tail of the queue. When a pointer reaches the end of the available store it returns to the head. The items in the circular buffer may themselves be pointers to the items in the queue.

City Direct A switched voice communications service offered by British Telecom International for companies that want to communicate between London and New York.

City-wide Centrex A Centrex feature which uses software to provide a multilocation customer all the appearance and conveniences of a single-location customer.

cladding The internal layer within a fiber optic cable, outside the conducting core and inside the opaque sheath, which assists in guiding light waves down the length of the conductor.

clamping voltages The "sustained" voltage held by a clamp circuit at some desired level.

CLASS Custom Local Area Signaling Services.

Class Of Office A classification scheme used to describe the relative position a telephone switching office occupies in the overall switching hierarchy. From the bottom of the hierarchy, they include Class 5 (end office); Class 4 (toll center); Class 3 (primary center); Class 2 (sectional center); and Class 1 (regional center).

Class Of Service (COS) In SNA, a designation of the path control network characteristics, such as path security, transmission priority, and bandwidth, that apply to a particular session. The end user designates class of service at session initiation by using a symbolic name that is mapped into a list of virtual routes, any one of which can be selected for the session to provide the requested level of service.

Class X Office Designation of a telephone switching facility in the overall telephone switching office hierarchy.

CLB Circuit Line Block.

CLEANUP In IBM's SNA, a network services request, sent by an SSCP to an LU, that causes a particular LU–LU session with that LU to be ended immediately without requiring the participation of either the other LU or its SSCP.

clear channel A communications path with full bandwidth available to the user. Control and other signals are typically transmitted on a separate channel.

clear data Data that is not enciphered. Synonymous with plaintext.

clear session A session in which only clear data is transmitted or received.

clearinghouse In Digital Equipment Corporation Network Architecture (DECnet), a collection of directory replicas stored together in one location.

Clear-To-Send (CTS) An RS-232 modem interface control signal (sent from the modem to the DTE on pin 5) which indicates that the attached DTE may begin transmitting; issued in response to the DTE's RTS. Called Ready-For-Sending in CCITT V.24.

Clear-To-Send (CTS) Delay The time required by a data set to inform a terminal device that it is ready to send or reply to information just received. Also called Modem Turnaround Delay.

CLI Change Level Indicator.

client In Digital Equipment Corporation Network Architecture (DECnet), the user of the service provided by a module or layer in the architecture.

CLIST In IBM's SNA, Command List.

CLNS ConnectionLess-mode Network Service.

clock 1. The timing signal used in synchronous transmission. 2. The source of such timing signals.

clock, external A clock or timing signal from another device. A modem can provide an external clock (clocking).

clock interrupt A type of interrupt which occurs at regular intervals and is used to initiate processes such as polling, which must happen regularly.

clock recovery The extraction of the clock signal which may accompany data. This recovery occurs only when the signal is received on a synchronous channel.

clocking Time synchronizing of communications information.

closedown In IBM's SNA, the deactivation of a device, program, or system.

closed architecture An architecture that is compatible only with hardware and software from a single vendor.

Closed User Group (CUG) In public data networks, a selected collection of terminal users that do not accept calls from sources not in their group. They are also often restricted from sending messages outside the group.

Closed-Circuit Television (CCTV) In LAN technology, one of the many services often found on broadband networks.

close-up A remote connectivity program marketed by Norton Lambert of Santa Barbara, CA. The program is remarketed by several vendors to provide off-site technicians with the ability to take control of a remote personal computer that is normally connected to a local area network or mainframe computer.

cluster A collection of two or more terminals or other devices in a single location.

Cluster Control Unit (CCU) In IBM 3270 systems, a device that controls the input/output operations of a group (cluster) of display stations. Also called terminal control unit.

cluster controller In IBM's SNA, a device that can control the input/output operations of more than one device connected to it. A cluster controller may be controlled by a program stored and executed in the unit, for example, the IBM 3601 Finance Communication Controller. Or it may be controlled entirely by hardware, for example, the IBM 3272 Control Unit.

CMA 1. Communications Managers Association. 2. Cellular Mobile Carrier.

CMC In IBM's SNA, Communication Management Configuration.

C-message noise Filtered measurement of noise similar to the response of the human ear to various frequencies over a telephone circuit.

CMIP Common Management Information Protocol.

CMOS Complementary Metal Oxide Semiconductor.

CMS Conversational Monitor System.

CNM Communications Network Management.

C-notched noise Measurement of metallic noise when a 1000 Hz tone is present through active components on a circuit. The tone is filtered or notched out during measurement.

C-Notched Noise Test The C-Notched Noise Test is used to determine the unwanted power present when a channel is carrying a normal signal. Thus, it actually measures the signal/noise ratio, which is most important with respect to data transmission. This test is a true measure of noise in circuits which have compandors.

CO Central Office.

co- and contra-directional interface The CCITT G.703 recommendation that specifies the relationship between the direction of the information and its associated timing signals. Co-directional specifies that the information and its timing are transmitted in the same direction while contra-directional specifies that the information and its timing are transmitted in opposite directions from one another.

COAX Coaxial cable.

coaxial cable A transmission medium noted for its wide bandwidth and for its low susceptibility to interference. Signals are transmitted inside a fully enclosed environment—an outer conductor or screen which surrounds the inner conductor. The conductors are commonly separated by a solid insulating material.

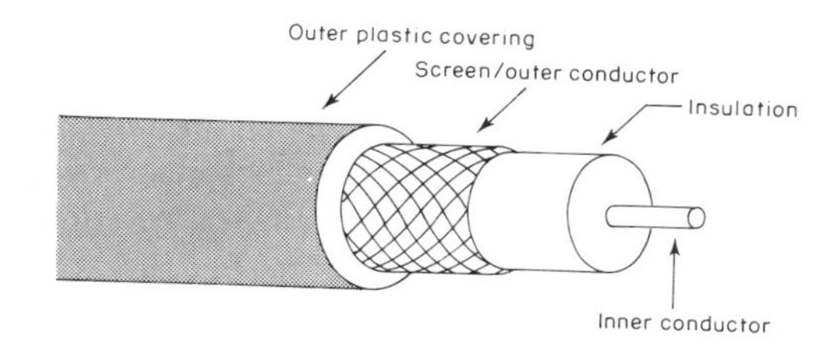

coaxial cable

coaxial converter In IBM 3270 systems, a protocol converter designed to be used between 3270 control units and attached to asynchronous devices; uses coaxial cable to connect to the control unit.

COBOL COmmon Business-Oriented Language.

code 1. A set of rules specifying the way in which data can be represented (e.g., the set of correspondences in the American Standard Code for Information Interchange) 2. In data communications, a system of rules and conventions specifying the way in which the signals representing data can be formed, transmitted, received, and processed. 3. The statements that make up a computer program.

code conversion The process of changing from one character coded representation into the corresponding character in a second code.

code level The number of bits used to represent characters.

CODEC An abbreviation of a coder/decoder. The coder is used to encode Pulse Amplitude Modulation (PAM) samples into PCM pulses while the decoder modifies the received PCM information and puts it in a form which can be understood by the receiver.

coding scheme Plan for changing human-usable language (letters and/or numbers) into computer usable form.

cohesion The cohesion of a connected network is the least number of lines which must be removed to separate the network into disconnected parts.

coin first A type of coin station which requires the presence of a coin in order to provide an off-hook indication to the local office.

COLAN Central Office Local Area Network.

cold boot Reloading a computer's operating system, by turning the electricity to the computer off and back on.

cold start A system restart that ignores previous data areas and accounting information in main storage and purges the queues upon starting.

collect connection A procedure whereby the charge for a virtual connection is billed to the destination Data Terminal Equipment (DTE).

collision 1. In LAN technology, the result of two stations attempting to use a shared transmission medium simultaneously. 2. In a half-duplex system, the result of both ends trying to transmit at the same time.

Collision Detection (CD) The ability of a transmitting node to detect simultaneous attempts on a shared medium.

colocation The practice of installing another organization's equipment at the central office of a Bell operating company (BOC) or an interexchange carrier (IX). Equipment is normally colocated to permit the BOC or IX to provide maintenance for the organization.

COM Computer Output Microfilm.

combined station In High-level Data Link Control (HDLC) protocol, a station capable of assuming either the role of a primary or a secondary station (balanced station).

COMCODE AT&T ordering code.

COMINT COMand INTerpreter.

Comité Consultatif Internationale de Télégraphique et Téléphonique (CCITT) An international consultative committee that sets international communications recommendations, which are frequently adopted as standards, and develops interface, modem, and data network recommendations. Membership includes PTTs, scientific and trade associations, and private companies. CCITT is part of the International Telecommunications Union (a United Nations treaty organization in Geneva). The following table is a selected list of CCITT recommendations.

NUMBER	GENERAL SUBJECT
	V series
V.10	Interchange circuits
V.11	Interchange circuits
V.21	300 bps dial modem
V.22	1200 bps dial or 2-wire leased line modem
V.22 bis	2400 bps dial 2-wire leased line modem
V.23	600/1200 bps dial modem
V.24	25-pin interface circuits
V.25	Dial parallel interface
V.25 bis	Dial serial interface
V.26	2400/1200 bps leased line modem
V.26 bis	2400/1200 bps dial modem
V.26 ter	2400 bps dial or 2-wire leased line modem
V.27	4800 bps leased line modem/manual equalizer
V.27 bis	4800 bps leased line modem/auto equalizer
V.27 ter	4800 bps dial modem
V.28	25-pin interface circuits
V.29	9600 bps leased line modem
V.32	9600 bps dial or 2-wire leased line modem
V.35	Wideband interface
	X series
X.3	PAD/terminal interface
X.21	DTE/DCE 15-pin interface
X.21 bis	DTE/DCE 25-pin interface
X.25	Packet switching
X.28	PAD/terminal interface
X.29	PAD/PDN interchange
X.32	Dial PDN access
X.121	PDN numbering plan

command A code that will cause a device to perform an electronic or mechanical action.

command file A file of commands that can be carried out by a computing device after initial setup.

Comand Interpreter (COMINT) Tandem's Guardian Command Interpreter program.

Command List (CLIST) In IBM's SNA (NCCF), a sequential list of commands and control statements that is assigned a name. When the name is invoked (as a command) the commands in the list are executed.

command port The console used to control and monitor a network or system. Also, the interface to which the console is connected.

command processor A software component that interprets control commands issued by an operator or user and invokes the requested function.

command set Commands that control an intelligent modem. Most modems support the Hayes AT command set, but many high-speed modems have their own enhanced command sets.

command state Operating mode wherein the modem will accept commands entered from a computer or terminal device. Most modems allow users to enter the command state while on-line without breaking the connection. In the AT command set, typing three "+" characters transfers the modem to command state.

common battery A dc power source in the Central Office that supplies power to switching equipment and to subscribers.

Common Business-Oriented Language (COBOL) COBOL is a common procedural language designed for commercial processing as developed and defined by a national committee of computer manufacturers and users. It is a specific language by which business-oriented data processing procedures may be precisely described in a standard form. COBOL is intended to communicate data processing procedures to users as well as providing a means to write business-oriented programs for computers.

common carrier A private data communications utility company or government organization that furnishes communications services to the general public and that is usually regulated by local, state, or federal agencies. Often, PTTs provide these services outside the USA, and telecos do inside.

common carrier principle The regulatory concept limiting the number of companies that provide needed services in a specific geographic area.

Common Channel Interoffice Signaling (CCIS) A technique by which the signaling information for a group of trunks is sent between switching offices over a separate voice channel. This technique uses time-division multiplexing.

Common Channel Signaling System (CCSS) A system where all signaling for a group of circuits is carried over a common channel.

Common Channel Signaling System No.7 The current international standard for out-of-band call control signaling on digital networks.

common communications carrier A firm engaged in the business of supplying communications lines for general public utility. Also called common carrier.

Common Control Switching Arrangement (CCSA) Switching facilities located in a common carrier's central offices. These facilities are used for the interconnection of corporate foreign exchange networks.

common logic redundancy See redundancy.

Common Management Information Protocol (CMIP) In Digital Equipment Corporation Network Architecture (DECnet), a management protocol which encompasses the Management Information Control and Exchange (MICE) and Management Event Notification (MEN) protocols.

communication adapter A hardware component inserted into a personal computer or a minicomputer to permit it to interface with a communication circuit.

Communication Control Unit (CCU) A nonprogrammed communications interface to a computer. This control unit will perform such functions as parity checking, block checking, and automatic answering of incoming calls.

communication controller IBM term for a front end processor.

Communication Identifier (CID) In IBM's VTAM, a key for locating the control blocks that represent a session. The key is created during the session-establishment procedure and deleted when the session ends.

communication line Deprecated term for telecommunication line and transmission line.

communication link The software and hardware, to include cables, connectors, converters, etc., required for two devices such as a computer and terminal to communicate.

Communication Macro Instructions In IBM's VTAM, the set of RPL-based macro instructions used to communicate during a session.

Communication Management Configuration (CMC) In IBM's SNA: 1. In VTAM, a technique for configuring a network that allows for the consolidation of many network management functions for the entire network in a single host processor. 2. A multiple-domain network configuration in which one of the hosts, called the communication management host, performs most of the controlling functions for the network, thus allowing the other hosts, called data hosts, to process applications. This is accomplished by configuring the network so that the communication management host owns most of the resources in the network that are not application programs. The resources that are not owned by the communication management host are the resources that are channel-attached stations of data hosts.

Communication Management Host In IBM's SNA, the host processor in a communication management configuration that does all network-control functions in the network except for the control of devices channel-attached to data hosts.

Communication Network Management (CNM) In IBM's SNA: 1. The process of designing, installing, operating, and managing the distribution of information and controls among end users of communication systems. 2. NCCF is an IBM program base that provides functions collectively called CNM. CNM programs include NLDM, NPDA, MVS/OCCF and VSE/OCCF.

Communication Network Management (CNM) application program In IBM's VTAM, an application program that is authorized to issue formatted management services request units containing physical-unit-related requests and to receive formatted management services request units containing information from physical units.

Communication Network Management (CNM) interface In IBM's SNA, the interface that allows an application program to send Forward request/response units to an access method and to receive Deliver request/response units from an access method. These request/response units contain network services request/response units (data and commands).

Communication Network Management (CNM) processor In IBM's SNA, a program that manages one of the functions of a communications system. A CNM processor is executed under control of NCCF and requires NCCF as a prerequisite program.

Communication Scanner Processor (CSP) In IBM's SNA, a processor in the 3725 communication controller that contains a microprocessor with control code. The code controls transmission of data over links attached to the CSP.

communication server An intelligent device normally located on the node of a local area network which performs communications functions.

communications The conveyance of information between two points, without alteration of the structure or content of the message.

Communications Act of 1934 This Act, passed by Congress in 1934, established a national telecommunications goal of high quality, universally available telephone service at reasonable cost. The act also established the FCC and transferred federal regulation of all interstate and foreign wire and radio communications to this commission. It requires that prices and regulations for service be just, reasonable, and not unduly discriminatory.

communications common carrier A government-regulated private company which provides the general public with telecommunications services and facilities.

communications control character A character used to control transmission over data networks. ASCII specifies 10 control characters that are used in byte-oriented protocols.

Communications Control Unit (CCU) In IBM 3270 systems, a communications computer, often a minicomputer, associated with a host mainframe computer. It may perform communications protocol control, message handling, code conversion, error control, and application functions.

communications line control Computer device which, with associated software, accepts data and performs necessary control functions.

communications link The physical medium connecting two systems.

communications monitor Software in mainframe that provides shared interface between applications programs and communications devices; includes communication access methods.

communications network The collection of transmission facilities, switches, and terminal devices that, when combined, provide the capability to communicate between two points.

communications processing The function of operating upon the information or upon control characters that precede, accompany, or follow the information to ensure its successful entry, transmission, and delivery.

communications protocol The means used to control the orderly exchange of information between stations on a data link or on a data communications network or system. Also called line discipline—or protocol, for short.

communications satellite An orbiting microwave repeater that relays signals between communications stations. Typically geosynchronous.

Communications Satellite Corp. (COMSAT) Private U.S. satellite carrier, established by Congress in 1962, for the coordination and construction of satellite communications and facilities for international and data communications (U.S. counterpart of Intelsat).

communications terminal Any device which generates electrical or tone signals that can be transmitted over a communications channel.

Community Antenna Television (CATV) A cable-based broadcast television distribution system typically covering an entire city or community based upon radio frequency (RF) transmission. Generally uses 75-ohm coaxial cable as the transmission medium.

compaction See compression.

companding Compressing/expanding. The process in which an analog signal is reduced in bandwidth to permit it to be transmitted over a smaller bandwidth channel and then reconstructed (expanded) to its original form.

compandor A combination of a compressor at one point in a communications path for reducing the volume range of signals, and an expandor at another, later, point for restoring the original volume range. Usually its purpose is to improve the ratio of the signal to the interference entering in the path between the compressor and expandor.

compatibility The ability to receive and process programs and data from one computer system on another computer system without modification.

compiler A computer program used to convert symbols meaningful to a human operator into codes meaningful to a computer.

Complementary Metal-Oxide Semiconductor (CMOS) A circuit technology that uses n-channel and p-channel MOS transistors. n-Channel and p-channel refer to the negatively charged electrons and the positively charged "holes" that make the circuit work. Such devices consume very little power.

complementing music-on-hold A PABX feature in which callers waiting after an announcement hear either a quiet tone or music-on-hold until an agent is assigned to the call.

composite The line side signal of a concentrator or multiplexer that includes all the multiplexed data.

composite link The line or circuit connecting a pair of multiplexers or concentrators; the circuit carrying multiplexed data.

composite loopback A diagnostic test that forms the loop at the line side (output) of a multiplexer.

Composite Signaling (CS) A direct current signaling system which requires a single line conductor for each signaling channel. The signaling system provides full-duplex operation.

compression Two types are available: data compression, which reduces the number of bits required to represent data (accomplished in many ways, including using special coding to represent strings of repeated characters or using fewer bits to represent the more frequently used characters); and analog compression, which reduces the bandwidth needed to transmit an analog signal. Also called compaction and companding.

compressor A device that performs analog compression.

compromise equalizer An equalizer set for best overall operation for a given range of line conditions. Often fixed, but may be manually adjustable.

Compuserve A subsidiary of H&R Block which operates a packet switching network and an information utility which provides subscribers with access to different databases and electronic mail.

computer An electronic system which, in accordance with its programming, will store and process information as well as perform high-speed mathematical or logical operations.

Computer Aided Design (CAD) Automated design operations. CAD allows users—via keyboard, light pen, and video display terminal—to manipulate design data in a computer's memory and generate varying video images or hard-copy printouts of an object.

Computer Aided Manufacturing (CAM) Automated manufacturing operations. CAM uses computers to control manufacturing processes.

computer branch exchange A computer-controlled PABX system.

computer conferencing A visual form of conference telephone call.

Computer Inquiry II More formally known as the Second Computer Inquiry, FCC Docket Number 20808, the final decision in 1980 resulted in a policy that resulted in competition for and deregulation of the telecommunications industry in the U.S.

Computer Inquiry III Computer Inquiry III removed the structural separation requirement between basic and enhanced services provided by Bell operating companies (BOCs) and AT&T. Adopted by the U.S. Federal Communications Commission in May 1986.

computer port The physical location where a communications channel interfaces to a computer.

Computerized Branch Exchange (CBX) A PBX that uses a computer with an electronic switching network.

Computer-to-PBX Interface (CPI) A specification originated by Northern Telecom, for a T1-based data communications interface between a mainframe computer and a Private Branch Exchange. Incompatible with both ISDN and AT&T's DMI specifications.

COMSAT Communications Satellite Corp.

concatenation 1. The linking of transmission channels or subnetworks end to end. 2. The linking of blocks of user data or protocol transmissions.

concentration Collection of data at an intermediate point from several low- and medium-speed lines for transmission across one high-speed line.

concentrator A device used to divide a data channel into two or more channels of average lower speed, dynamically allocating space according to the demand in order to maximize data throughput at all times. Also called an intelligent TDM, STDM, or statistical multiplexer.

concurrent Pertaining to the occurrence of two or more events or activities within the same specified interval of time.

Conditional End Bracket (CEB) In IBM's SNA, the value (binary 1) of the conditional end bracket indicator in the request header (RH) of the last request of the last chain of a bracket. The value denotes the end of the bracket.

conditioning The "tuning" or addition of equipment to improve the transmission characteristics or quality of a leased voice-grade line so that it meets specifications for data transmission.

COMMON CARRIER CLASSIFICATION OF LINE PARAMETERS

C conditioning applies only to frequency response and delay distortion characteristics.

Actual measurements are at 4 Hz higher, such as 1004, etc.

The + gain or − loss is with respect to a 1004 Hz reference

D1 Point to point channels

D2 Two or three point channels

All D conditioned channels must meet the following:

1. Signal to C-notched noise = 28 dB
2. Signal to second harmonic distortion = 35 dB
3. Signal to third harmonic distortion = 40 dB

conference call A telephone call among three or more locations.

Conference Européenne des Postes et Télécommunications (CEPT) A regulatory body formed in 1959 which holds plenary sessions every two years. The resolutions concerning technical standards from these sessions apply only to its 26 nation members. Because the rollcall of CEPT comprises of many telecommunication administrators from CCITT, both organizations influence each other.

configuration An arrangement of parts to achieve some purpose. For example: network configuration or node configuration; hardware configuration—the equipment to be used and the way it is to be connected; software configuration—a procedure performed to prepare a software program for operation or define a work station's resources to the file server.

Configuration Control Program (CCP) In IBM's SNA, an SSP interactive application program by which configuration definitions for the IBM 3710 Network Controller can be created, modified, and maintained.

Configuration Report Program (CRP) In IBM's SNA, an SSP utility program that creates a configuration report listing network resources and resource attributes for networks with NCP, EP, PEP, or VTAM.

Configuration Restart In IBM's ACF/VTAM, the recovery facility that can be used after a failure or deactivation of a major node, ACF/VTAM, or the host processor to restore the domain to its status at the time of the failure or deactivation.

configuration services In IBM's SNA, one of the types of network services in the system services

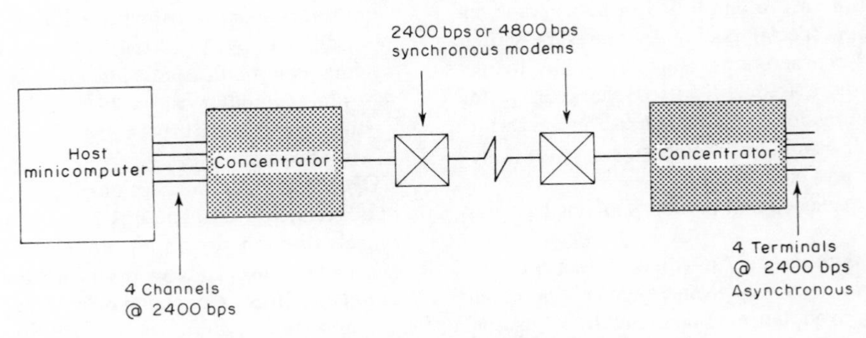

Typical concentrator configuration

concentrator

Bandwidth parameter limits

Channel condition	Frequency response relative to 1004 Hz		Envelope delay distortion	
	Frequency range	Variation in dB	Frequency range	Variation microseconds
Basic	500-2500 300-3000	-2 to +8 -3 to +12	800-2600	1750
C1	1000-2400 300-2700 300-3000	-1 to +3 -2 to +6 -3 to +12	1000-2400 800-2600	1000 1750
C2	500-2800 300-3000	-1 to +3 -2 to +6	1000-2600 600-2600 500-2800	500 1500 3000
C4	500-3000 300-3200	-2 to +3 -2 to +6	1000-2600 800-2800 600-3000 500-3000	300 500 1500 3000
C5	500-2800 300-3000	-0.5 to +1.5 -1 to +3	1000-2600 600-2600 500-2800	100 300 600

conditioning (basic and C-level)

control point (SSCP) and in the physical unit (PU). Configuration services activate, deactivate, and maintain the status of physical units, links, and link stations. They also shut down and restart network elements and modify path-control routing tables and address-transformation tables.

configuration table A table of program statements used to describe the topology, equipment characteristics, and operating parameters of a device in a network. The table enables the device to interface with other devices in the network.

congestion Congestion occurs when a network, or part of a network, is overloaded and has insufficient communications resources for the volume of traffic.

congestion avoidance A mechanism used to adjust the load on the network to prevent congestion.

connect time 1. A measure of system usage: the interval during which the user was on-line for a session. 2. The interval during which a request for a connection is being completed.

connected In IBM's ACF/VTAM, pertaining to a PU or LU that has an active physical path to the host processor containing the SSCP that controls the PU or LU.

connecting arrangement Interface equipment that is required by the common carrier when connecting customer-provided equipment to the public network.

connecting block Flame retardant plastic blocks containing quick clips for terminating cable and wire without the removal of insulation from the conductor. In addition, they provide an electrically tight connection between the cable and the cross-connect wire.

connection 1. An established data communications path. 2. The process of establishing that path. 3. A point of attachment for that path.

Connection Control In Digital Equipment Corporation Network Architecture (DECnet), the Session Control function concerned with the system-dependent functions related to creating, maintaining, and destroying Transport Connections.

connection point manager In IBM's SNA, a component of the transmission control layer that: (1) performs session-level pacing of normal-flow requests, (2) checks sequence numbers of received request units, (3) verifies that request units do not exceed the maximum permissible size, (4) routes incoming request units to their destinations within the half-session, and (5) enciphers and deciphers FMD request units when crytography is selected. The connection point manager coordinates the normal and expedited flows for one half-session. *Note*: The sending connection point manager within a half-session builds the request/response header (RH) for the outgoing request/response units, and the receiving connection point manager interprets the request/response headers that precede incoming request/response units.

Connectionless-Mode Network Service In Digital Equipment Corporation Network Architecture (DECnet), a network service which operates according to a datagram model. Each message is routed and delivered to its destination independently of any other. The Network Layer of DNA provides this type of service.

connectionless service Transport of a single quantum of information not set-up by a signaling or administrative procedure (i.e., a packetgram).

Connection-Mode Network Service (CONS) In Digital Equipment Corporation Network Architecture (DECnet), a network service which operates according to a connection oriented model. Before data can be exchanged, a connection must first be established.

connectivity The connections of communications lines in a network.

connector A physical interface, such as an RJ plug/jack or DP25P/S plug/socket.

CONS Connection-mode Network Service.

console The device used by the operator, system manager, or maintenance technician to monitor or control computer, system, or network performance.

constant carrier A type of modem that supports only full-duplex handshaking. CD is always on.

Consultative Committee on International Telephony and Telegraphy (CCITT) An international consultative committee that sets international communications recommendations, which are frequently adopted as standards, and develops interface, modem, and data network recommendations such as V.22, V.27, and V.25. Membership includes PTTs, scientific and trade associations, and private companies. CCITT is part of the International Telecommunications Union (a United Nations treaty organization in Geneva). The CCITT meets every four years (last in 1988) to settle, revise, and publish current standards. Each meeting's version of the complete set of standards is published as a multi-volume document identified by color (e.g. the 1980 CCITT Yellow Book; the 1984 CCITT Red Book). See Comité Consultatif Internationale de Télégraphique et Téléphonique for selected list of CCITT recommendations.

contact bounce An imperfection in relay switching resulting from the physical properties of relay contacts which may initially move apart when closed prior to forming a complete closure. A coating of mercury can be used as a conductive coating that will bridge the contacts together during the bouncing period.

contended access In LAN technology, a shared access method that allows stations to use the medium on a first-come, first-served basis.

contending port A programmable port type which can initiate a connection only to a preprogrammed port or group of ports.

contention The facility provided by the dial network or a data PABX which allows multiple terminals to compete on a first-come, first-served basis for a smaller number of computer ports.

contention delay Time spent waiting to use a facility due to sharing with other users.

contention group A number of ports functioning as a group. Port connection request are addressed to the group, and the call is routed to any available port within the contention group.

contiguous ports Ports occurring in unbroken numeric sequence.

Continue-Any Mode In IBM's ACF/VTAM, a state into which a session is placed that allows its input to satisfy a Receive request issued in any-mode. While this state exists, input on the session can also satisfy Receive requests issued in specific-mode.

Continue-Specific Mode In IBM's ACF/VTAM, a state into which a session is placed that allows its input to satisfy only Receive requests issued in specific-mode.

continuous ARQ A protocol in which error checking and correction procedures consist of the receiving station notifying the sending station that blocks or frames of data were received in error. Continuous ARQ allows the sending station to send more than one block or frame before stopping to wait for acknowledgment.

Continuous Variable-Slope Delta Modulation (CVSD) A method of digitizing analog voice signals that uses 16 000 to 64 000 bps bandwidth depending on the sampling rate. CVSD reduces the bandwidth needed for a voice channel from 64 Kbps to 32 Kbps or less.

continuously variable Capable of having one of an infinite number of values, differing from each other by an arbitrary small amount. Usually used to describe analog signals or analog transmission.

Continuously Variable Slope Delta (CVSD) A method of converting an analog (voice) signal into a digital stream by sampling the signal and encoding the delta or change that occurred since the previous sample.

control center A site from which a network administrator can exercise distributed systems administration and control functions.

control character A non-printing character used to start, stop, or modify a function. The carriage return (CR) is an example of a control character.

control codes Special nonprinting codes which cause the terminal or computer to perform specific electronic or mechanical actions (such as setting tabs, etc.).

control console A computer device used for control of system operation.

control line timing Clock signals between a modem and communications control unit.

control mode The state that all terminals on a line must be in to allow line control actions, or terminal selection to occur. When all terminals on a line are in the control mode, characters on the line are viewed as control characters performing line discipline, e.g., polling or addressing.

Control Program (CP) In an IBM environment, the Control Program in a virtual memory system manages the real resources of the computer to include memory and direct access storage devices.

control programs Control programs contain many routines that would otherwise have to be put into each individual program. Such routines include those for handling error conditions, interruptions from the console, or interruptions from a communications terminal.

control signal An interface signal used to announce, start, stop, or modify a function. For example, CD is an RS-232 control signal that announces the presence of a carrier.

EXAMPLE OF RS-232 CONTROL SIGNALS

Pin	*Control signal*	*From*	*To*
4	Request-To-Send (RTS)	DTE	DCE
5	Clear-To-Send (CTS)	DCE	DTE
6	Data Set Ready (DSR)	DCE	DTE
8	Carrier Detect (CD)	DCE	DTE
20	Data Terminal Ready (DTR)	DTE	DCE
22	Ring Indicator (RI)	DCE	DTE

control statement In IBM's SNA (NCCF), a statement in a command list that controls the processing sequence of the command list or allows the command list to send messages to the operator and receive input from the operator.

control station A station on a network which supervises network control procedures such as polling, selecting, and recovery. The control station is also responsible for establishing order on the line in the event of contention, or any other abnormal situation that may arise between stations on the network.

Control Switching Points (CSP) The regional, sectional, and primary switching centers for telephone dialing.

control unit In an IBM host system, equipment coordinating the operation of an input/output device and the CPU.

controlled A terminal and communications system design in which a central host or switching system exerts control over terminals to prevent them from transmitting when the host or switch is unable to service the traffic.

controlling application program In IBM's ACF/VTAM, an application program with which a logical unit (other than a secondary application program) is automatically put in session whenever the logical unit is available.

controlling logical unit In IBM's VTAM, a logical unit with which a secondary logical unit (other than an application program) is automatically put in session whenever the secondary logical unit is available. A controlling logical unit can be either an application program or a device-type logical unit.

Conversant A registered trademark of AT&T used for a series of voice-response systems. Callers interact with the system by pressing keys on a touch-tone phone or by speaking words or numbers.

conversational mode A mode of communications between terminals. In this mode, an entry by one terminal elicits a reply by the other terminal.

Conversational Monitor System (CMS) An IBM operating system that is used to manage a virtual machine.

converted command In IBM's SNA, an intermediate form of a character-coded command produced by ACF/VTAM through use of an unformatted system services definition table. The format of a converted commmand is fixed; the unformatted system services definition table must be constructed in such a manner that the character-coded command (as entered by a logical unit) is converted into the predefined, converted command format.

copper wire Literally, the wire formed from copper which carries the electrical signals which comprise telephone transmission. Frequently used as slang for coaxial cable.

core The region of a fiber optic waveguide through which light is transmitted—typically 8 to 12 micrometres in diameter for single-mode fiber, and from 50 to 200 micrometres for multimode fiber.

Cornet An ISD networking upgrade for the Siemens Saturn family of digital voice/data business communications systems.

Corporation for Open Systems (COS) An American consortium of companies, founded in 1986, dedicated to the development of worldwide standards through interoperability and conformance testing of OSI products.

COS 1. Call Originate Status. 2. Class of Service. 3. Corporation for Open Systems.

COSAC COmmunications SAns Connections.

COST In Digital Equipment Corporation Network Architecture (DECnet), a metric used by the routing algorithm. Each link is assigned a cost, and the routing algorithm selects paths with minimum cost.

COT Central Office Terminal.

country code Second set of digits to place an international call following the international access code. Country codes are listed in the following table.

COUNTRY	COUNTRY CODE
Algeria	213
American Samoa	684
Andorra	33
Anguilla	(809)
Antigua (including Barbuda)	(809)
Argentina	54
Aruba	297
Ascension Island	247
Australia (including Tasmania)	61
Austria	43
Bahamas	(809)
Bahrain	973
Bangladesh	880
Barbados	(809)
Belgium	32
Belize	501
Benin	229
Bermuda	(809)
Bolivia	591
Brazil	55
British Virgin Islands	(809)
Brunei	673
Bulgaria	359
Cameroon	237
Canada	Use area code
Cayman Islands	(809)
Chile	56
China	86
Colombia	57
Costa Rica	506
Cyprus	357
Czechoslovakia	42
Denmark (including Faeroe Islands)	45
Dominica	(809)
Dominican Republic	(809)
East Germany	37
Ecuador	593
Egypt	20
El Salvador	503
Ethiopia	251
Fiji Islands	679
Finland	358
France	33
French Antilles (including Martinique, St.Barthelemy, and St. Martin)	596
French Guiana	594
French Polynesia	689
Gabon	241
Gambia	220
Gibraltar	350
Greece	30
Greenland	299
Grenada (including Carriacou)	(809)
Guadeloupe	590
Guam	671
Guantanamo Bay	53
Guatemala	502
Guyana	592
Haiti	509
Honduras	504
Hong Kong	852
Hungary	36
Iceland	354
India	91
Indonesia	62
Iran	98
Ireland	353
Israel	972
Italy	39
Ivory Coast	225
Jamaica	(809)
Japan	81
Jordan	962
Kenya	254

Korea	82
Kuwait	965
Lesotho	266
Liberia	231
Libya	218
Liechtenstein	41
Luxemburg	352
Macao	853
Malawi	265
Malaysia	60
Malta	356
Marshall Islands	692
Mexico	52
Micronesia	691
Monaco	33
Montserrat	(809)
Morocco	212
Namibia	264
Netherlands	31
Netherlands Antilles	599
Nevis	(809)
New Caledonia	687
New Zealand	64
Nicaragua	505
Nigeria	234
Norway	47
Oman	968
Pakistan	92
Panama	507
Papua New Guinea	675
Paraguay	595
Peru	51
Philippines	63
Poland	48
Portugal (including Madeira Islands)	351
Qatar	974
Romania	40
Saipan (including Rota and Tinian)	670
San Marino	39
Saudi Arabia	966
Senegal	221
Singapore	65
South Africa	27
Spain (including Balearic Islands, Canary Islands, Ceuta, and Melilla)	34
Sri Lanka	94
St. Kitts	(809)
St. Lucia	(809)
St. Pierre/Miquelon	508
St. Vincent and the Grenadines	(809)
Surinam	597
Swaziland	268
Sweden	46
Switzerland	41
Taiwan	886
Tanzania	255
Thailand	66
Togo	228
Trinidad and Tobago	(809)
Tunisia	216
Turkey	90
Turks Islands and Caicos Islands	(809)
Uganda	256
United Arab Emirates	971
United Kingdom (England, N. Ireland, Scotland, Wales)	44
Uruguay	598
Vatican City	39
Venezuela	58
West Germany	49
Yemen Arab Republic	967
Yugoslavia	38
Zaire	243
Zambia	260
Zimbabwe	263

Note: () indicates area code.

coupler A device which interconnects a circuit to a terminal.

coupling loss The price in power a light pulse pays when it jumps between two optical devices such as a laser and a fiber lightguide.

CP 1. Command Port. 2. Control Program.

CPE Customer Premises Equipment.

CPH Characters Per Hour.

CPI Computer to PBX Interface.

CPM Control Program for Microcomputers.

CPMS Central Processing Unit Master Switching Subsystem.

CPNI Customer Proprietary Network Information.

CPS Characters Per Second.

CPU Central Processing Unit.

CQMS Circuit Quality Monitoring System.

CR Carriage Return.

crash (*verb*) Of hardware or software, to stop functioning properly.

CRC Cyclic Redundancy Check.

CRC-6 The North American standard for Cyclic Redundancy Checking on 1.544 Mbps circuits that use the Extended Superframe Format. The 6-bit CRC value, calculated on the previous Superframe, is transmitted in the Framing Bits of the 4th, 8th, 12th, 16th, 20th, and 24th frames of an Extended

Superframe. The CRC-6 is used as a measure of signal quality over time, not as a check on the validity of individual Superframes.

CRC-8 The European standard for Cyclic Redundancy Checking on 2.048 Mbps circuits. The CRC-8 value, calculated on the previous Multiframe, is transmitted in the Framing Bits of the 1st, 3rd, 5th, 7th, 9th, 11th, 13th, and 15th frames of the 16-frame Multiframe. The CRC-8 is used as a measure of signal quality over time, not as a check on the validity of individual Multiframes.

credit In Digital Equipment Corporation Network Architecture (DECnet), a flow control mechanism whereby the receiver of data tells the transmitter how many messages it is prepared to receive at a given point.

credit window In Digital Equipment Corporation Network Architecture (DECnet), the credit window identifies the range of message numbers which the receiver is prepared to receive at a given point.

critical angle The minimum angle at which a light wave striking the cladding in an optical fiber will be reflected back into the core.

crossbar A circuit switch having several horizontal and vertical paths that are electromagnetically or electronically interconnected.

crossbar switch An electromechanical switch used in older telephone central offices for switching and connecting circuits.

crossbar system A type of line switching which uses crossbar switches.

cross-connect 1. A piece of hardware used to interconnect multiplexers with line terminating equipment and other multiplexers. 2. Distribution system equipment where communication circuits are administered (that is, added or rearranged using jumper wires or patch cords). In a wire cross connect, jumper wires or patch cords are used to make circuit connections. In an optical cross-connect, fiber patch cords are used. The cross-connect is located in an equipment room, backbone closet, or satellite closet.

cross-connect field Wire terminations grouped to provide cross connect capability. The groups are identified by color-coded sections of backboards mounted on the wall in equipment rooms, backbone closets, or satellite closets, or by designation strips placed on the wiring block or unit. The color coding identifies the type of circuit that terminates at the field.

cross-connection The wire connections running between terminals on the two sides of a distribution frame, or between binding posts in a terminal.

cross-domain (*adjective*) In IBM's SNA, pertaining to control of resources involving more than one domain.

cross-domain keys In IBM's SNA, a pair of cryptographic keys used by a system services control point (SSCP) to encipher the session cryptography key that is sent to another SSCP and to decipher the session cryptography key that is received from the other SSCP during initiation of cross-domain LU–LU sessions that use session-level cryptography. Synonymous with cross keys.

cross-domain link In IBM's SNA: 1. A subarea link connecting two subareas that are in different domains. 2. A link physically connecting two domains.

cross-domain LU–LU session In IBM's SNA, a session between logical units (LUs) in different domains.

Cross-Domain Resource (CDRSC) In IBM's SNA, a resource owned by a Cross-Domain Resource Manager (CDRM) in another domain but known by the CDRM in this domain by network name and associated Cross-Domain Resource Manager.

Cross-Domain Resource Manager (CDRM) In IBM's VTAM, the function in the system services control point (SSCP) that controls initiation and termination of cross-domain sessions.

crossed pinning Configuration that allows two DTE devices or two DCE devices to communicate. Also called cross-over cable and null-modem cable.

cross-keys Synonym for cross-domain keys.

cross-modulation Interference caused by two or more carriers in a transmission system that interact through non-linearities in the system.

cross-network (*adjective*) In IBM's SNA, pertaining to control or resources involving more than one SNA network.

cross-network LU–LU session In IBM's SNA, a session between logical units (LUs) in different networks.

cross-network session In IBM's SNA, an LU–LU or SSCP–SSCP session whose path traverses more than one SNA network.

cross-over cable A cable which permits a terminal

to be directly connected to a computer port (DTE to DTE) or a tail circuit modem to a modem port (DCE to DCE). Also called null-modem cable.

cross-subarea In IBM's SNA, pertaining to control or resources involving more than one subarea node.

cross-subarea link In IBM's SNA, a link between two adjacent subarea nodes.

Crosstalk A communications program for use on IBM PCs and compatible computers from the Crosstalk Communications Division of Digital Communications Associates, Inc., of Roswell, GA. The program includes numerous file transfer protocols and allows up to 15 concurrent communications sessions.

crosstalk The unwanted transfer of a signal from one circuit, called the disturbing circuit, to another, called the disturbed circuit.

crosstalk, far-end Crosstalk which travels along the disturbed circuit in the same direction as the signals in that circuit. To determine the far-end crosstalk between two pairs, 1 and 2, signals are transmitted on pair 1 at station A, and the level of crosstalk is measured on pair 2 at station B.

crosstalk, near-end Crosstalk which is propagated in a distributed channel in the direction opposite to the direction of propagation of the current in the distributing channel. Ordinarily, the terminal of the disturbed channel, at which the near-end crosstalk is present, is near or coincides with the energized terminal of the disturbing channel.

CRP In IBM's SNA, Configuration Report Program.

CRQ Call ReQuest.

CRT Cathode Ray Tube.

CRV In IBM's SNA, CRyptography Verification.

cryptographic Pertaining to the transformation of data to conceal its meaning.

cryptographic algorithm A set of rules that specify the mathematical steps required to encipher and decipher data.

cryptographic key In systems using the Data Encryption Standard (DES), a 64-bit value (containing 56 independent bits and 8 parity bits) provided as input to the algorithm in determining the output of the algorithm.

cryptographic session In IBM's SNA products, an LU–LU session in which a function management data (FMD) request may be enciphered before it is transmitted and deciphered after it is received.

cryptographic session key In IBM's SNA, deprecated term for session cryptography key.

cryptography The process of hiding information from unauthorized disclosure by conversion through a secret algorithm or key.

Cryptography Verification Request (CRV) In IBM's SNA, a request unit sent by the primary logical unit (PLU) to the secondary logical unit (SLU) as part of cryptographic session establishment, to allow the SLU to verify that the PLU is using the correct cryptographic session key.

CS Composite Signaling.

CSA Carrier Serving Area.

CSDC Circuit Switched Digital Capability.

CSIRONET Commonwealth Scientific & Industrial Research Organization Network.

CSMA Carrier Sense Multiple Access.

CSMA/CA Carrier Sense Multiple Access with Collision Avoidance.

CSMA/CD Carrier Sense Multiple Access with Collision Detection.

CSNET Computer Science NETwork.

CSP 1. Control Switching Points. 2. Communication Scanner Processor (IBM's SNA).

CSR Customer Station Rearrangement.

CSU Channel Service Unit.

CT2 A second generation cordless phone technology being developed in the United Kingdom.

CTAK Cipher Text Auto Key.

CTDB Terminal Descriptor Block (Common).

CTS Clear To Send.

CTTU Central Trunk Test Unit.

C-type conditioning Conditioning used to control attenuation distortion and envelope delay distortion.

CU Control Unit.

CUG Closed User Group.

current The amount of electrical charge flowing past a specified circuit point per unit time, measured in amperes.

current loop 1. (Single-current signaling, used in U.S.) Method of interconnecting Teletype terminals and transmitting signals that represent a mark by current on the line and a space by the absence of current. 2. (Double-current signaling, used everywhere else.) A mark is represented by current in one direction and a space by current in the other direction.

cursor A movable underline, rectangular-shaped block of light, or an alternating block of reversed video on the screen of a display device, usually indicating where the next character will be entered.

custom chip A microchip designed for a specific job or function; e.g., the echo canceler chip.

Custom Local Area Signaling Services (CLASS) Number translation services provided by a communications carrier based upon the availability of common channel interoffice signaling. Examples of CLASS include call-forwarding and caller identification.

customer The person, firm, corporation, or other entity that orders services or products and is responsible for payment of charges and for compliance with carrier tariff regulations. Except for duly authorized and regulated common carriers, no one may be a customer who does not have a communications requirement of his own for service. Also referred to as a subscriber.

Customer Controlled Reconfiguration (CCR) A feature offered on AT&T T1 circuits that allows customers to change the channel assignments (e.g. destinations of individual channels) on a T1 circuit by placing a data call to the central office on a separate line.

Customer Information Control System (CICS) IBM communications monitor program product that provides an interface between the operating system access method and applications programs to allow remote or local display terminal interaction with a data base in the central processor.

Customer Information Control System/Virtual Storage (CICS/VS) IBM host application program/operating system.

Customer Premises Equipment (CPE) Devices at the customer location for interfacing between public transmission facilities and other equipment such as telephones, terminals, multiplexers, and computers. Devices classified as CPE can be supplied by third party vendors as well as the telephone companies.

Customer Proprietary Network Information (CPNI) Information to include a company's calling patterns, billing, network design and use of network services that telephone companies keep in their existing customer data bases.

customer station rearrangement A feature that allows business customers to change, display, and verify data affecting their telephone service from an on-premises terminal.

Customer-Provided Equipment (CPE) Any device connected to a common carrier facility which is not provided by the common carrier.

CUT Control Unit Terminal.

cut down A method of securing a wire to a wiring terminal. The insulated wire is placed in the terminal groove and pushed down with a special tool. As the wire is seated the terminal cuts through the insulation to make an electrical connection, and the tool's spring-loaded blade trims the wire flush with the terminal. Also called "punch down."

cutset A minimal set of elements of a connected graph (nodes, lines or both) which, when removed from the graph, disconnects it. It must be a minimal set, which means that any proper subset of the cutset does not disconnect the graph. The graph is 'disconnected' when it is separated into at least two parts without connecting links between them. Cutsets are used to measure cohesion and connectivity.

CVSD Continuous Variable Slope Delta Modulation.

CXR Carrier A communications signal used to indicate the intention to transmit data on a line. Also called Carrier Detect.

Cybernet Network of Control Data Corp.

cycle 1. A complete sequence of a wave pattern that recurs at regular intervals. 2. One iteration or loop through a set of logical points.

cycles per second (CPS) Measure of frequency.

cyclic access to store Access to a store (for reading or writing) in which successive addresses are accessed in turn returning eventually to the first address.

Cyclic Redundancy Check (CRC) An error-detection scheme in which the block check character is the remainder after dividing all the serialized bits in a transmission block by a predetermined binary number, or a polynomial based on the transmitted data.

D

D4 An AT&T specified frame format that designates every 193rd bit position in an AT&T supplied T1 facility as reserved for D4 allows continuous monitoring and non-destructive diagnostic framing to be implemented by the carrier.

D4 channel bank The current generation of central office T1 multiplexing equipment. Capable of handling D4 and Extended Superframe formats.

D-1 conditioning AT&T conditioning for two-point services not arranged for switching and for two-point services arranged for through switching, which controls the signal to C-notched noise ratio and intermodulation distortion.

D-2 conditioning AT&T conditioning for two-point switched services, which controls the signal to C-notched noise ratio and intermodulation distortion.

D-3 conditioning AT&T conditioning for SCAN System B access lines terminated in a four-wire telephone arrangement to control the signal to C-notched noise ratio and intermodulation distortion.

D-5 conditioning AT&T conditioning for polled multipoint data services not arranged for switching, which controls the signal to C-notched noise ratio and intermodulation distortion.

D4 framing A framing technique for T1 circuits using D4 channel banks. D4 framing uses a 12-frame Superframe with bit-robbing for AB signaling in each channel octet of the 6th and 12th frames of each superframe. D4 framing is now being superseded by the Extended Superframe Format.

D-channel (delta channel) The Common Channel Signaling channel of an ISDN circuit. In the Basic Rate service, the D-channel occupies 16 Kbps of 144 Kbps bandwidth, carries signaling for two B-channels, and can be used for low-speed, end user packet data as well as for signaling. In the Primary Rate service, the D-channel occupies 64 Kbps of 1.544 Mbps (or 2.048 Mbps) of bandwidth, and carries signaling for 24 (or 30) B-channels. In both ISDN services, the D-channel uses the LAP-D link protocol.

DA Data Available.

DA-15 A 15-pin D-type physical connector used in T1 multiplexing equipment as an alternative to WECO 310 or Bantam Connector.

DAA Data Access Arrangement.

D/A conversion Digital-to-Analog (D/A) conversion.

DAB Demand Assigned Bus.

DACS 1. Data Acquisition and Control System. 2. Digital Access Cross-connect System.

DACS/CCR Digital Access Cross-connect System/Customer Controlled Rerouting.

DAF Dedicated Access Facility.

Daini-Denden A Tokyo-based alternative long distance communications carrier.

daisy chaining The physical connection of cables that allows the cascading of signals between devices.

DAL Data Access Line.

DAP Data Access Protocol.

DARPA Defence Advanced Research Projects Agency.

DAS Digit AnalysiS.

DASD Data Access Storage Device.

data Any material which is represented in a formalized manner so that it can be stored, manipulated, and transmitted by machine.

Data Access Arrangement (DAA) DCE furnished or approved by a common carrier that permits privately owned DCE or DTE to be attached to the common carrier's network. All modems now built for the public telephone network have integral DAAs.

data acquisition A system for measurement and recording of data from physical entities and devices.

Data Acquisition and Control System (DACS) System designed to handle a variety of real-time applications, process control, and high-speed data acquisition. Each system is individually tailored with modularity allowing satisfaction of specific system requirements.

data bank A centralized collection of computer information usually accessible by dial-up.

data base A collection of data stored electronically in a predefined format and according to an established set of rules, often called a schema.

data blocks Logical units of information that are being sent over computer I/O channels.

data call queueing standby A PABX feature that places a data caller into a "standby" queue for shared facilities, such as computer ports and modem pools.

Data Carrier Detect (DCD) A signal sent from a data set which informs the terminal that a carrier waveform is being received. The signal may also be called carrier detected, carrier found, carrier on, etc.

data channel The communication path along which data can be transmitted.

data character A character containing alphanumeric information mainly used in a communications link between terminals and/or computers.

data circuit Communication facility permitting transmission of information in digital or analog form.

Data Circuit-terminating Equipment (DCE) Equipment that provides the functions required to establish, maintain, and terminate a connection, the signal conversion, and coding for communications between data terminal equipment (DTE). Also called data communications equipment.

data class of service A PABX feature that allows the communications manager to customize service by controlling access to data facilities, devices, and features.

data collection Procedure in which data from various sources is accumulated at one location (in a file or queue) before being processed.

data communication service A specified user information transfer capability provided by a data communication system to two or more end users.

data communication session A coordinated sequence of user and system activities whose purpose is to cause digital user information present at one or more source users to be transported and delivered to one or more destination users. A normal data communications session between a user pair comprises: (1) an access function, (2) a user information transfer function, (3) a disengagement function for each user. A data communication session is formally defined by a data communication session profile.

data communication session profile The exact sequence of user/system interface signals by which data communication service is provided in a typical (successful) instance. A complete data communication session profile should also include any possible blocking (service refusal) sequences for a particular set of users.

data communication subsystem A group of data communication system elements terminated within the end user interfaces.

data communication system A collection of transmission facilities and associated switches, data terminals, and protocols that provide data communication service between two or more end users. The data communication system includes all functional and physical elements that participate in transferring information between end users. The system element that interfaces with the end user is a data terminal or a computer operating system. A computer operating system normally serves as the first point of contact for application programs requiring data communication service.

data communication user 1. An end user of a data communication system. 2. An aggregate user of a data communication subsystem.

data communications 1. The processes, equipment, and/or facilities used to transport signals from one data processing device at one location to another data processing device at another location. 2. In IBM's SNA, the transmission and reception of data.

Data Communications Equipment (DCE) The equipment that provides the functions required to establish, maintain, and terminate a connection as well as the signal conversion required for communications between the DTE and the telephone line or data circuit. Synonymous with modem in common usage. Also called data circuit-terminating equipment.

data compression A method of reducing the amount of transmitted data by applying an algorithm to the basic data at the point of transmission. A decompression algorithm expands the data back at the receiving end into its original format.

data dictionary A list of all the data names and data elements in a system.

data encrypting key A cryptographic key used to

encipher and decipher data transmitted in a cryptographic session.

Data Encryption Standard (DES) A cryptographic algorithm endorsed by the National Bureau of Standards (NBS) to encrypt data using a 56 bit key. Specified in the Federal Information Processing Standard Publication (FIPS PUB) 46 dated 15 January, 1977.

data entry Introducing data into a data processing or information processing system for input.

Data Flow Control (DFC) In IBM's SNA, a request/response unit (RU) category used for requests and responses exchanged between the data flow control layer in one half-session and the data flow control layer in the session partner.

Data Flow Control (DFC) layer In IBM's SNA, the layer within a half-session that (1) controls whether the half-session can send, receive, or concurrently send and receive request units (RUs), (2) groups related RU's into RU chains, (3) delimits transactions via the bracket protocol, (4) controls the interlocking of requests and responses in accordance with control modes specified at session activation, (5) generates sequence numbers, and (6) correlates requests and responses.

data flow control protocol In IBM's SNA, the sequencing rules for requests and responses by which network addressable units in a session coordinate and control data transfer and other operations.

data host In IBM's SNA communication management configuration, a host that is dedicated to processing applications and does not control network resources, except for its channel-attached or communication adapter-attached devices.

data hot line A feature on a PABX that allows the user to place a data call to a particular terminal automatically.

data integrity A measure of data communications performance, indicating a sparsity (or, ideally, the absence) of undetected errors.

data integrity point The generic name given to the point in an IBM 3800 model 3 printing process at which the data is known to be secure. Also called the stacker.

data least cost routing A PABX feature that automatically selects the most economical route (WATS, FX, tie-line, switched network) for each outgoing call.

data line group A feature on a PABX that allows a single number to find a free port in a group of numbers.

data link A serial communications path between nodes or devices without any intermediate switching nodes.

data link connector A type of single fiber optical connector used to connect fiber cable to a host computer or other equipment.

Data Link Control (DLC) The combination of software and hardware that manages the transmission of data over the communications line.

Data Link Control (DLC) layer In IBM's SNA, the layer that consists of the link stations that schedule data transfer over a link between two nodes and perform error control for the link. Examples of data link control are SDLC for serial-by-bit link connection and data link control for the IBM System/370 channel.

data link control protocol A set of rules used by two nodes on a data link to accomplish an orderly exchange of information. Synonymous with line control.

data link layer The second layer in the OSI model. It takes data from the network layer and passes it on to the physical layer. It is responsible for transmission and reception of packets, datagram service, local addressing, and error detection (but not error correction).

data medium 1. The material in which or on which a specific physical variable may represent data. 2. The physical quantity that may be varied to represent data.

data message detail recording A PABX feature which provides a record of all external and internal data calls. Information recorded can include the calling party's extension number, the destination number, time of call, and call duration.

data mode A condition of a modem or a DSU with respect to the transmitter in which its Data Set Ready and Request to Send circuits are on and it is ready to send data.

data network A system consisting of a number of terminal points that are able to access one another through a series of communication lines and switching arrangements.

Data Network Identification Code (DNIC) A four-digit number assigned to public data networks as well as specific services on some of those networks. See the following table.

DATA NETWORK IDENTIFICATION CODE (DNIC)

Country	*Network*	*DNIC*
Australia	Austpac	5052
Australia	Midas	5053
Austria	Radio Austria	2329
Belgium	DCS	2062
Brazil	Interdata	7240
Canada	Datapac	3020
Canada	Globedat	3025
Canada	Infoswitch	3029
Denmark	Datapak	2382
Eire	Eirpac	2724
Finland	Finnpak	2442
France	NTI	2081
France	Transpac	2080
French Polynesia	Tompac	5470
Gabon	Gabopac	6282
Germany (FDR)	Datex P	2624
Great Britain	IPSS	2341
Great Britain	PSS	2342
Greece	Helpac	2022
Guadeloupe	Dompac	3400
Hong Kong	Intelpac	4542
Hong Kong	Datapac	4545
Israel	Isranet	4251
Italy	Itapac	2222
Ivory Coast	Sytranpac	6122
Japan DDX P	4401	
Japan	Venus P	4408
Luxembourg	Luxpac	2704
Martinique	Dompac	3400
Netherlands	DABAS	2044
Netherlands	Datanet 1	2041
New Zealand	Pacnet	5301
Norway	Datapak	2422
Singapore	Telepac	5252
South Africa	Saponet	6550
Spain	TIDA	2141
Spain	Iberpac	2145
Sweden	Datapak	2402
Switzerland	Datalink	2289
Switzerland	Telepac	2284
USA	Autonet	3126
USA	Compuserve	3132
USA	DBS (WUI)	3104
USA	Datapak	3119
USA	LSDS (RCA)	3113
USA	Marknet	3136
USA	Telenet	3110
USA	Tymnet	3106
USA	UDTS (ITT)	3103
USA	Uninet	3125
USA	Wutco	3101

data networking A capability that will allow users to combine separate databases, telecommunication systems, and specialized computer operations into a single integrated system, so that data communication can be handled as easily as voice messages.

Data Over Voice (DOV) Technology used to transmit data and voice simultaneously over twisted-pair copper wire. Primarily used with local Centrex services or special customer premises PBXs. An FDM technique which combines data and voice on the same line by assigning a portion of the unused bandwidth to the data. Usually implemented on the twisted pair cables used for in-house telephone system wiring.

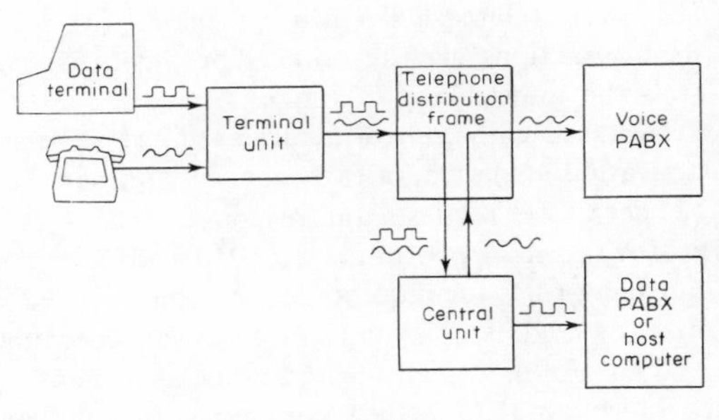

Data Over Voice (DOV)

Data Packets Per Second (DPPS) The rate at which data packets are processed through a communications processor.

Data PBX A switch that allows a user's circuit to select other circuits for the purpose of establishing a connection. Only digital transmission, and not analog voice, is switched.

Data Qualifier (DQ) packet Defined in X.25 for signaling overhead information across an X.25 virtual circuit. In Telenet, the DQ bit is set to 1 to signal that an X.29 Packet Assembler/Disassembler (PAD) message is contained in the data field of the X.25 data packet. High-level Data Link Control (HDLC) and Synchronous Data Link Control

(SDLC) also use the DQ packet.

Data Quality Monitor (DQM) A device used to measure data bias distortion above or below a threshold.

Data Race A company located in San Antonio, TX which specializes in the manufacture of high-speed modems designed for use on the public switched telephone network.

data rate, data signaling rate A measure of how quickly data is transmitted, expressed in bps. Also commonly, but often incorrectly, expressed in baud. Synonymous with speed.

Data Service Unit (DSU) A device used to provide a digital data services interface. Located on the user's premises, the DSU interfaces directly with the data terminal equipment. The DSU provides loop equalization, remote and local testing capabilities, and the logic and timing necessary to provide a standard EIA/CCITT interface. May have an integrated Channel Service Unit (CSU). Also called digital service unit.

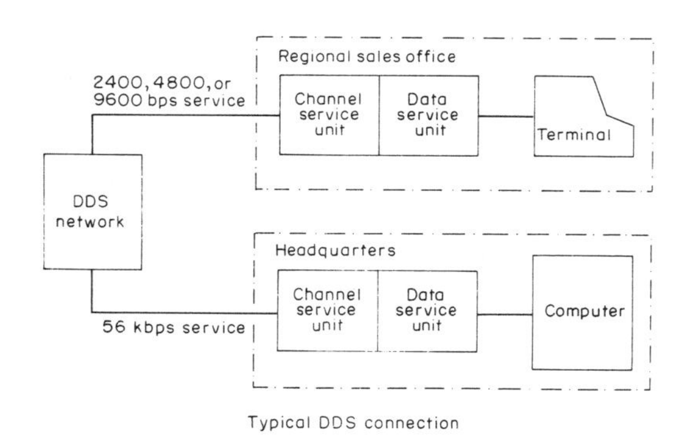

Data Service Unit (DSU)

data set A software term for a certain type of data file. A synonym for modem (coined by AT&T).

Data Set Ready (DSR) An RS-232 modem interface control signal (sent from the modem to the DTE on pin 6) which indicates that the modem is connected to the telephone circuit. Usually a prerequisite to the DTE issuing RTS.

data set separator pages Those pages of printed output that delimit data sets.

data signaling rate Used in communications to define the rate at which signal elements are transmitted or received over a transmission path by data terminal equipment. The data signaling rate is expressed in bits per second (bps) and baud.

data sink A device that can accept data signals from a transmission device.

data stream The collection of characters and data bits transmitted through a channel.

data structure A system of relationships between items of data. To express these relationships when a data structure is stored, lists or other systems using pointers etc. may be used.

data switch A system that connects network lines to a specific input/output port of a computer or front end processor (FEP).

data terminal A device associated with a computer system for data input and output that may be at a location remote from the computer system, thus requiring data transmission.

Data Terminal Equipment (DTE) In data communications notation, a user device such as a terminal or computer, connected to a circuit.

Data Terminal Ready (DTR) An RS-232 modem interface control signal (sent from the DTE to the modem on pin 20) which indicates that the DTE is ready for data transmission and which requests that the modem be connected to the telephone circuit.

data traffic reset state In IBM's VTAM, the state usually entered after Bind Session, if Cryptography Verification is used, and after Clear, but prior to Start Data Traffic. While a session is in this state, requests and responses for data and data flow control cannot be sent. Only certain session control requests can be sent.

data transfer mode In packet switching, data packets carrying customer data are exchanged in this virtual circuit mode. A virtual circuit is in data transfer mode after a call is successfully established. It remains in data transfer mode until it is cleared or until an escape to command mode is accomplished.

data transfer rate The average number of bits, characters, or blocks per unit of time transferred from a data source to a data link.

data transmission The technology of transmitting and receiving information over communication channels.

data types In IBM's SNA (NPDA), a concept to describe the organization of data displays. Data types are defined as alerts, events, and statistics. Data types are combined with resource types and display types to describe NPDA display organization.

Data Under Voice A technique used for combining data and voice transmission which exploits the unused part of the transmission channel's bandwidth for data transmission.

datacommonality Used by General DataComm, Inc., to describe a unique packaging technique which provides the following benefits: (1) high density modular packaging, (2) a broad array of versatile data sets and accessories, (3) system flexibility and ease of expansion, (4) low power consumption and heat dissipation, (5) quick and simple installation, (6) at-a-glance monitoring of system operation, (7) convenient, low cost maintenance and (8) high reliability.

Datalynx/3174 A protocol converter manufactured by Local Data Corporation (now Andrew Corp.) that provides as many as 32 serial ports allowing ASCII terminals, PCs, minicomputers, microcomputers, and printers, to emulate IBM full screen 327x, 318x, 319x terminal, and 328x printer devices.

Datalynx/3274 A protocol converter manufactured by Local Data Corporation (now Andrew Corp.) that provides up to nine serial ports, allowing asynchronous terminals, PCs, minicomputers, microcomputers, printers, etc. to emulate IBM 3278 models 1–5 terminals and 327x printer devices.

Datamizer A family of data compression products manufactured by Symplex Communications Corporation of Ann Arbor, Michigan.

datagram In packet switched networks, an abbreviated, connectionless, single-packet message from one station to another. Rarely, if ever, implemented on current PDNs.

Dataphone Both a service mark and a trademark of AT&T. As a service mark, it indicates the transmission of data over the telephone network. As a trademark, it identifies the communications equipment furnished by AT&T for data communication service.

Dataphone Digital Service (DDS) AT&T's private line service, filed in 1974, for transmitting data over a digital system. The digital transmission system transmits electrical signals directly, instead of translating the signals into tones of varied frequencies as with the traditional analog transmission system. The digital technique provides more efficient use of transmission facilities, resulting in lower error rates and costs than analog systems. In 1987, AT&T introduced switched 56 Kbps Dataphone Digital Service.

Dataphone Digital Service with Secondary Channel A private line service offered by AT&T and several Bell operating companies (BOCs) that allows 64 Kbps clear channel data with a secondary channel that provides end-to-end supervisory, diagnostic and control functions. Also referred to as DDS II.

Dataspeed An AT&T marketing term for a family of medium-speed paper tape transmitting and receiving units.

Datel A British Telecom service which provides a means for transmitting data over the public telephone network in the United Kingdom and from the United Kingdom to 70 countries worldwide.

Datex-P The packet switching network operated by Deutsche Bundespost in West Germany.

dB, db Decibel.

dBa (dBrn adjusted) A unit of noise measurement for a circuit having F1A-line or HA-1-receiver weighting. 0 dBa is noise power of 3.16 picowatts (−85 dBm), equivalent to 0 dBrn (−90 dBm or 1 picowatt with C-message weighting) adjusted to F1A weighting.

dBa0 The power level of noise expected at a particular location in a circuit, called "dBrn adjusted reference level".

dBa (F1A) The noise level measured on a line by a noise measuring set having F1A-line weighting.

dBa (HA-1) The noise level measured across the receiver of a 302-type or similar telephone handset measured by a test set having HA-1-receiver weighting.

DBAAU Dial Backup Auto Answer Unit.

D-BIT The delivery confirmation bit in a X.25 packet that is used to indicate whether or not the DTE wishes to receive an end-to-end acknowledgment of delivery.

dBm A decibel referenced to 1 milliwatt. Absolute measure of signal power where 0 dBm is equal to one milliwatt. In the telephone industry, dBm is based on 600 ohms impedance and 1000 hertz frequency. 0 dBm is 1 milliwatt at 1000 Hz terminated by 600 ohms impedance.

dBm0 The abbreviation commonly used to indicate the signal magnitude in dBm at the 0TLP. The Transmission Level Point (TLP) is the ratio (in dB) of the power of a signal at any particular point to the power of the same signal at the reference point:

TLP (dB) + power at 0TLP
= measured power at TLP (dBm)

Example: The receive TLP of a data circuit is −3 dB. Data is normally transmitted at −13 dB below 0TLP or −13 dBm0. Using this information, what is the expected measured receive level on our circuit? If we used the formula:

−3 + (−13 dBm0) = −16 dBm

DBMS Database Management System.

DB25P EIA designation for the male plug used to connect communications cables which fit into a DB25S socket plug.

dBrn The abbreviation for decibel reference noise. In their desire to make things easier for the technician, the early telco engineers created a separate scale for noise that generated larger positive numbers for larger amounts of noise. The zero difference used in describing noise is −90 dBm or 0 dBrn. The relationship between these two scales is shown in the following table. It is important to remember that dBrn is a measurement of noise. It is also important to remember the relationship, for example 60 dBrn is equal to −30 dBm.

dBrn	*dBm*
90	0
80	−10
70	−20
60	−30
50	−40
40	−50
30	−60
20	−70
10	−80
0	−90

dBrnC The abbreviation for decibel reference noise C-message weighted). Noise contains numerous irregular waveforms which have a wide range of frequencies and powers. Although any noise superimposed upon a conversation has a degrading effect, experiments have shown that this effect is greatest at the mid-range of the voice frequency band. To obtain a useful measurement of the interfering effect of noise, the various frequencies contributing to the overall noise are weighted in accordance to their relative interference effect. This weighting is accomplished through a weighting network or filter. The filter which closely emulates these characteristics is called a C-message weighted filter.

dBrnC0 The abbreviation for decibel reference noise C-message weighted zero transmission level. Noise may also be referenced to the 0TLP. For example, the noise measured at a −3 TLP is 33 dBrnC. To determine the noise at the 0TLP:

(TLP) − (actual measured noise)
= (noise at 0 TLP) −3 − 33 dBrnc
= 36 dBrnC0

DBS Direct Broadcast Satellite.

DB25S EIA designation for the female socket used to connect communications cables.

DBSS Direct Broadcast Satellite Service.

DBU Dial Back Up.

DC Device Control.

dc Direct Current.

dc current loop See current loop.

dc continuity A circuit that appears to be a continuous circuit. A metallic circuit composed of wire not interrupted by amplifiers, transformers, etc.

DCA Distributed Communications Architecture.

DCD Data Carrier Detect.

DCE Data Circuit-terminating Equipment.

DCL Digilog Command Language.

DCLU Digital Carrier Line Unit.

DCM 1. Digital Circuit Multiplexing. 2. Digital Circuit Multiplication. 3. Dynamic Connection Management.

D-conditioning Service offering from telephone companies to control the signal-to-noise ratio and harmonic or non-linear distortion. It is currently used for transmission above 9600 bps with complex modems on voice grade private lines.

DCPSK Differentially Coherent Phase Shift Keying.

DCTU Directly Connected Test Unit.

DDCMP Digital Data Communications Message Protocol.

DDD Direct Distance Dialing.

DDD network The long-distance Direct Distance Dialing (DDD) telephone network.

DDN Defense Data Network.

DDP Distributive Data Processing.

DDR Dual Dial Restore.

DDS Dataphone Digital Service.

DDS II Dataphone Digital Service with Secondary Channel.

DDS-SC Dataphone Digital Service with Secondary Channel.

DDT/DDS Digital Data Throughput/Digital Dataphone Service.

dead front An insulated surface that eliminates any exposure of metal parts. Eliminates inadvertent disruptions and noise.

dead letter message A notification sent to the originator of an electronic mail message indicating that the message could not be delivered.

deadly embrace A state of a system of cooperating processes in which it is logically impossible for the activity of some or all of these processes to continue. A deadly embrace may result, for example, when the existence of a critical section is not recognized.

deallocate To release a resource that is assigned to a specific task.

DEC Digital Equipment Corporation.

decentralized Pertaining to processing in which intelligence is located at several remote sites of the same processing system.

decentralized network An information processing system in which processing and data base storage are distributed among two or more locations.

decibel (dB) A tenth of a bel. A unit of measuring relative strength of a signal parameter such as power, voltage, etc. The number of decibels is ten times the logarithm (base 10) of the ratio of the measured quantity of the reference level. The reference level must always be indicated, such as 1 milliwatt for power ratio.

decimal A digital system that has ten states, 0 through 9.

decipher In IBM's VTAM, to convert enciphered data into clear data. Synonymous with decrypt.

DECnet Trademark for DEC's communications network architecture that permits interconnection of DEC computers using DDCMP.

DECNET E-net DEC Engineering Network.

decoding Changing a digital signal into analog form or into another type of digital signal.

decrypt In IBM's VTAM, to convert encrypted data into clear data. Synonym for decipher.

decryption The reversal of encryption. Encryption and decryption are carried out by mathematically applying a key (a 64-bit number) to an algorithm to scramble and descramble the data.

DED Dynamically Established Data Link.

dedicated Committed to one specific use, such as a dedicated port on a computer to a specified terminal or microcomputer.

Dedicated Access Facility (DAF) Connects customer-premise devices to a public network via leased access channels.

dedicated circuit A circuit offered on a full-period basis, as opposed to on an on-demand switched basis. A leased circuit.

dedicated communications link A link that is continuously available to the user. Also called private or leased.

dedicated connection A non-switched circuit that is always available to the user.

dedicated line A communications line that is not dialed. Also called a leased line or private line.

dedicated network A network confined to the use of one customer. Also called private network.

dedicated port Port which is assigned to communicate with one other port.

default 1. Condition which exists from POWER ON or RESET if no instructions to the contrary are given to a device. 2. A value or option assumed by a computer or a network program, when no other value is specified.

default drive The disk or diskette drive currently in use by a work station.

default server The server in a local area network to which your default drive is mapped.

default SSCP list In IBM's VTAM, a list of SSCPs, either in VTAM's network or in another network, that can be used when no predefined CDRSC or name translation function is provided specifying an LU's owning CDRM. This list is filed as a part of an adjacent SSCP table in the VTAM definition library.

default SSCP selection In IBM's VTAM, a function that selects a set of one or more SSCPs to which a session request can be routed when there is no predefined CDRSC or name translation function provided that specifies an LU's owning CDRM.

Defense Advanced Research Projects Agency (DARPA) A funding agency for computer networking experiments performed on ARPANET.

definite response In SNA, a value in the form-of-response-requested field of the request header. The value directs the receiver of the request to return a response unconditionally, whether positive or negative, to that request.

definition statement IBM's SNA: 1. In VTAM, the means of describing an element of the network. 2. In NCP, types of instructions that define a resource to the NCP.

degradation Deterioration in the quality or speed of data transmission, caused as more users access a computer or computer network.

degree In the application of graph theory to networks, the degree of a node is the number of lines which connect to it.

DEL The ASCII DELETE code used in some instances to delete transmitted characters or to exit modes of operation.

delay The waiting time between two events in a communications system.

delay, absolute The length of time taken for a signal to travel from one point to another in a communications system. Absolute delay is dependent on the length, frequency, and the medium of transmission.

delay distortion A distortion occurring on a communication line that is due to the different propagation speeds of signals at different frequencies. Some frequencies travel more slowly than others in a given transmission medium and therefore arrive at the receiver at different times. Delay distortion is measured in microseconds of delay relative to the delay at 1700 Hz. This type of distortion can have a serious effect on data transmissions because the data bits going down a transmission line cover a wide bandwidth. Also known as envelope delay.

delay equalizer A corrective device which is designed to make the phase delay or envelope delay of a circuit or system substantially constant over a desired frequency range.

delay modulation A modulation scheme that uses different forms of delay in a signal element. Frequently used in radio, microwave, and fiber optic systems.

delay vector Associated with one node of a packet switched network, the delay vector has as its elements the estimated transit times of packets starting there and destined for other nodes in the network. Nodes may send copies of their delay vector to their neighbors as part of an adaptive routing scheme.

delayed-request mode In SNA, an operational mode in which the sender may continue sending request units on the normal flow after sending a definite-response request chain on that flow, without waiting to receive the response to that chain.

delayed-response mode In SNA, an operational mode in which the receiver of normal-flow request units can return responses to the sender in a sequence different from that in which the corresponding request units (RUs) were sent. *Note*: An exception is the response to the DFC request CHASE. All resonses to normal-flow request units received before CHASE must be sent before the response to CHASE is sent.

delimit The process of structuring data with specific boundaries, such as control characters which define the beginning and end of a block of data.

delimiter A special character or group of characters or word that enables a computer to recognize the beginning or end of a portion of a program or segment of data.

delivery time Time from start of transmission at the transmitting terminal to reception at the receiving terminal.

delta A multi-node network topology that consists of three multiplexer nodes each interconnected with one another with at least one aggregate link.

delta channel See D-channel.

Delta Modulation (DM) A signal-difference sampling technique which uses a one bit if the amplitude difference between two samples is increasing and a zero bit if it is decreasing.

delta routing A method of routing in which a central routing controller receives information from nodes and issues routing instructions, but leaves a degree of discretion to individual nodes.

demand multiplexing A form of time division multiplexing in which the allocation of time to subchannels is made according to their need to carry data. A subchannel with no data to carry is given no time slot. Other names are dynamic multiplexing and statistical multiplexing.

DEMARC Rate DEMARCation point.

demarcation The physical point where a regulated service is provided to a user, e.g., the point where the wires associated with a dedicated circuit terminate in a connection strip or connector block.

demarcation strip A terminal board used to provide a physical point of interconnection between communications equipment on the user's premises and the transmission facilities of the communications line vendor.

DEMARK Demarcation.

demodulation The process of retrieving data from a carrier. The reverse of modulation.

demultiplexing The process of breaking a composite signal into its component channels; the reverse of multiplexing.

D-encapsulant A pourable, fast-setting, all-weather, two part, reenterable splice encapsulant. It is used to encapsulate splices on aircore and waterproof cables as well as the interface between aircore and waterproof cables.

Department Of Defense (DOD) Part of U.S. government executive branch that handles military matters, including data communications. Responsible for some LAN-associated protocols and standards, such as TCP/IP, as well as selected FIPS.

DEPIC Dual Expanded Plastic Insulated Conductors, consisting of a dual plastic coating made up of a foamed core within a thin, solid skin.

DES Data Encryption Standard.

designated gateway SSCP In IBM's VTAM, a gateway SSCP designated to perform all the gateway control functions during LU–LU session setup.

Designet A network design tool of BBN Communication Corp. of Cambridge, MA.

despooling The process of reading records off the spool into main storage. During the despooling process, the physical track addresses of the spool records are determined.

destination A logical location to which data is moved (a data receiver). A destination can be a terminal, a person, or a program.

destination field A field contained in a message header which contains the address of the station to which a message is being directed.

destination group See rotary.

Destination Logical Unit (DLU) In IBM's VTAM, the logical unit to which data is to be sent.

Destination Service Access Point (DSAP) In Digital Equipment Corporation Network Architecture (DECnet), the one byte field in an LLC frame on a LAN that identifies the receiving data link client protocol.

destination user The user to whom a data communication system is to deliver a particular user information block (or unit).

destination user information bits The binary representation of user information transferred from a data communications system to a destination user. When the user information is output as non-binary symbols (e.g., alphanumeric characters), the destination user information bits are the bits on which a final decoding is performed to generate the delivered symbols.

detector A transducer that provides an electrical output signal in response to an incident optical signal. The current is dependent on the amount of light received and the type of device.

deterministic Pertaining to an access method that requires each station on a local area network to wait its turn to transmit.

deterministic network A network having the characteristic of predictable access delay.

Deutsche Bundepost The PTT of the Federal Republic of Germany (FRG).

device A piece of computer equipment, such as a display or a printer, that performs a specific task.

Device Control (DC) A category of control characters primarily intended for turning on or off a subordinate device. (DC1, DC2, DC3, and DC4) also (X-ON and X-OFF).

device control character In IBM's VTAM, (ISO) a control character used for the control of ancillary devices associated with a data processing system or data communication system, for example, for switching such devices on or off.

device driver A software component that controls all data transfers to or from a peripheral or communications device.

device partitioning A pool of devices (called a fence) to be used exclusively by a set of jobs in a specific job class allowing an installation to tailor its device usage to its anticipated workload.

device-type logical unit In IBM's VTAM, a logical unit that has a session limit of one and usually acts as the secondary end of a session. It is typically an SNA terminal (such as a logical unit for a 3270 terminal or a logical unit for a 3790 application program). It could be the primary end of a session, for example, the logical unit representing the Network Routing Facility (NRF) logical unit.

DFC Data Flow Control (IBM's VTAM).

DFI Digital Facilities Interface.

DFN Deutsche Forschungsnetz.

DFSYN response In IBM's VTAM, a data flow synchronous (DFSYN) response which is a normal-flow response that is treated as a normal-flow request so that it may be received in sequence with normal-flow requests.

DFT Distributed Function Terminal.

DFTAC Distributing Frame Test Access Controller.

DIA Document Interface Architecture.

diagnostic programs Computer programs used to check equipment malfunctions and pinpoint faulty components. Can be used manually or can be automatically called in by supervisory programs.

diagnostic unit A device used on a conventional computer to detect faults in the system. Separate unit diagnostics will check such items as arithmetic circuitry, transfer instructions, each input–output unit, and so on.

diagnostics Programs or procedures used to test a piece of equipment, a communications link or network, or any similar system.

dial backup The process of using the public switched telephone network to reestablish communications if a leased line facility becomes inoperative.

Dial Backup Auto Answer Unit (DBAAU) An ancillary device for automatic phone answering.

dial line (dial-up line) A line that requires signaling, such as tone or pulse, to alarm a potential receiving station that a call is pending.

dial network A network that is shared among many users, any one of whom can establish communication between desired points by use of a dial or pushbutton telephone. Synonymous with public switched telephone network.

dial pulse A current interruption in the dc loop of a calling telephone. It is produced by the breaking and making of the dial pulse contacts of a calling telephone when a digit is dialed. The loop current is interrupted once for each unit of value of the digit.

dial tone A tone indicating that automatic switching equipment is ready to receive dial signals.

dial train The series of pulses or tones that contain the information necessary to route a call over the public switched network.

dial transparency The dial transparency option (EIA passthrough) allows a multiplexer to pass RS-232C interface signals from a remote, dial-up modem to the central-site computer as if the remote modem were connected directly to the computer.

DIAL X.25 Packet mode communications carried out by dial-in X.25 Data Terminal Equipments (DTEs) that are temporarily connected to the network (as distinct from a hard-wired connection).

dial-back security A security technique whereby the remote-access system will limit access to users who have the proper password and log-on and are dialing from a prearranged location. After providing the proper log-on sequence, the system will hang up and dial a number that was previously stored in the system from which the user was supposed to be originating the call.

Dialed Number Identification Service (DNIS) A service offered by MCI which identifies specific 800 number categories to 800 number subscribers.

dial-in The ability of a modem to answer an incoming call from a remote terminal.

dialing Establishing a connection through common communications lines over the switched telephone network.

Dial-it 900 Service An AT&T mass announcement service for larger volumes of calls over short timeframes, used to provide a pre-recorded or live message or to register an audience's opinion. Can be designed to have a certain frequency of calls forwarded to a live operator or another message to award a prize, speak to a celebrity, collect a donation, etc. Callers ordinarily pay for the calls.

dialog A comprehensive on-line information service that has over 300 databases covering such subjects as business, science, medicine, law, energy, agriculture, and the humanities. For additional information, telephone 1-800-334-2564.

dial-out The ability of a modem to call a remote device.

dial-tone first service A coin telephone service that permits callers to reach the operator and to dial directory assistance or 911 without charge.

dial-up The use of a dial or pushbutton telephone to

initiate a station-to-station telephone call.

dial-up line, dial-in line, dial line A circuit or connection on the public telephone network.

dial-up network The use of a dial or pushbutton instrument, or an automatic machine, to initiate a call on the public switched telephone network.

dibit A group of two bits. In 4-phase, phase modulation such as DPSK, each possible value of a dibit is encoded as a unique carrier phase shift; the four possible values of a dibit are 00, 01, 10, and 11. The following table illustrates one common assignment of dibit values to phase shifts.

DPSK PHASE SHIFTS

Dibit value	*Phase shift*
00	45 degrees
01	135 degrees
10	315 degrees
11	225 degrees

dielectric An electrically non-conductive material through which signals can be induced by magnetic flux.

differential modulation A type of modulation in which the absolute state of the carrier for the current signal element is dependent on the state after the previous signal element.

Differential Phase Shift Keying (DPSK) Modulation technique used in Bell 201 modem.

diffraction The deviation of a wavefront from the path predicted by geometric optics when a wavefront is restricted by an opening or an edge of an object. *Note*: Diffraction is usually most noticeable for openings of the order of a wavelength. However, diffraction may still be important for apertures many orders of magnitude larger than the wavelength.

Digilog A company located in Montgomeryville, PA, which specializes in the manufacture of protocol analyzers.

Digilog Command Language (DCL) A programming language used to prepare setup and testing programs for a Digilog protocol analyzer.

digit One of the numerals in a number system.

digital Discretely variable as opposed to continuously variable. Data characters are coded in discrete, separate pulses or signal levels.

Digital Access and Cross-connect System/ Customer Controlled Rerouting (DACS/CCR) Allows DACS routing functionality to be under the control of the T1 user.

Digital Access Cross Connect Switch (DACS) An electronic switch which is capable of switching data of either the DS-0 or DS-1 levels. DACS allows T1 carrier facilities or any of its subchannels to be switched or cross-connected to another T1 carrier.

digital adaptive equalization A repeating pattern is transmitted by a modem on one side of a circuit which enables a modem receiver at the opposite end of the circuit to measure and build up compensation to correct line distortion.

digital–analog converter A device that converts a representative number sequence into a signal that is a function of a continuous waveform.

Digital Business System (DBS) A term used by Panasonic Co. for a midrange digital hybrid product line.

Digital Carrier Line Unit (DCLU) Provides integrated (i.e., no central office terminal) access to an AT&T 5ESS switch for a SLC 96 carrier in a pre-ISDN environment. The DCLU-5 extends that integrated access to ISDN service.

Digital Circuit Multiplexing (DCM) A proprietary speech compression technique developed by ECI Telecom, Inc., which increases the voice capacity over the TAT-8 transatlantic cable from 10 000 to 50 000 voice channels.

Digital Circuit Multiplication (DCM) A method for increasing the effective capacity of primary-rate and higher-level PCM hierarchies based upon speech coding at 64 Kbps.

digital city A city in which a Principle Telephone Company Central Office is located and serves a specific geographical area for Digital Data Service (DDS).

digital computer A programmable electronic device that handles information in the form of discrete binary numbers. The user's program and data are stored in the memory of a computer.

Digital Cross-connect (DSX) A Bell standard transmission interface which specifies signal levels, signal format, and connectivity.

Digital Cross-connect, Level 1 (DSX-1) A Bell standard transmission interface which specifies signal levels, format, and connectivity for DS-1.

digital data Information represented by a code consisting of a sequence of discrete elements.

Digital Data Communications Message Protocol (DDCMP) A communications protocol used in DEC computer-to-computer communications.

digital data network A network specifically designed for the transmission of data, wherever possible in digital form, as distinct from analog networks such as telephone systems, on which data transmission is an exception.

Digital Data Service (DDS) The Bell System developed a digital transmission network to provide higher data rates with fewer errors and lower costs. Known as Digital Data Service (DDS), this network provides full-duplex, private line service at synchronous data rates of 2400, 4800, 9600, and 56 000 bps, or 1.544 Mbps as well as switched point-to-point transmission at 56 000 bps. DDS uses bipolar pulse code modulation which is a way of transmitting a binary stream of data. A zero is sent as "no pulse" and a one is sent as a "pulse." This service is employed over a combination of local distribution facilities, intermediate length lines (T1 and T2 carriers), and long haul facilities. Regenerative repeaters, spaced every few thousand feet in the local loops, refresh the digital pulse.

Digital Data Throughput (DDT/DDS) A service offered by carriers for terminating dataphone digital service circuits at a central office via a T1 connection. DDS circuits are 56 Kbps data circuits offered by telcos. Data rates of 2.4 Kbps, 4.8 Kbps, 9.6 Kbps, and 56 Kbps can be used on DDS circuits.

Digital Equipment Corporation (DEC) A leading manufacturer of minicomputers and related hardware and software.

digital error Digital transmission where a zero is interpreted as one or vice versa.

digital hierarchy A series of multiplexing schemes over digital circuits in which signals within the bandwidth of any circuit at a given bit rate can be multiplexed into the bandwidth of a circuit at the next higher rate. The current North American Digital Hierarchy incorporates the following multiplexing levels: DS-0, 64 Kbps; DS-1, 1.544 Mbps, octect interleaved, carries 24 DS-0 circuits; DS-1C, 3.152 Mbps, bit interleaved, carries two DS-1 circuits or 48 DS-0 circuits; DS-2, 6.312 Mbps, bit interleaved, carries two DS-1C circuits, doue DS-1 circuits, or 96 DS-0 circuits; DS-3, 44.746 Mbps, bit interleaved, carries seven DS-2 circuits, 14 DS-1C circuits, 28 DS-1 circuits, or 672 DS-0 circuits; and DS-4, 274.176 Mbps, bit interleaved, carries six DS-3 circuits, 42 DS-2 circuits, 84 DS-1C circuits, 168 DS-1 circuits, or 2016 DS-0 circuits.

DIGITAL HIERARCHY: The six Levels of Transmission in the current North American Digital Hierarchy

Circuit type	*Bit rate*	*Equivalent number of circuits*				
		DS-0	DS-1	DS-1C	DS-2	DS3
DS-0	64 K	1	—	—	—	—
DS-1	1.544 M	24	1	—	—	—
DS-1C	3.152 M	48	2	1	—	
DS-2	6.312 M	96	4	2	1	—
DS-3	44.736 M	672	28	14	7	1
DS-4	274.176 M	2016	168	84	42	6

digital loopback Technique for testing the digital circuitry of a communications device. May be initiated locally, or remotely via communications circuits. Device tested will transmit back a received test message, after decoding. Results are compared with original transmission.

digital loopback testing In data communications, a

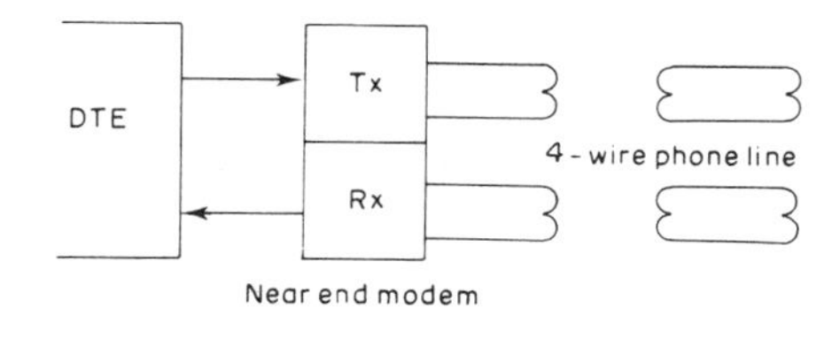

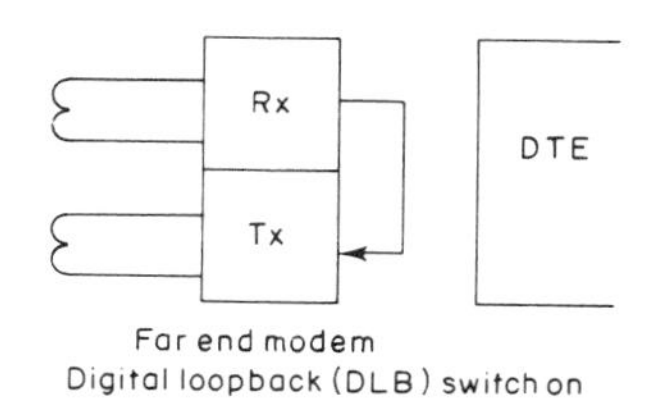

digital loopback testing

technique whereby a test device transmits a known pattern through a local modem and a communications channel to a remote modem. The test pattern is demodulated, remodulated, then internally looped back in the remote modem and transmitted back to the test device. Any bit errors induced in the loopback path will be detected and counted.

(1) Tests both modems and phone lines from the near end. (2) Loopback is initiated from the near end if the modem is capable, otherwise far-end operator intervention is required.

digital milliwatt A pulse code modulated signal equivalent to an analog signal of 1 kHz at 1 mW. Used as a reference pattern for testing digital circuits. A digital milliwatt signal fulfills the one's density requirements for digital transmission through repeaters.

Digital Multiplexed Interface (DMI) A specification, originated by AT&T, for a T1 based data communications interface between a mainframe computer and a private branch exchange. Unlike Northern Telecom's CPI, the DMI interface conforms to current ISDN specifications.

Digital Multiplexed Interface Bit-Oriented Signaling (DMI-BOS) A simple common channel signaling protocol for AT&T's Digital Multiplexed Interface. Designed as an interim step between channel-associated signaling and the more complex DMI Message-Oriented Signaling (DMI-MOS, based on Common Channel Signaling System No. 7), DMI-BOS uses bits in the signaling channel to emulate the E and M wires of an analog telephone.

Digital Multiplexed Interface Message-Oriented Signaling (DMI-MOS) A common channel signaling protocol for mature installations of AT&T's Digital Multiplexed Interface. Based on Common Channel Signaling System No. 7.

digital multiplexers Digital multiplexers form the interface between digital transmission facilities of different rates. They combine digital signals from several lines in the same level of the hierarchy into a single pulse stream suitable for connection to a facility of the next higher level in the hierarchy.

digital PBX (private branch exchange) A private, on premises, digital telephone switching exchange, owned and managed by the end user.

digital repeaters Devices inserted in a transmission medium to regenerate a digital signal sent over the medium.

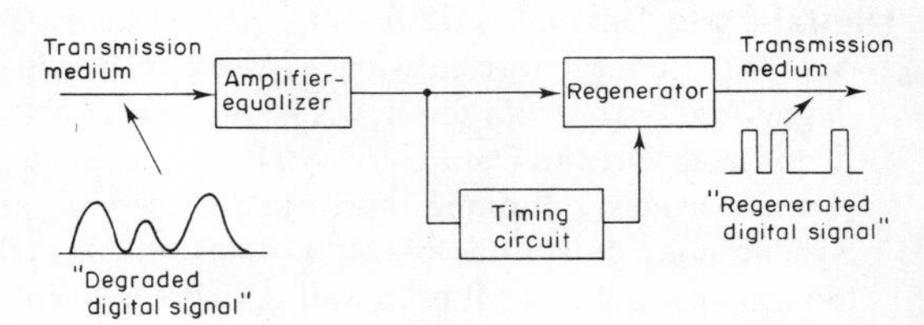

digital repeaters

digital routing A software-defined network (SDN) service feature offered by AT&T which enables an SDN user to transmit digital data at 56 Kbps between locations.

Digital Service, Europe A digital circuit of 2.048 Mbps, the European standard bit rate for DS-1 (T1) transmissions.

Digital Service, Level n A general descriptor for a level of the North American Digital Hierarchy.

Digital Service Area (DSA) A metropolitan area that is served by digital service.

Digital Service unit (DSU) Connects DTE to a digital transmission line such as AT&T's DDS. Also called data service unit.

Digital Service Unit/Channel Service Unit (DSU/CSU) A combined digital service unit and channel service unit. The DSU portion converts the digital signals from the data terminal equipment

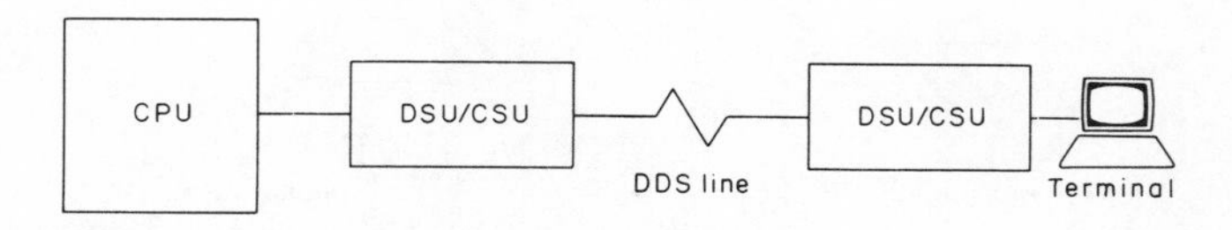

Digital Service Unit/Channel Service Unit (DSU/CSU)

to the form required for operation with the digital network, while the CSU portion is the circuitry designed as the interface to the digital lines.

digital signal A discrete or discontinuous signal, one whose various signals are discrete intervals apart.

Digital Signal Processing (DSP) Refers to the technology that enables a modem to convert digital signals to analog signals and analog signals to digital signals.

digital signaling The transmission of digital data over a communications medium by means of electrical signaling directly driven by the digital data stream, as opposed to analog signaling in which an indirect representation is used.

Digital Speech Interpolation (DSI) A voice compression technique which takes advantage of the pauses inherent in human speech to multiplex additional voice conversations onto the same transmission link.

digital station terminal A path for digital transmission furnished within the serving area of a digital city between the principal telephone company central office and a station.

digital switch A solid-state device that routes information in a digital form. The term is also used to describe a digital switching system.

digital switching The process of establishing and maintaining a connection, under program control, where digital information is routed between input and output. Generally, a "virtual circuit."

digital system cross-connect frame (DSX) A bay or panel to which T1 lines and DFI circuit packs are wired that permits cross-connections by patch cords and plugs. A DSX panel is used in small office applications where only a few digital trunks are to be installed.

Digital Termination System (DTS) An FCC-proposed system that integrates cellular radio for local access and microwave or satellite for long distance.

Digital to Analog (D/A) The conversion of a digital (discrete) signal to an analog (continuous) signal.

digital transmission A transmission in which signal elements are in the form of discrete pulses of constant rather than variable (analog) amplitude.

Digital Trunk and Line Unit (DTLU) Provides system access for T1 carrier lines used for interoffice trunks or remote switching module umbilicals.

digitize To convert information from its nominal state into a digital code.

digroup Represents the information content of 24 pulse code modulated (PCM) channels.

diode A one-way device that controls the flow of electrical charge. The diode lets current flow in one direction.

DIP Dual In-line Package/Pins.

dip switches Switches for opening and closing leads between two devices. When a lead is opened, a jumper wire can be used to cross or tie it to another lead.

direct access 1. The process of direct interaction between an on-line user program and a terminal, with the user program controlling requests for data and the immediate output of data. 2. A method for reading or writing a record in a file by supplying its key or index value.

direct access memory A method of computer storage that allows a particular address to be accessed independently of the location of that address. Therefore, the items stored in the memory can be accessed in the same amount of time, regardless of location.

direct activation In IBM's VTAM, the activation of a resource as a result of an activation command specifically naming the resource.

Direct Broadcast Satellite Service (DBSS) A TV transmission system in which video is broadcast via satellite directly to subscribers.

direct conduit method A ceiling distribution method in which cables are run in conduit from a satellite location directly to the desired information outlets.

direct current (dc) An electrical current that is used for digital signaling.

direct current loop A data transmission interface technique usually used with teleprinters, which transmit a digital signal usually at 20 milliamperes dc. Also called current loop.

direct deactivation In IBM's VTAM, the deactivation of a resource as a result of a deactivation command specifically naming the resource.

direct delivery A type of delivery in which an electronic message is sent directly to a device.

Direct Distance Dialing (DDD) A telephone service in North America which enables users to call their subscribers outside their local area without operator assistance. In the United Kingdom and some other countries, this service is known as STD (subscriber trunk dialing).

Direct Inward Dialing (DID) A feature of a PBX and some Centrex systems that allows outside callers to call an extension directly without local operator assistance.

Direct Memory Access (DMA) A method of transferring information to or from a computer memory device independently of the central processing unit.

Direct Outward Dialing (DOD) A feature of a PBX and Centrex systems that allows an inside caller to call an external number without local operator assistance.

directional antenna An antenna which limits the beamwidth of a received radio signal to the direction from which the radio signals arrive.

directive In Digital Equipment Corporation Network Architecture (DECnet), a management request sent by a director to an entity.

Director In Digital Equipment Corporation Network Architecture (DECnet), the management software used by a network manager.

directory 1. A structure containing one or more files and a description of those files. 2. In DECnet, a container in the Naming Service which stores a set of names.

directory routing A message-routing system which uses a directory at each switching node. Each directory contains the preferred outgoing link for each destination.

dirty loop A two-wire telephone loop that has impairments that results in a normal voice connection becoming noisy.

disable To turn off a device or prevent certain interrupts from occurring in a processing unit (such as a network communications card). Interrupts may be disabled by several means to include setting a switch or a jumper, or issuing a command.

disabled In IBM's VTAM, pertaining to an LU that has indicated to its SSCP that it is temporarily not ready to establish LU–LU sessions. An Initiate request for a session with a disabled LU can specify that the session be queued by the SSCP until the LU becomes enabled. The LU can separately indicate whether this applies to its ability to act as a primary logical unit (PLU) or a secondary logical unit (SLU).

disc See disk.

disconnect The process of disengaging or releasing a circuit between two stations.

disconnect sequence In a switching network, a sequence of characters used to terminate a call connection and exit the network.

disconnect signal A signal transmitted from one end of a switched circuit that the established connection should be broken. The signal is sent to the other end of the circuit.

disconnection The termination of a physical connection.

discrete access In LAN technology, an access method used in star LANs. Each station has a separate (discrete) connection through which it makes use of the LAN's switching capability.

discretely variable Capable of one of a limited number of values. Usually used to describe digital signals or digital data transmission.

discrimination In frequency modulation systems, the process at a receiver of detecting frequency changes of a carrier signal and deriving the information originally used to cause the frequency shifts. Equivalent to frequency demodulation.

disengagement confirmation The event that verifies that a particular user's participation in an established data communications session has been terminated. For that user, it completes the disengagement function and stops the counting of disengagement time. Disengagement confirmation occurs, for a particular user, when (1) disengagement of that user has been requested, and (2) the user is able to initiate a new access attempt. Most data communication systems notify the user that a new access attempt can be initiated by issuing an explicit disengagement confirmation signal. In cases where no disengagement confirmation signal is issued, the user may initiate a new access attempt to confirm disengagement.

disengagement request The event that notifies the system of a user's desire to terminate an established data communication system. It is complementary to the access request in most systems. Each disengagement request begins the disengagement function and starts the counting of disengagement time for one or both users. Disengagement is normally requested simultaneously for both users in connection-oriented sessions, and independently for each end user in connectionless sessions.

disk An electromagnetic storage medium for digital data.

Disk Operating System (DOS) A program or set of programs that instruct a disk-based computing system to schedule/supervise work, manage computer resources, and operate/control peripheral devices.

disk server A node on a local area network that allows other nodes to create and store files on its disks.

diskless work station A device connected to a local area network that does not have a local disk. The diskless work station uses the disk of a server connected to the network.

DISOSS DIStributed Office Support System.

dispersed network A communications network consisting of more than one processor, where processing may take place in many locations. This network may have both host and satellite processors.

dispersion In an optical system, dispersion is the spreading of photons in a light pulse as they speed down a lightguide. This "smears" the pulses and reduces the rate at which they can be transmitted. Most common are: (1) chromatic dispersion (different wavelengths of light within a pulse travel at different speeds); (2) modal dispersion (some parts of the pulse follow longer paths through the lightguide than other parts).

display In IBM's SNA (NPDA), refers to the title page and all following pages of a display. It is also referred to as a screen or panel.

display converter In IBM 3270 systems, a coaxial converter that allows asynchronous display terminals to emulate IBM 3278 Display Stations.

display station, display terminal A device consisting of a keyboard and video or CRT display. In the IBM 3270 Information Display System, a 3278 is an example of a display station. An ASCII CRT terminal is an example of a display terminal.

display types In IBM's SNA (NPDA), a concept to describe the organization of data displays. Display types are defined as total, most recent, user action, and detail. Display types are also referred to as display levels. Display types are combined with resource types and data types to describe NPDA display organization.

display unit, video A device that provides a visual representation of data.

Display-System Protocol (DSP) Packet Assembler/Disassembler (PAD) designed to support 3270 devices. 3270 Binary Synchronous Communications (BSC) DSP supports 3270 BSC devices. Systems Network Architecture (SNA) DSP supports 3270 Synchronous Data Link Control (SDLC) devices.

distance sensitivity A method of charging for circuit services based on the distance traveled, as opposed to distance insensitivity, in which other bases than distance such as duration or data volume are used.

distortion The unwanted changes in signal or signal shape that occurs during transmission between two points.

distributed adaptive routing A method of routing in which the decisions are made on the basis of exchange of information between the nodes of a network.

distributed architecture In LAN technology, a LAN that uses a shared communications medium. Used on bus or ring LANs. Uses shared access methods.

distributed computing The name of the trend to move computing resources such as minicomputers, microcomputers, or personal computers to individual work stations.

distributed management In Digital Equipment Corporation Network Architecture (DECnet), a form of management where network managers and directors are dispersed across many systems.

distributed network A communications network made up of two or more processing centers. The processing may be distributed between each center and each processor must be in communication with all others.

Distributed Office Systems Support (DISOSS) One of IBM's strategic Systems Application Architecture (SAA) products for document library storage and corporate electronic messaging.

distributed processing An arrangement that allows separate computers to share work on the same application program. Often erroneously used to mean distributed computing.

distributed system management model The framework within which DNA network management is designed.

distributing frame A structure for terminating permanent wires of a telephone Central Office, private branch exchange, and for permitting the easy change of connections between them by means of cross-connecting wires.

distributing rings Rings used to provide support for wires in building apparatus and distribution closets.

distribution block Centralized connection equipment where telephone or data terminal wiring is terminated and cross-connections are made.

distribution list Contains references to subsets of electronic mail users that can be used for broadcast addressing. A distribution list is used to send the same message simultaneously to a group of users.

distribution switch An X.25 processor, used to route and switch data packets through a Packet Switching Network (PSN), and which can be equipped with Packet Assembler/Disassembler (PAD) software to interface terminal and host devices to the network.

diversity A technique employing two or more channels over which a single transmission is carried, thus improving reliability, security, etc.

divestiture The breakup of AT&T by the Federal Court based on an antitrust agreement reached between AT&T and the U.S. Department of Justice. Effective 1 January, 1984.

DLB Digital Loopback.

DLC Data Link Control.

DLCF Data Link Control Field.

DLI Dual Link Interface.

DLO Data Line Occupied.

DLTU Digital Line Trunk Unit.

DLU Destination Logical Unit (IBM's SNA).

DM Delta Modulation.

DMA Direct Memory Access.

DMI Digital Multiplexed Interface.

DMI-BOS Digital Multiplexed Interface Bit-Oriented Signaling.

DMI-MOS Digital Multiplexed Interface Message-Oriented Signaling.

DNA Digital Network Architecture.

DNIC Data Network Identification Code.

document content architecture An IBM architecture, part of SNA which defines a standard for the transmission and storage of documents over networks to include voice, data, and video.

document interface architecture An IBM architecture which is included in SNA for controlling the exchange of document information between dissimilar editing systems.

documentation A written description of a program that includes its name, purpose, how it works, and, frequently, operating instructions. Good documentation reduces both labor and ulcers for those who later use and debug a program.

DOD Department of Defense (U.S.).

domain In IBM's SNA, a host-based System Services Control Point (SSCP) and the physical units (PUs), logical units (LUs), links, link stations, and associated resources that the host SSCP has the capability to control.

domain operator In IBM's SNA multiple domain network, the person or program that controls the operation of the resources controlled by one System Services Control Point (SSCP).

domain specific part In Digital Equipment Corporation Network Architecture (DECnet), the part of an NSAP address assigned by the addressing authority identified by the IDP.

dormant state A state of inactivity in which there is no current request for a task.

DOS Disk Operating System.

double-current transmission, polar direct-current system A form of binary telegraph transmission in which positive and negative direct currents denote the significant conditions.

DOV Data Over Voice.

Dow Jones news/retrieval service An on-line information service that specializes in providing subscribers with business news to include stock prices, annual report information, mergers, and other news of interest to investors. For additional information, telephone 1-800-257-5114.

down time Period when all or part of a system or network is not available to end users due to failure or maintenance.

downline loading The process of sending configuration parameters, operating software, or related data from a central source to individual stations.

downlink The transmission path from a satellite to a ground station.

down-loop A fixed pattern (repeating 100) which forces a receiving CSU to loopback the signal to its transmit side.

downstream In IBM's SNA, in the direction of data flow from the host to the end user.

downstream device For the IBM 3710 Network Controller, a device located in a network, such that the 3710 is positioned between the device and a host. A display terminal downstream from the 3710 is an example of a downstream device.

downstream line In IBM's SNA, a telecommunication line attaching a downstream device to an IBM 3710 Network Controller.

Downstream Load Utility (DSLU) In IBM's SNA, a program product that uses the communication network management (CNM) interface to support the load requirements of certain type 2 physical units, such as the IBM 3644 Automatic Data Unit and the IBM 8775 Display Terminal.

DP Data Processing.

DP Dial Port.

DPDU Data Link Protocol Data Unit.

DPPS Data Packets Per Second.

DPSK Differential Phase Shift Keying.

DQ packet Data Qualifier packet.

DR Dynamic Reconfiguration (IBM's SNA).

Draft Proposal An ISO standard document that has been registered and numbered but not yet given final approval.

DRAM module Dynamic Random Access Memory module.

DRDS Dynamic Reconfiguration Data Set (IBM's SNA).

driver 1. A software module that controls an input/output port or external device. 2. Short for line driver.

drop Individual connections (sometimes called nodes) on a multipoint (also called multidrop) circuit.

drop, subscriber's The line from a telephone cable to a subscriber's building.

drop cable In local area networks, a short cable connecting a network device with the network cable.

drop-and-insert A T1 multiplexing technique similar to Channel Bypass, used when some channels in a DS1 stream must be demultiplexed at an intermediate node. With drop-and-insert, all channels are demultiplexed, those destined for the intermediate node are dropped, traffic from the intermediate node is added, and a new T1 stream is created to carry all channels to the final destination. With bypass, only those channels destined for the intermediate node are demultiplexed—ongoing traffic remains in the T1 signal, and new traffic bound from the intermediate node to the final destination takes the place of the dropped traffic.

drop-and-insert emulation A technique for T1 protocol analysis, in which the protocol analyzer terminates one channel of the DS-1 bit stream in one or both directions, and inserts test data into that channel while maintaining the DS-1 line protocol (useful for testing terminal equipment through a T1 multiplexer or PBX).

dropouts Cause of errors and loss of synchronization with telephone data transmission; where signal level unexpectedly drops at least 12 dB for more than 4 milliseconds. AT&T standards suggest only two dropouts per 15-minute period.

dropped channels In a multinode multiplexer network dropped channels are channels which are terminated at a node.

DRS Data Rate Selector.

dry T1 T1 with an unpowered interface.

DS-0 1. Digital System, level 0. An individual 64 Kbps data or digitized voice channel. Voice digitization is 8-bit PCM. 2. Digital Signal zero, 64 Kbps.

DS-1 Digital Signal Level 1. The type of signal carried by a T1 facility, normally at 1.544 Mbps.

DS-1C Digital Signal Level 1C. The type of signal carried by a 3.152 Mbps digital transmission facility. A DS1C circuit carries the equivalent of 2 DS1 circuits, or 448 DS0 circuits.

DS-2 Digital Signal Level 2. The type of signal carried by a 6.312 Mbps T2 digital transmission facility. A DS2 circuit carries the equivalent of two DS1C circuits, four DS1 circuits, or 96 DS0 circuits.

DS-3 1. Digital Service, Level 3. A digital circuit of 44.736 Mbps. A DS3 circuit carries the equivalent of 7 DS2 circuits, 14 DS1C circuits, 28 DS1 circuits, or 672 DS0 circuits. DS3 circuits are now the largest units of digital transmission available for private use. 2. Digital Signal Level 3C, 44.736 Mbps digital signal.

DS-4 1. Digital Service, Level 4. A digital circuit of 274.176 Mbps, currently the largest unit of transmission in the North American Digital Hierarchy. A DS4 circuit carries the equivalent of 6 DS3 circuits, 42 DS2 circuits, 84 DS1C circuits, 168 DS-1 circuits, or 4032 DS-0 circuits. *Note*: Do not confuse the DS-4 transmission level with D4 framing, a technique used only on DS-1 circuits. 2. Digital Signal Level 4C, 274.176 Mbps digital signal.

DSA Digital Service Area.
DSC Direct Satellite Communications.
DSE 1. Distributive System Environment. 2. Digital Service Europe.
DSI Digital Speech Interpolation.
DSIRnet Government network in New Zealand.
DSN Digital Service, level N.
DSP 1. Digital Signal Processing. 2. Domain Specific Part.
DSR Data Set Ready.
DSU Data Service Unit.
DSU/CSU Digital Service Unit/Channel Service Unit.
DSX Digital Cross Connect.
DSX-1 Digital signal cross-connect level 1. The set of parameters used where DS-1 digital signals are cross-connected (T1 carrier).
DTE Data Terminal Equipment.
DTLU Digital Trunk and Line Unit.
DTMF Dual Tone Multi-Frequency.
DTMF/MF compelled operation The method of using DTMF/MF tones to communicate commands and responses to and from a master controlling unit.
DTR Data Terminal Ready.
DTS 1. Digital Tandom Switch. 2. Digital Termination System.
D-type conditioning Conditioning used to control the signal to C-notched noise ratio and intermodulation distortion.
dual cable A type of broadband cable system in which separate cables are used for transmission and reception.
dual chassis Two interconnected chassis in a cabinet with one network address.
Dual Dial Restoral (DDR) An option that provides public switched telephone network backup to the modem's normal leased line channel in the event of failure of that channel.
Dual Fiber Cable A type of optical fiber cable that has two single fiber cables enclosed in an extruded overjacket of polyvinyl chloride, with a ripcord for peeling back the overjacket to access the fibers.
Dual In-line Pins (DIP) Term used to describe the pin arrangement on an integrated circuit (IC) or a multiple (electric) switch.
Dual-Tone Multiple-Frequency (DTMF) Term used to describe the audio signaling frequencies on Touch-Tone, pushbutton telephones. The telephone handset generates a composite audio signal made of the superposition of two tones selected by line-and-column addressing of a keyboard. This scheme is shown in the illustration. These frequencies were chosen in such a way that neither their harmonics nor their intermodulation products fall in one of the tone bands. Separation between tones is typically 10%.

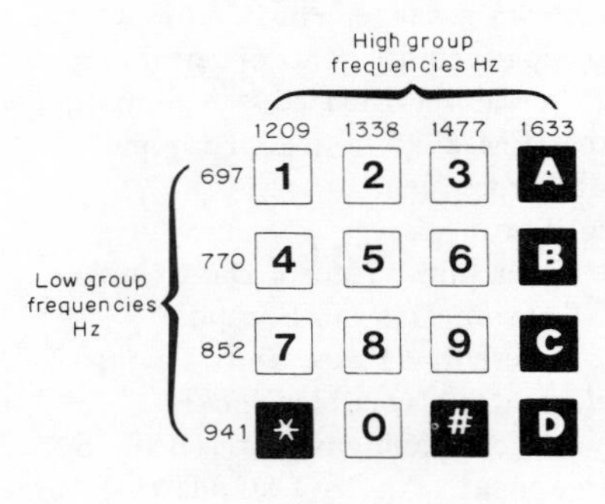

Dual-Tone Multiple-Frequency (DTMF)

The fourth column of buttons (1633 Hz tone) is not usually present in today's telephone sets. However, it is available for future use. A 4 × 4 matrix of buttons is presently used in some military phones and is suitable for implementation in telephone sets intended for multiple functions such as financial transaction terminals, credit checking machines, and some PBX or CBX systems.

The dual-tone signal is heard in the ear-piece and is sent over the telephone lines to the private branch exchange (PBX) or directly to the Central Office, where it must be decoded into the digit that it represents. The decoder circuitry—or receiver—will convert the DTMF signal to a binary format or to a 2-of-8 format. For the traditional dial-pulse equipment, a further conversion is needed to achieve a complete Touch-Tone to dial-pulse interface, allowing telephone companies to offer subscribers Touch-Tone service in areas where the equipment is not compatible. Such an application is termed a tone-to-pulse converter.
duct system The tubular enclosures that hold wires and cables, enabling them to pass through the floors or ceilings of a building.

dumb terminal A term used to describe a Teletype or Teletype-compatible terminal. The dumb terminal is an asynchronous terminal that may operate at speeds as high as 9600 bps or higher. The dumb terminal is an ASCII terminal that, although it may be "intelligent" in many of the functions it provides, uses no communications protocol.

duobinary signalling A method of transmitting asynchronous binary waveform in which neighboring signal segments influence one another in a controlled manner.

duplex A facility which permits transmission in both directions simultaneously (sometimes referred to as full duplex, contrasted with half-duplex).

duplex chassis Two independent chassis in a cabinet with two network addresses.

duplex circuit A circuit used for transmission in both directions at the same time. It may be called full duplex to distinguish it from half duplex.

duplex signaling (DX) A signaling system which occupies the same cable pair as a voice path, yet does not require filters.

duplex transmission Simultaneous two-way independent transmission in both directions. Also called full-duplex.

duplexing An outdated term referring to the use of duplicate computers, files or circuitry, so that in the event of one component failing, an alternative one can enable the system to carry on its work.

duty cycle The relationship between the time a device is on and the time it is off. For example, a terminal that is powered on for one hour a day is said to have a low duty cycle.

DUV Data Under Voice.

DX Duplex Transmission.

Dyad A Siemens trademark for the vendor's series of digital telephones that are used with the firm's Saturn digital voice/data business communications systems.

dynamic allocation A technique in which the resources assigned are determined by criteria applied at the moment of need.

dynamic assignment In Digital Equipment Corporation Network Architecture (DECnet), the use of a Dynamically Established Data Link by the Network layer, in such a way that a connection is only made when data is required to be transferred, and the subnetwork address to which the connection is made is determined by the destination NSAP address.

Dynamic Connection Management (DCM) In Digital Equipment Corporation Network Architecture (DECnet), the use of a Dynamically Established Link by the Network layer, in such a way that a connection is only made when data is required to be transferred.

dynamic LPDA In IBM's SNA, a function enabling the NCCF application to set/query the LPDA status for a link/station.

dynamic multiplexing Time-division multiplexing in which the allocation of time to constituent channels is based on the demands of these channels.

Dynamic Random Access Memory (DRAM) A type of computer memory in which information can be stored and retrieved in miscellaneous order, but which must be maintained or "refreshed" by a periodic electrical charge if the memory is not "read out" or used immediately.

Dynamic Reconfiguration (DR) In IBM's VTAM, the process of changing the network configuration (peripheral PUs and LUs) associated with a boundary node, without regenerating the boundary node's complete configuration tables.

Dynamic Reconfiguration Data Set (DRDS) In IBM's VTAM, a data set used for storing definition data that can be applied to a generated communication controller configuration at the operator's request.

dynamic tables A configuration in which tables can be downline-loaded into a network device while the device is in operation.

dynamic threshold alteration In IBM's SNA, an NCP function to allow an NCCF operator to dynamically change the traffic count and temporary error threshold values associated with SDLC and BSC devices.

dynamic threshold query In IBM's SNA, an NCP function to allow an NCCF operator to query the current settings of a traffic count or temporary error threshold value associated with an SDLC or BSC device.

dynamically established data link In Digital Equipment Corporation Network Architecture (DECnet), a connection-oriented subnetwork used by the Network layer.

D&I Drop-and-Insert.

E

E78 A software program marketed by Digital Communications Associates which enables a personal computer to emulate an IBM 3278 or 3279 terminal when an appropriate adapter card is installed in the PC.

EADAS Engineering and Administrative Data Acquisition System.

EAEO Equal Access End Office.

EARN European Academic Research Network.

earth stations Ground terminals that use antennas and other related electronic equipment designed to transmit, receive, and process satellite communications.

EAS Extended Area Service.

Eastern Mediterranean Optical System (EMOS) A fiber optic undersea cable which is designed to link the European Continent with Sicily, Greece, Turkey, and Israel.

Easylink An electronic mail service marketed by Western Union.

Easyplex An electronic mail system offered by CompuServe Information Services Division.

Easytalk A trademark of Nady Systems, Inc., of Oakland, CA, as well as a series of telephone headsets.

EBCDIC Extended Binary Coded Decimal Interchange Code. The EBCDIC code implemented for the IBM 3270 information display system is illustrated.

ECB Event Control Block (IBM's SNA).

ECC Error Correcting Code.

ECD Equipment Configuration Database.

echo Distortion caused when a signal being reflected back to the originating station is perceptively delayed in time.

echo canceller A chip that identifies and erases echoes occurring in satellite and terrestrial long-distance phone calls of about 2000 miles and beyond. Essentially, it creates a mirror image of an echo which then blots up the real echo when it comes along.

echo cancellation The technique used in high speed modems which isolates and removes unwanted signal energy resulting from echoes of the transmitted signal.

echo canceller testing The testing of an echo canceller's ability to attenuate unwanted echo signals, signal distortion, and verify disable operations.

echo check A method of checking data transmission accuracy whereby the received data are returned to the sending end for comparison with the original data.

echo distortion A telephone line impairment caused by electrical reflections at distant points where line impedances are dissimilar.

echo modulation A method of producing a shaped pulse in which a main pulse with "echoes" is generated and put through a simple band-limiting filter. The "echoes" are pulses of controlled amplitude occurring both before and after the main pulse.

echo plug A standard DB25 connector wired to echo transmitted data and modem signals back to received-data and modem-signal inputs for diagnostic purposes. Also called a loop back plug.

echo suppressor A device used by telcos or PTTs that blocks the receive side of the line during the time that the transmit side is in use, preventing energy from being reflected back (echoed) to the transmitter. Because of impedance irregularities in the terminating networks, a small amount of the energy transmitted in one direction over the switched network is often reflected at the receiving end back toward the originating end. This reflection is commonly referred to as a talker echo. If the talker echo encounters another impedance irregularity at the originating end, a second echo called a listener echo is produced.

Most modems are designed to ignore talker echoes. On the other hand, listener echoes can interfere directly with receiving data if the time delay is sufficient and the echo is of sufficient amplitude.

EBCDIC code implemented for the IBM 3270 information display system.

	Bits 0, 1 →	00				01				10				11			
	Bits 2, 3 →	00	01	10	11	00	01	10	11	00	01	10	11	00	01	10	11
Bits 4567	Hex 1 ↓ / Hex 0 →	0	1	2	3	4	5	6	7	8	9	A	B	C	D	E	F
0000	0	NUL	DLE			SP	&	·									0
0001	1	SOH	SBA					/		a	j			A	J		1
0010	2	STX	EUA		SYN					b	k	s		B	K	S	2
0011	3	ETX	IC							c	l	t		C	L	T	3
0100	4									d	m	u		D	M	U	4
0101	5	PT	NL							e	n	v		E	N	V	5
0110	6			ETB						f	o	w		F	O	W	6
0111	7			ESC	EOT					g	p	x		G	P	X	7
1000	8									h	q	y		H	Q	Y	8
1001	9		EM							i	r	z		I	R	Z	9
1010	A					¢	!	¦	:								
1011	B					.	$	,	#								
1100	C		DUP		RA	<	*	%	@								
1101	D		SF	ENQ	NAK	(	)	_	'								
1110	E		FM			+	;	>	=								
1111	F		ITB		SUB	|	¬	?	"								

EBCDIC character set

As a consequence, echo suppressors are selectively employed in the switched network. The echo suppressor blocks the reflections.

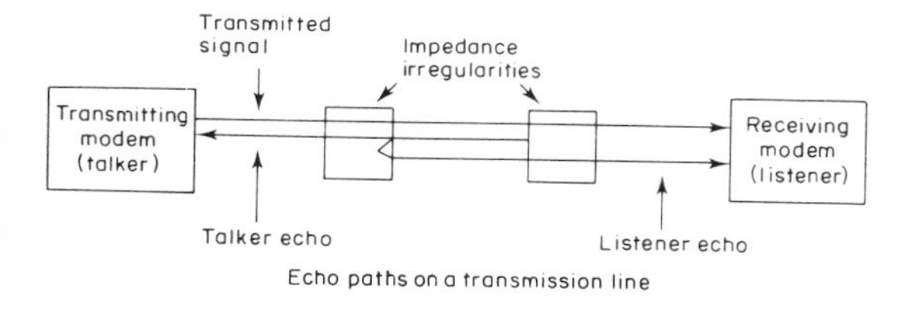

Echo paths on a transmission line

echo suppressor

Characteristics of the echo suppressors are:

(1) There is only one echo suppressor in each transmission path and it is between regional centers.

(2) It is normally used where the echo signal path delay is greater than 45 milliseconds.

(3) The echo suppressor has a turnaround time of 100 milliseconds.

(4) Echo suppressors are disabled with a signal operating in the range of 2010 to 2240 Hz and the signal must be present for at least 400 milliseconds.

(5) Echo suppressors are enabled when no power is present, in the 300 to 3000 Hz range for a period greater than 100 milliseconds.

echo testing A method of checking data transmission for accuracy by returning received data to the sending end for comparison with the original data. (1) Immediate Echo Data is retransmitted character by character as soon as it is received by the echoing device. (2) Buffered Echo Data is captured and stored by the echoing device until a predefined message end character is received. The accumulated data is then retransmitted to the original transmitter.

Echoplex A method of checking data integrity by returning characters to the sending station for verification of data integrity. Also called echo check.

ECL Emitter Coupled Logic.

ECMA European Computer Manufacturers Association.

ECN Engineering Change Notice.

ECSA Exchange Carriers Standards Association.

ECSC Exchange Carriers Standards Commission.

EDI Electronic Data Interchange.

EDP Electronic Data Processing.

EDTV Enhanced Definition TV.

EEPROM Electrically Erasable Programmable Read-Only Memory.

EFS Error Free Seconds.

EFT Electronic Funds Transfer.

EHF Extremely High Frequency.

EIA Electronic Industries Association.

EIA interface A set of signal characteristics (time duration, voltage, and current) specified by the Electronic Industries Association and accepted by the American National Standards Institute (ANSI).

EIA standard equipment rack unit The distance between mounting holes (center to center) in front vertical rails of an EIA standard equipment rack/cabinet. One unit equals 1.75 inches.

EIA standards

RS-232-C	An EIA-specified physical interface, with associated electrical signaling, between data circuit-terminating equipment (DCE) and data terminal equipment (DTE). The most commonly employed interface between computers and modems.
RS-422-A	Electrical characteristics of balanced-voltage digital interface circuits.
RS-423-A	Electrical characteristics of unbalanced-voltage digital interface circuits.
RS-485	The recommended standard of the Electronic Industry Association that specifies the electrical characteristics of generators and receivers for use in balanced digital multipoint systems.
RS-499	General-purpose 37-position and 9-position interface for data terminal equipment and data circuit-terminating equipment employing serial binary data interchange.
RS-499-1	Addendum 1 to RS-499

EIN European Infomatics Network.

EKRAN satellite The second Soviet geostationary communications satellite that was launched in 1976. This satellite was the world's first direct television broadcast satellite.

ELA Extended Line Adapter.

elastic store A storage buffer designed to accept data under one clock and deliver it under another, to accommodate short-term instabilities (jitter) in either clock. Elastic stores are extensively used in multiplexing and clock recovery applications.

Electrically Erasable Programmable Read Only Memory (EEPROM) A non-volatile memory device in which the contents can be changed by special circuitry incorporated in the parent system or device.

Electrically Programmable Read Only Memory (EPROM) Memory which is initially programmed by the manufacturer. Changes can be made to conform to customer requirements at their premises.

Electromagnetic Compatibility (EMC) The ability of suitable designed electronic equipment to operate in the same location without mutual interference.

Electromagnetic Interference (EMI) The interference in signal transmission or reception caused by the radiation of electrical and magnetic fields.

electromagnetic pulse An intense, short-duration burst of electromagnetic waves generated by a nuclear explosion.

electromagnetic wave A wave of energy consisting of interrelated electric and magnetic waves. Both light and radio are examples of electromagnetic waves.

electron An elementary particle made of a tiny charge of negative electricity.

electronic banking Any banking transaction which takes place electronically, including automatic teller machines, bank-to-bank wire transfers, etc.

electronic funds transfer The use of communications to simultaneously debit an account in one location and credit an account in another, thus "transferring" funds.

Electronic Industries Association (EIA) An American group whose members include manufacturers, that formulates technical standards and marketing data. The EIA often represents the telecommunications industry in governmental concerns. U.S. contributions to the CCITT are made through the EIA.

electronic mail A store and forward service for text messages from one computer terminal or system to another. The text is stored for the recipient until that person logs into the system to retrieve messages.

Electronic Order Exchange (EOE) A data communications approach to inter-company transactions between buyers and sellers. EOE can be employed to transmit purchase orders, price and product listings, and order-related information.

electronic patch panel Programming mechanism that enables the user to change dedicated port routing using program commands.

Electronic Switched Network (ESN) A comprehensive private networking package developed by Northern Telecom which connects geographically dispersed PBXs into a single, unified, private communications network.

Electronic Switching System (ESS) 1. A type of telephone switching system which uses a special purpose stored program digital computer to direct and control the switching operation. ESS permits the provision of custom calling services such as speed dialing, call transfer, three-way calling, etc. 2. AT&T manufactured, stored-program-control, central-office switches. Common ESSs include Nos 1, 1A, 4 and 5 switches.

electronic wire transfer In banking, a means by which funds from an account in one bank are transferred to another via telephone.

Electrostatic Discharge (ESD) A surge of electric current from an accidental source of stored electric charge. This is usually an accidental event resulting from a user moving around in a dry environment.

element In IBM's SNA: 1. A field in the network address. 2. The particular resource within a subarea identified by the element field.

element address In IBM's SNA, a value in the element address field of the network address identifying a specific resource within a subarea.

EM End of Medium.

embedded signalling A technique for transmitting signalling information on the same channel as the customer information.

EMC ElectroMagnetic Compatibility.

Emergency Power System (EPS) Consists of an engine and a generator set. This system, designed to start within 30 seconds of a commercial power failure, carries the electrical load for the duration of the commercial power failure.

EMI ElectroMagnetic Interference.

EMI/RFI ElectroMagnetic Interference/Radio Frequency Interference

Emitter Coupled Logic (ECL) A type of transistor circuit characterized by fast operation and high heat dissipation.

EMOS Eastern Mediterranean Optical System.

EMPI Electrical Pulse Interference.

empty slot ring In LAN technology, a ring LAN in which a free packet circulates past (or, more precisely, through) every station. A bit in the packet's header indicates whether it contains any messages (if it contains messages, it also contains source and destination addresses).

emulation A device or computer program, or combination that acts like another device or computer program. For example, a protocol analyzer can emulate DTE or DCE.

emulation mode In IBM's SNA, the function of a network control program that enables it to perform activities equivalent to those performed by a transmission control unit.

Emulation Program (EP) In IBM's SNA, a control program that allows a local 3705 Communications Controller to emulate the function of an IBM 2701 Data Adapter Unit, an IBM 2702 Transmission Control, or an IBM 2703 Transmission Control.

enable To turn on a device or place it in a state which will allow certain interrupts to occur in a processing unit. Interrupts are usually enabled by issuing a command or setting a switch or jumper.

enabled In IBM's ACF/VTAM, pertaining to an LU that has indicated to its SSCP that it is now ready to establish LU–LU sessions. The LU can separately indicate whether this applies to its ability to act as a PLU or an SLU.

encipher In IBM's SNA: 1. To scramble data or convert it, prior to transmission, to a secret code that masks the meaning of the data to any unauthorized recipient. 2. In ACF/VTAM, to convert clear data into enciphered data.

enciphered data In IBM's SNA, data that is intended to be illegible to all except those who legitimately possess the means to reproduce the clear data. Synonymous with ciphertext.

encode To code.

encoder A portion of a multibit encoded modem. The encoder groups bits for modulation.

encoding/decoding The process of reforming information into a format suitable for transmission.

encrypt In IBM's VTAM, to convert clear data into enciphered data. Synonym for encipher.

encryption The process of altering information in such a way that it is unintelligible to unauthorized parties. The process involves transforming a clear message into ciphertext to conceal its meaning.

end bracket In IBM's SNA, the value (binary 1) of the end bracket indicator in the request header (RH) of the first request of the last chain of a bracket. The value denotes the end of the bracket.

end distortion Communications signal distortion usually caused by speed error. It is the percentage difference in time between two consecutive mark to space transitions as compared to the ideal time.

End distortion = ((ideal pulse width − actual pulse width)/ideal pulse width) * 100%

If the received speed is too fast the result is positive. If the received speed is too slow the result is negative.

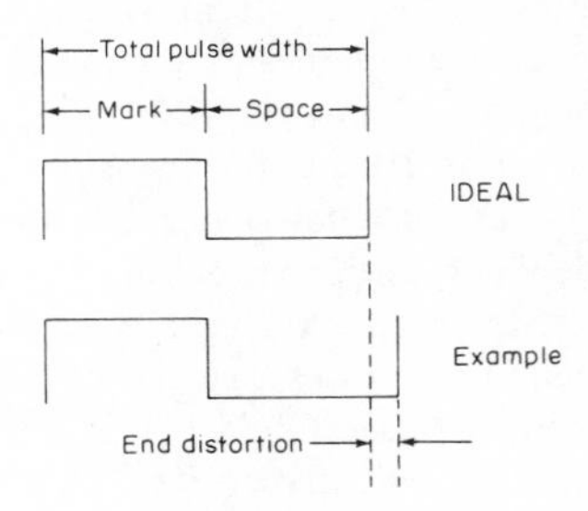

end distortion

end of address The character used to separate the address or routing part of a message from the subsequent parts of a message. Sometimes called end-of-routing symbol.

end of medium (EM) A control character that may be used to identify the physical end of data recorded on a medium. Also (end of file).

End Of Text (ETX) Indicates end of message. If multiple transmission blocks are contained in the message in Binary Synchronous Communications, ETX terminates the last block.

End Of Transmission (EOT) An ASCII code that means "end of transmission" (EOT). Used in the EOT/ACK handshaking protocol. The computer sends an EOT at the end of each transmission to the terminal. When the terminal is ready to receive more data, it transmits an acknowledge (ACK) back to the computer.

End Of Transmission Block (ETB) A transmission control character used to indicate the end of a transmission block of data.

end office A Class 5 telephone central office; local subscribers are connected to an end office.

end system In Digital Equipment Corporation Network Architecture (DECnet), a system that transmits and receives NPDUs from other systems, but that does not relay NPDUs between other systems.

end user 1. In IBM's SNA, the ultimate source or destination of application data flowing through an SNA network. An end user may be an application program or a terminal operator. 2. An individual who utilizes an information system during the performance of his or her normal duties. End users can include, but are not limited to, bank tellers, clerks, factory workers, engineers, and managers.

end-to-end responsibility The principle that assigns to communications common carriers complete responsibility for all equipment and facilities involved in providing a telecommunications service from one end of the connection to the other. Also called end-to-end accountability.

endpoint A named source and/or destination activity, which can send and/or receive information through the distributed system. Generally synonymous with mailbox.

Engineering Change Notice (ECN) A document that identifies and authorizes a hardware change and explains why the change is necessary. ECNs are used to maintain effective review and control of engineering changes.

Enhanced Definition TV (EDTV) An improved quality of television signal that is compatible with existing transmission and receiving equipment but has a lower quality than HDTV (high definition TV).

Enhanced Interactive Network Optimization System (E-INOS) A network design tool product of AT&T Communications of Bedminster, NJ.

Engineering Change Order (ECO) Any design change that will require revision to specifications, drawings, documents, or configurations.

ENQ ENQuiry.

enquiry (ENQ) A control character (control E in ASCII) used as a request to obtain identification or status.

ENQ/ACK protocol A Hewlett-Packard communications protocol. The HP3000 computer follows each transmission block with ENQ to determine if the destination terminal is ready to receive more data. The destination terminal indicates its readiness by responding with ACK.

entity A network component that can be managed using network management.

entity attribute A piece of management information maintained by an entity. DNA defines Identification, Characteristic, Status, and Counter attributes.

entity hierarchy The logical structure of manageable components in a system.

entity instance An occurrence of entity.

entity name The name of an entity consisting of a global name part and a local name part.

entry point The location in an IBM SNA network from which network management and session data is transported to a host over the same link as the one to which the device is attached.

envelope, analog In an amplitude modulated signal the waveform has maxima and minima. The location of these maxima and minima can be joined by two smooth curves which form the envelope of the waveform.

envelope, digital A group of bits in a specific format which usually has a data field as well as qualifier addresses.

envelope delay Characteristics of a circuit which result in some frequencies arriving ahead of other frequencies, even though the frequencies were transmitted together.

EOA End Of Address.

EOB End Of Block; a control character.

EOM End Of Message.

EON End Of Number.

EOT End Of Transmission.

EP Emulation Program (IBM's SNA).

EPROM Erasable Programmable Read Only Memory.

EPS Emergency Power System.

equal access The requirement due to the divestiture of AT&T for the Bell Operating Companies (BOCs) to provide interexchange carriers other than AT&T with the same level of access to their Central Office switches as that provided to AT&T. This permits, as an example, telephone company subscribers to dial long distance using their preferred interexchange carrier without having to dial extra digits.

Equal Access End Office (EAEO) An EES switch used to provide carrier access to collocated stations.

equalization A method of compensating for the distortion introduced during data transmission over telephone channels. Usually a combination of adjustable coils, capacitors, and resistors are used to compensate for the difference in attenuation and delay at different frequencies.

equalizer A device used by modems to compensate for distortions caused by telephone line conditions.

equalizer, adaptive An equalizer that can constantly change its equalization to compensate for changes on the line. Also called automatic equalizer.

equalizer, automatic An equalizer circuit that can adjust itself to compensate for distortion on the line. Also called adaptive equalizer.

equalizer, fixed An equalizer circuit that is permanently set.

equalizer, manual An equalizer that must be adjusted by the user.

equipment room The room in which voice and data common equipment is housed, protected, and maintained, and where circuit administration is done using the trunk and distribution cross-connects.

equipment wiring subsystem The cable and distribution components in an equipment room that interconnect system common equipment, other associated equipment, and cross-connects.

equivalent four-wire system A transmission system using frequency division to obtain full-duplex operation over only one pair of wires.

ER Explicit Route (IBM's SNA).

Erasable Programmable Read-Only Memory (EPROM) A non-volatile semiconductor PROM that can have its current contents cleared (usually through exposure to ultraviolet light) and then accept new contents for storage.

erlang A standard unit of measurement for communications traffic, equal to 36 CCS. Used for throughput and capacity planning, the erlang represents the full-time use of a circuit.

error Any inconsistency between the true state of what is transmitted and the information that is received.

error burst A sudden increase in errors over a short period of time, as compared to the period immediately preceding and following the occurrence.

error control An arrangement that detects and may correct errors. In some systems, refinements are added that correct the detected errors, either by operation on the received data or by retransmission from the source.

error controller A device that provides error control, usually installed in pairs between the modem and DTE at each end of a data link.

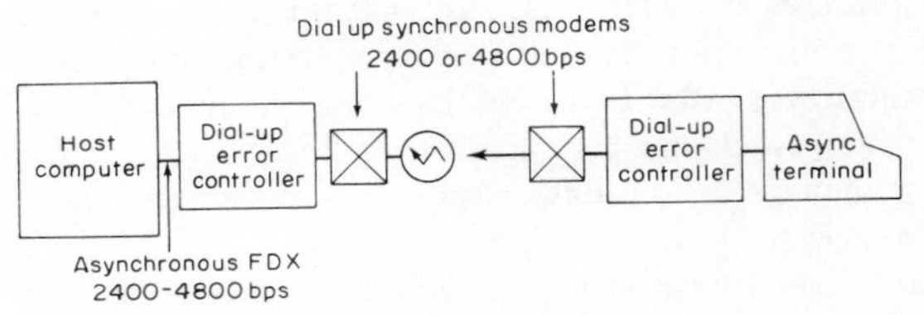

error controller

error correction An arrangement that restores data integrity in received data, either by manipulating the received data or by requesting retransmission from the source.

error detection An arrangement that senses flaws in received data by examining parity bits, verifying block check characters, or using other techniques.

error detection with retransmission Error handling technique where the receiver can detect errors, then request retransmission of the message from the sender. Also called ARQ.

error injection In BERT testing, the process of inserting known violations of the test pattern into the transmitted bit stream, either on command (on low-speed circuits) or at a predetermined rate (on high-speed circuits, such as T1).

error injection rate In BERT testing, the rate at which the test set can be programmed to inject errors into the transmitted test pattern.

error rate A measure of data integrity, given as the fraction of bits which are flawed. Often expressed as a negative power of 10, as in 1E−6 (a rate of one error in every one million bits).

Error-Correcting Code (ECC) An error-detecting code that incorporates additional signaling elements and enables errors to be detected and corrected at the receiving end. Also called forward error correction (FEC).

error-detecting telegraph code A telegraph code in which each telegraph signal conforms to specific rules of construction, so that departures from this construction in the received signals can be automatically detected.

Error-Free Second (EFS) A second of activity on a communications circuit during which no bit errors occur.

ES Errored-Second.

ESC Escape Character.

Escape Character (ESC) A code meaning "escape" which is used to control various electronic and mechanical functions of the terminal or the software program currently in use.

escape sequence In a switching network, a sequence of characters entered during a call to access commands without disconnecting the call.

ESD ElectroStatic Discharge.

ESF Extended Superframe Format.

ESPRIT European Strategic Program for Research and Development in Information Technology.

ESS Electronic Switching System.

essential facilities In packet switched networks, essential facilities are standard network facilities which are on all networks.

ESTAE Extended Specify Task Abnormal Exit (IBM's SNA).

ET Exchange Termination.

ETB End of Transmitted Block.

Ethernet In LAN technology, a *de facto* standard, developed first by Xerox and then sponsored by Xerox, Intel, and DEC. An Ethernet LAN uses coaxial cables and CSMA/CD. Ethernet is similar to an IEEE 802.3 LAN (they can share the same cable and can communicate with each other).

Ethernet cable A generic term used to refer to a type of 50 ohm, coaxial cable.

ETS Electronic Tandem Switch.

ETX End of TeXt character.

EUnet European Unix network.

European Academic Research Network (EARN) A network which connects over 550 computer systems in 24 countries. EARN was founded in 1983 to further research collaboration among academics in Europe.

European Computer Manufacturers Association (ECMA) A Western European trade organization that issues its own standards and belongs to ISO. Membership include western European computer suppliers and manufacturers.

European framing The international standard framing format for 2.048 Mbps circuits. In European framing, each frame contains 256 bits, and each multiframe contains 16 frames, divided into two sub-multiframes of 8 frames each. The first 8 bits of each frame are reserved; the remaining 248 bits contain 31 octet channels. Within the multiframe, frames are numbered 0 through 15, the second through eighth bits of frames 0, 2, 4, 6, 8, 10, 12, and 14 of each multiframe carry the Frame Synchronization Pattern (0011011). The first bit of each frame is reserved for national use, and on networks crossing international borders, its value is fixed at 1. In most national networks, the first bit of each frame alternates among three patterns; the first bits of frames 0, 2, 4, 6, 8, 10, 12, and 14 carry a CRC-8 computed over the previous multiframe; the first bits of frames 1, 3, 5, 7, 9, and 11 carry the pattern 001011; and the first bits of frames 13 and 15 are spare bits reserved further for national use. In frames not carrying the framing synchronization pattern (i.e. frames 1, 3, 5, 7, 9, 11, 13, and 15), the value of the second bit is fixed at 1 to avoid confusion with the framing synchronization pattern, and bit 3 carries a remote alarm indication (0 if no alarm, 1 if alarm condition is indicated); bits 4 through 8 of these frames are spare bits reserved for national use.

European Informatics Network (EIN) A European project to facilitate research into networks and promote agreement on standards. Its first task was to construct a packet switched subnet and this became operational in 1976. The five centers on the network are in London, Paris, Zurich, Milan and Ispra and there are also 'secondary centers' which join in the research on network protocols. The transport protocol of the EIN is the basis of the one adopted by IFIP Working Group 6.1.

Euroset Plus A trademark of Siemens for that firm's quality telephones designed for home and business use.

EVD EVent Dispatcher.

even parity check (odd parity check) This is a check which tests whether the number of digits in a group of binary digits is even (even parity check) or odd (odd parity check).

event An occurrence of a normal or abnormal condition detected by an entity and of interest to network management.

Event Control Block (ECB) In IBM's SNA, a control block used to represent the status of an event.

Event Dispatcher (EVD) The intermediary between event sources and event sinks.

event driven reconfiguration The ability of an unattended multiplexer automatically to install a predetermined alternate configuration in response to a predetermined event, such as a link or node failure.

event sink An entity which is a consumer of event reports.

event source An entity that detects events and generates event reports.

exception message A message displayed when an abnormal network condition exists or an illegal command is given.

Exception Request (EXR) In IBM's SNA, a request that replaces another message unit in which an error has been detected. *Note*: The exception request contains a 4-byte sense field that identifies the error in the original message unit and, except for some path errors, is sent to the destination of the original message unit. If possible, the sense data is returned in a negative response to the originator of the replaced message unit.

exception response In IBM's SNA, a value in the form-of-response-requested field of a request header. The receiver is requested to return a response only if the request is unacceptable as received or cannot be processed; that is, a negative response, but not a positive one, may be returned.

excess zeros More consecutive zeros received than are permitted for the selected transmission coding technique. For AMI encoded DS1 signals, 15 or more consecutive zeros are defined as excessive. For B8ZS encoded signals, 8 or more zeros are defined as excessive.

excessive bipolar violations An error condition on an AT&T T1 circuit defined as the occurrence of 1544 bipolar violations in a period of 1000 seconds that includes no period of 85 or more seconds without a bipolar violation.

exchange A unit established by a common carrier for the administration of communications services in a specified geographical area such as a city. It consists of one or more central offices together with the equipment used in providing the communications services. Frequently used as a synonym for Central Office.

exchange, classes of Class 1—Regional Center; Class 2—Sectional Center; Class 3—Primary Center; Class 4—Toll Center; Class 5—End Office. (ATT definition.)

exchange, private automatic (PAX) A dial telephone exchange that provides private telephone service to an organization and that does not allow calls to be transmitted to or from the public telephone network.

exchange, private automatic branch (PABX) A private automatic telephone exchange that provides for the transmission of calls to and from the public telephone network.

exchange, private branch (PBX) A manual exchange connected to the public telephone network on the user's premises and operated by an attendant supplied by the user. PBX is today commonly used to refer to an automatic exchange.

exchange, trunk An exchange devoted primarily to interconnecting trunks.

exchange area A geographical area that has a single uniform set of telephone charges.

Exchange Carriers Standards Association (ESCA) An ANSI standards group that existed prior to divestiture of AT&T and which now consists of Regional Bell Operating Companies (RBOCs) and independent telephone companies.

Exchange Carriers Standards Commission (ECSC) Committee organized to carry out national integration formerly assumed by AT&T.

exchange service A service permitting interconnection of any two customers' stations through the use of the exchange system.

EXEC In an IBM environment, an EXEC file contains a sequence of CMS commands that can be invoked by typing the name of the file and pressing the ENTER key.

Exit List (EXLST) In IBM's VSAM and VTAM, a control block that contains the addresses of routines that receive control when specified events occur during execution, for example, routines that handle session-establishment request processing or I/O errors.

exit routine In IBM's SNA, any of several types of special-purpose user-written routines.

EXLST exit routine In IBM's VTAM, a routine whose address has been placed in an exit list (EXLST) control block. The addresses are placed there with the EXLST macro instruction, and the routines are named according to their corresponding operand—hence DFASY exit routine, TPEND exit routine, RELREQ exit routine, and so forth. All exit list routines are coded by the VTAM application programmer.

expandor A device that reverses the effect of analog compression.

expedited flow In IBM's SNA, a data flow designated in the transmission header (TH) that is used to carry network control, session control, and various data flow control request/response units (RUs). The expedited flow is separate from the normal flow (which carries primarily end user data) and can be used for commands that affect the normal flow. *Note*: The normal and expedited flows move in both the primary-to-secondary and secondary-to-primary directions. Requests and responses on a given flow (normal or expedited) usually are processed sequentially within the path, but the expedited flow traffic may be moved ahead of the normal-flow traffic within the path at queuing points in the half-sessions and for half-session support in boundary functions.

explicit access In LAN technology, a shared access method that allows stations to use the transmission medium individually for a specific time period. Every station is guaranteed a turn, but every station must also wait for its turn.

Explicit Route (ER) In IBM's SNA, the path control network components, including a specific set of one or more transmission groups, that connect two subarea nodes. An explicit route is identified by an origin subarea address, a destination subarea address, an explicit route number, and a reverse explicit route number.

explicit route length In IBM's SNA, the number of transmission groups in an explicit route.

exponential backoff The practice of waiting a random period of time after a collision is detected on a CSMA/CD local area network prior to attempting to transmit again on the network.

exposed Cable, wire, strand, etc., which are subject to disturbance by lightning, possible contact, or induction from electrical currents in excess of 300 volts to ground, or ground potential rises from nearby power generating stations, substations, or higher voltage industrial transformers. A building is considered exposed or unexposed according to the classification of the outside plant which serves it.

Express Transfer Protocol (XTP) A protocol which corresponds to the network and transports layers of the International Standards Organization Open Systems Interconnection (OSI) network model and which is designed to be implemented in silicon to maximize the throughput of high-speed local networks.

EXR EXception Request (IBM's SNA).

extended addressing In bit-oriented protocol, a facility allowing a larger address to be used. IBM SNA adds two high-order bits to the basic address.

Extended Area Service (EAS) A service that permits calls to a designated area beyond the local exchange.

Extended Binary Coded Decimal (EBCD) A code using six bits plus one parity bit (IBM).

Extended Binary Coded Decimal Interchange Code (EBCDIC) An 8-bit character code used primarily in IBM equipment. The code provides for 256 different bit patterns.

extended distance cables Data communication cables which operate at longer distances than standard cables.

extended network addressing In IBM's SNA, the network addressing system that splits the address into a subarea and an element portion. The subarea portion of the address uses a fixed number of bits to address host processors or communication controllers. The element portion uses the remaining bits to permit processors or controllers to address resources.

extended numbering A feature of CCITT Recommendation X.25 that provides an extension of frame sequence numbering. With extended numbering, for example, frame sequence numbers range from 0 through 127, allowing for larger window sizes to compensate for satellite circuit delays. Also known as Modulo 128.

extended sequence numbering An option in HDLC which permits sequence numbers up to 127 to be used rather than up to seven.

Extended Specify Task Abnormal Exit (ESTAE) In IBM'S VTAM, an MVS macro instruction that provides recovery capability and gives control to the user-specified exit routine for processing, diagnosing an ABEND, or specifying a retry address.

Extended Superframe Format (ESF) A framing format, specified by AT&T, and used on North American T1 circuits, in which each Superframe contains 24 frames, and in which the framing bits of sequential frames carry four signaling patterns (Frame Synchronization, Signaling Synchronization, CRC-6, and Link Data Channel (LDC). The Extended Superframe Format is gradually replacing the 12-frame, D4 format, in North American telephony. The Interview 7000 Series can test both ESF- and D4-framed T1 circuits.

extent A group of contiguous allocated sectors on a disk.

external domain In IBM's SNA, the part of the network that is controlled by an SSCP other than the SSCP that controls this part.

external gateways Gateways into electronic-mail systems on computers not included in the LAN.

external modem A standalone modem, as opposed to a modem board that is inserted into the system unit of a personal computer.

external procedure A routine that is assembled or compiled separately from the program that called it.

external transmit clock An interface timing signal provided by a DTE device, which synchronizes the transfer of transmit data.

external writer In OS/VS2, a program that supports the ability to write SYSOUT data in ways and to devices not supported by the job entry subsystem.

Extremely High Frequency (**EHF**) That portion of the electromagnetic spectrum consisting of frequencies in the microwave range of approximately 30 to 300 GHz.

eye diagram The diagram produced by superimposing many received waveforms, taken with a wide range of bit patterns (e.g., a pseudo-random bit sequence). A useful eye diagram shows the waveforms at the point at which regeneration takes place. The "eye" is the hole in the pattern which is necessary for successful regeneration.

E & M signalling A two-state voltage signalling technique commonly used between PBXs to signal on-hook and off-hook conditions. Commonly referred to as ear and mouth.

F

facility 1. In general, a feature or capability offered by a system, item of hardware, or software. 2. In telco environments, line and equipment used to furnish a completed circuit. 3. In packet switched networks, non-standard facilities selected for a given national network which may or may not be found on other networks.

Facility Restriction Level (FRL) Definition of the calling privileges associated with a line; for example, intragroup calling only in the warehouse, but unrestricted calling in the boardroom.

facsimile (fax) The transmission of image via communications channels by means of a device which scans the original document and transforms the image into coded signals. Facsimile device operations typically follow one of four CCITT standards for information representation and transmission:

Group 1— analog with page transmission in four or six minutes

Group 2— analog with page transmission in two or three minutes

Group 3— digital with page transmission in less than one minute

Group 4— digital with page transmission in less than 10 seconds

fading A loss in transmission intensity caused by changes in the transmission medium. Typically encountered in microwave and radio transmission.

fail safe Describing a circuit or device which fails in such a way as to maintain circuit continuity or prevent damage.

fail soft The ability of a system to detect component failures and modify its processing quickly and temporarily to prevent irretrievable loss of data or equipment.

fail softly When a piece of equipment fails, a fail softly program lets the system fall back to a degraded mode of operation rather than let it fail catastrophically and give no response to its users.

failed second A second during which a digital circuit is in a failed signal state.

failed signal state An error condition on a digital circuit, defined as the occurrence of 10 consecutive Severely Errored Seconds, and considered to be cleared after 10 consecutive seconds without a Severely Errored Second. Each second of a failed signal state is called a failed second.

F1A-line weighting A telephone noise measuring class of weighting used to measure noise on a line terminated by a 302-type or similar telephone handset.

fallback, double Fallback in which two separate equipment failures have to be contended with.

fallback data rate A feature in a modem where, if a communications line deteriorates to the point where reception is difficult, the modem can be switched to transmit at a lower speed.

fallback procedures Predefined operations (manual or automatic) invoked when a fault or failure is detected in a system.

fanning strip A narrow strip having smooth holes or slots through which cross-connecting can be brought for orderly termination on a connecting block, terminal block, or terminal strip.

far-end crosstalk Crosstalk that travels along a circuit in the same direction as the signals in the circuit.

FAS Flexible Access System.

fast select In packet switched networks, a calling method which allows the user to expedite the transmission of a limited amount of information (usually 128 bytes). The information is sent along with the call request packet; therefore, the information arrives faster than in other call methods (which send the information in the packets that follow the call request packet).

fastrun In IBM's SNA, one of several options available with NCP/EP Definition Facility (NDF) that indicates only the syntax is to be checked in generation definition statements.

FAX Facsimile.

F-bits (framing bits) In multiplexed digital transmission, the leading bit of each frame, not included in any octet channel of user data. On any T1 circuit, framing bits occupy 8 Kbps bandwidth. In D4 framing, the framing bits of successive frames carry the Frame Synchronization Pattern and the Signaling Synchronization Pattern. In ESF, the framing bits of sequential frames carry four signaling patterns—Frame Synchronization, Signaling Synchronization, CRC-6, and Link Data Channel (LDC).

FCB File Control Block.

FCC Federal Communications Commission.

FCS Frame Check Sequence.

FCT Frame Creation Terminal.

FD Full-Duplex.

FDDI Fiber Distributed Data Interface.

FDM Frequency Division Multiplexing.

FDP Field Developed Program.

FDX Full-DupleX.

FE Format Effector.

Fe A common abbreviation for Extended Superframe Format, a framing format, specified by AT&T, and used on North American T1 circuits, in which each Superframe contains 24 frames, and in which the framing bits of sequential frames carry four signaling patterns—Frame Synchronization, Signaling Synchronization, CRC-6, and Link Data Channel (LDC). The Extended Superframe Format is gradually replacing the 12-frame, D4 format, in North American telephony.

FE bits (framing bits extended) Framing bits in Extended Superframe Format.

Feature Interactive Verification Environment (FIVE) An AT&T test facility which can be used by telephone companies and their customers to test, verify and evaluate the AT&T 5ESS digital central office switch.

FEC Forward Error Correction.

FEDAC A trademark of Frederick Engineering, Inc. of Columbia, MD, as well as a dataline monitor/protocol analyzer on an IBM PC adapter card which permits on-site and unattended remote operations.

Federal Communications Commission (FCC) A U.S. government board of seven presidential appointees established by the Communications Act of 1934 that has the power to regulate all USA interstate communications systems as well as all international communications systems that originate or terminate in the USA.

Federal Information Processing Standards (FIPS) The standards formulated by Telecommunications Standards program of the Federal Government—usually in close coordination with industry—and published by the National Bureau of Standards.

FED-STD FEDeral STandarD.

FED-STD-1001 Synchronous high-speed data signaling between data terminal equipment (DTE) and data communications equipment (DCE).

FED-STD-1002 Time and frequency reference information in telecommunications systems.

FED-STD-1003 A synchronous bit-oriented data-link control procedure; Advanced Data Communications Procedures (ADCCP).

FED-STD-1005 Coding and modulation requirements for 2400 bps modems.

FED-STD-1006 Coding and modulation requirements for 4800 bps modems.

FED-STD-1007 Coding and modulation requirements for duplex 9600 bps modems.

FED-STD-1008 Coding and modulation requirements for duplex 600 bps and 1200 bps modems.

feedback information In IBM's VTAM, information that is placed in certain RPL fields when an RPL-based macro instruction is completed.

feedline A coaxial cable or waveguide connecting a receiver or transmitter to an earth station antenna.

Feline A general-purpose protocol analyzer constructed on an adapter card that is installed in an IBM PC or compatible computer. FELINE is manufactured by Frederick Engineering, Inc., of Columbia, MD.

FELINE-LT A trademark of Frederick Engineering of Columbia, MD as well as a dataline monitor/protocol analyzer constructed on an adapter card that can be inserted into lap-top computers.

femtosecond A quadrillionth, or million billionth of a second (10^{-15}s).

FEP Front-End Processor.

FF Form Feed.

F1F2 A modem that operates over a half-duplex line to produce two subchannels at two different frequencies for low-speed, full-duplex operation.

FFTSO File Transfer Time Sharing Option.

FGND Frame GrouND.

fiber bandwidth The lowest frequency at which the

magnitude of the transfer function of an optical waveguide has fallen to half the zero frequency value; in other words, the frequency at which the signal loss has increased by 3 dB (decibels). Since the bandwidth of an optical waveguide is approximately reciprocal to its length (mode mixing), the bandwidth–length product (MHz–km) is often quoted as a quality characteristic.

fiber buffer Material used to protect an optical fiber or cable from physical damage, providing mechanical isolation or protection. Fabrication techniques include both tight jacket, or loose tube, buffering, as well as multiple buffer layers.

fiber distributed data interface A specialized local area network concept developed under American National Standards Institute (ANSI) auspices for use in computer-to-computer exchanges via fiber optic links operating at data rates up to 100 Mbps. The standard specifies multimode fiber 50/125, 62.5/125, or 85/125 core-cladding, a LED or laser light source, and 2 kilometers for unrepeated data transmission at 40 Mbps.

fiber loss Attenuation of light signal in optical fiber transmission.

fiber optic cable A transmission medium composed of small strands of glass, each of which provides a path for light rays which act as a carrier.

fiber optic facility A transmission facility in which information is transmitted as light pulses over fiberglass threads. Fiber optic facilities are immune to electrical interference, require very little power and can transmit at very high rates of speed.

fiber optic waveguides Thin filaments of glass or other transparent materials through which light beams can be transmitted for long distances through multiple internal reflections.

fiber (optical) Any filament or fiber, made of dielectric materials, that guides light.

fiber optics A signal-conducting medium that conveys light waves through transparent fibers. These fibers act as waveguides for light waves.

fiber SLC carrier system A system developed by AT&T Bell Laboratories that now carries up to 96 voice circuits on a pair of fiber lightguides between a central office and a remote terminal. The first installation was in Chester Heights, Pennsylvania, in 1982.

FIC First-In-Chain (IBM's SNA).

FID Format IDentification (IBM's SNA).

FIDB Facility Interface Data Bus.

field A group of bits that describes a specified characteristic; displayed on a reserved area of a CRT or located in a specific part of a record.

field-formatted In IBM's SNA, pertaining to a request or response that is encoded into fields, each having a specified format such as binary codes, bit-significant flags, and symbolic names.

field-formatted request In IBM's SNA, a request that is encoded into fields, each having a specified format such as binary codes, binary counts, bit-significant flags, and symbolic names. A format indicator in the request/response header (RH) for a request is set to zero.

Field Programmable Logic Array (FPLA) A logic element that can replace a conventional Read-Only Memory (ROM), provided that only some of the possible words in the storage matrix and address decoder are required. Both the address decoder and the storage matrix are programmed to obtain the maximum reduction in chip size. Devices in which the programming can be carried out by the user are termed FPLA.

Field Replaceable Unit (FRU) Same as service parts.

field-wire line Simple pairs of insulated wire twisted together. Field-wire lines are typically used in military applications for emergency and temporary short distance connections due to their high transmission loss.

FIFO (first in, first out) A method used to process an item in a queue according to which item has been in the queue longest.

FIGS FIGures Shift.

figures shift (FIGS) 1. A physical shift in a terminal using Baudot Code that enables the printing of numbers and symbols. 2. The character that causes the shift.

file A collection of records. A file might include all the names and addresses of a company's employees.

file attribute Any of the set of file characteristics which determines its accessibility and degree of protection from other than its current owner. Common file attributes are read, write, modify, and share.

File Control Block (FCB) A data structure that controls a user's access to a file.

file name A name assigned to a collection of related

data records or to a peripheral or communications device.

file organization A method that establishes the relationship between a record and its location in a file.

File Separator (FS) A control character used to separate and qualify data logically; normally delimits a data item called a file.

file server A device or station, usually in a local network, that provides file and storage services to other stations on the network.

file server protocol In LAN technology, a communications protocol that allows application programs to share files.

File Transfer, Access, and Management (FTAM) An ISO application-layer standard for network file transfer and remote file access.

File Transfer Customer Monitor System (FTCMS) A term used to describe third party software that permits file transfers between a personal computer and an IBM mainframe's CMS subsystem.

File Transfer Time Sharing Option (FTTSO) A term used to describe third party software that permits file transfers between a personal computer and an IBM mainframe time sharing subsystem.

filter An arrangement of electronic components designed to pass signals in one or several frequency bands and to attenuate signals in other frequency bands.

final route The last-choice route in an automatic switching system.

final-form document An electronic document that can only be printed or displayed.

FIPS Federal Information Processing Standards.

FIPS PUB 1-1 Code for information interchange.

FIPS PUB 7 Implementation of the code for information interchange and related standards.

FIPS PUB 15 Subset of the standard code for information interchange.

FIPS PUB 16-1 Bit sequencing of the code for information interchange in serial-by-bit data transmission.

FIPS PUB 17-1 Character structure and character parity sense for serial-by-bit data communications in the code for information interchange.

FIPS PUB 22-1 Specifies synchronous signaling rates between data terminal equipment (DTE) and data circuit-terminating equipment (DCE).

FIPS PUB 37 Synchronous high-speed data signaling rates between data terminal equipment (DTE) and data circuit-terminating equipment (DCE).

FIPS PUB 46 Specifies the 64-bit key, Data Encryption Standard.

FIPS PUB 71 Advanced Data Communications Control Procedures (ADCCP).

FIPS PUB 78 Guidelines for implementing Advanced Data Communications Control Procedures (ADCCP).

Fireberd A trademark of Telecommunications Techniques Corporation. A widely used BERT tester, capable of testing high-speed digital circuits such as T1 lines.

firmware A computer program or software stored permanently in PROM or ROM or semipermanently in EPROM or EEPROM.

First-In-Chain (FIC) In IBM's SNA, a request unit (RU) whose request header (RH) begin chain indicator is on and whose RH end chain indicator is off.

first-in/first-out operation A PABX feature which answers the longest waiting call first.

first-party maintenance End-users of equipment doing self-maintenance.

first speaker In IBM's SNA, the LU–LU half-session defined at session activation as: (1) able to begin a bracket without requesting permission from the other LU–LU half-session to do so, and (2) winning contention if both half-sessions attempt to begin a bracket simultaneously.

FIU Facilities Interface Unit.

FIVE Feature Interactive Verification Environment.

five-pair rubber cable A cable which contains ten conductors arranged in five pairs.

Fixed Loss Loop (FLL) A classification of FCC registered modems that limits output to 4 dB. In the fixed loss loop jack, it is required that the modem

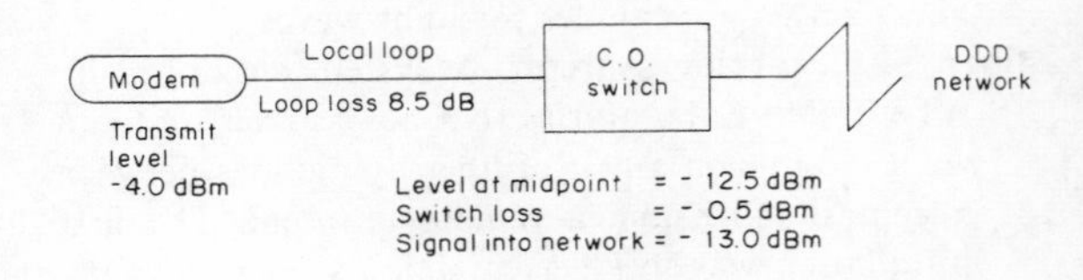

Fixed Loss Loop (FLL)

does not exceed a transmit level of −4 dBm. Once the loop loss of the subscriber line is measured, a resistor is selected to "build out" the loss of the loop so that the data signal hits the telephone company's central office at a predescribed level. This process is designed so that data hits the message telephone network 13 dB below the normal voice level.
The voice TLP at the input to the Message Telephone System network is 0 dBm. If data is 13 dB below voice, then we would expect the data level to be −13 dBm at this point. First, the installer must measure the insertion loss of the circuit from the customer's location to the telephone company central office. Let's assume that the installer measures a loss of 4 dB.

We already know that the modem transmits at the −4 dBm. With a little addition:

−4 dBm (modem's transmit level)
−4 dB (loop loss)

−8 dBm (total measured signal at the central office)

Notice that the objective was to be at a −13 dB at the central office. In order to accomplish this, the installer selects a resistive pad (attenuator) to reduce the signal in the data jack to meet the requirements. Now we have:

−4 dBM (modem's transmit level)
−5 dB (loss of resistive pad in data jack)
−4 dB (loss of loop)

−13 dBm (total measured signal at the central office)

fixed routing A method of routing in which the behavior is predetermined, taking no account of changes in traffic or of network component outages.

fixed-length record A record stored in a file in which all of the records are the same length.

flag A bit pattern of six consecutive "1" bits with a prefix and suffix "0" bit (character representation 01111110) used in many bit-oriented protocols to mark the beginning and end of a frame.

flat rate service Service where the user is entitled to an unlimited number of telephone calls within a specified local area for a fixed rate.

flat weighting A type of telephone noise weighting in which measurements are conducted using an input filter to obtain a flat amplitude–frequency response over a specified frequency range.

Fleet Satellite Communications System (FLTSATCOM) The U.S. Navy's network of UHF communications satellites that links aircraft, ships, submarines, ground stations, the U.S. Air Force Strategic Air Command, and national command authorities.

Flexible Access System (FAS) A fiber optic transmission system that spans the United Kingdom. Designed for completion in the mid-1990s, FAS is an overlay network that is part of the UK's public network. FAS allows large numbers of leased lines to be controlled by software, permitting business customers to mix connections of various circuits to the network as requirements change.

flexible routing A software-defined network (SDN) service feature offered by AT&T which lets the user reroute calls automatically to another location or be answered with a recorded announcement that instructs the caller to dial another site.

Flexgel Trademark name for AT&T developed filling compound used in waterproof cable. Waterproof material consisting of extended thermoplastic rubber (ETPR).

FLL Fixed Loss Loop.

flooding A packet routing method which replicates packets and sends them to all nodes, thus ensuring that the actual destination is reached.

flooding, network layer A means of propagating a message throughout a network. The message is transmitted to each neighboring system except the one from which it was received.

floppy disk A flexible plastic $5\frac{1}{4}$-inch disk coated with magnetic material and used to store data. Newer $3\frac{1}{2}$-inch diskettes are encased in a hard plastic case.

flow control The procedure for controlling the transfer of messages or characters between two points in a data network—such as between a protocol converter and a printer–to prevent loss of data when the receiving device's buffer begins to reach its capacity. Also called pacing.

flowchart A chart to represent the flow of data or instructions through a process, including decision points and loopbacks where appropriate. Often used for designing and documentating computer programs.

FLTSATCOM FLeeT SATellite COMmunications System.

flyback buffering delay The time delay provided by the statistical multiplexer to prevent loss of data during mechanical functions of terminal equipment such as printer carriage return and line feed. The unit of time delay is in character-times.

FM 1. Function Management (IBM's SNA). 2. Frequency Modulation.

FM subcarrier One-way data transmission using signals modulated in the unused portions of the FM radio frequency spectrum.

FMD Function Management Data (IBM's SNA).

FMH Function Management Header (IBM's SNA).

FOC Fiber Optic Communications.

focal point The location in an IBM SNA network that provides central network management for the domain.

footprints Compatible configuration of two mating pieces of apparatus. Specifically, an exclusion feature which is molded into the base of a protector unit which matches a raised detail in the protector.

foreign attachment Equipment attached to a common carrier facility which is not provided by the common carrier.

foreign exchange (FX) line A communications common carrier line in which a termination in one central office is assigned a number belonging to a remote central office.

foreign exchange service A service which connects a customer's telephone to a telephone company central office normally not serving the customer's location.

Form Feed (FF) A printer control character used to skip to the top of the next page (or form).

format A message structure that may include the code set and protocol characters.

Format Effectors (FE) Category of control characters mainly intended for the control of the layout and positioning of information on printing and/or display devices. Samples of FE characters are: back space (BS), carriage return (CR), form feed (FF), horizontal tab (HT), line feed (LF), and vertical tab (VT).

format identification (FID) field In IBM's SNA, a field in each transmission header (TH) that indicates the format of the TH; that is, the presence or absence of certain fields. TH formats differ in accordance with the types of nodes between which they pass. *Note*: There are six FID types:

FID0, used for traffic involving non-SNA devices between adjacent subarea nodes when either or both nodes do not support explicit route and virtual route protocols.

FID1, used for traffic between adjacent subarea nodes when either or both nodes do not support explicit route and virtual route protocols.

FID2, used for traffic between a subarea node and an adjacent PU type 2 peripheral node.

FID3, used for traffic between a subarea node and an adjacent PU type 1 peripheral node.

FID4, used for traffic between adjacent subarea nodes when both nodes support explicit route and virtual route protocols.

FIDF, used for certain commands (for example, for transmission group control) sent between adjacent subarea nodes when both nodes support explicit route and virtual route protocols.

formatted system services In IBM's SNA, a portion of VTAM that provides certain system services as a result of receiving a field-formatted command, such as an Initiate or Terminate command.

Forms Control Buffer (FCB) A buffer that is used to store vertical formatting information for printing, each position corresponding to a line on the form.

FORmula TRANslator (FORTRAN) A compiler language developed by the IBM Corporation. FORTRAN was originally conceived for use in scientific problems; however, it has now been adopted for commercial application as well. FORTRAN programs consist of sets of equation statements.

FORTRAN FORmula TRANslator.

fortuitous distortion Distortion resulting from causes generally subject to random laws (accidental irregularities in the operation of the apparatus and of the moving parts, disturbances affecting the transmission channel, etc.).

forward channel The communications path carrying data from the call initiator to the called party. The opposite of reverse channel. Also main channel.

Forward Error Correction (FEC) Technique allowing the receiver to correct errors occurring in a transmission channel without requiring retransmission of the data.

four-wire circuit A circuit using two pairs of conductors, one pair for the "transmit" channel and the other pair for the "receive" channel.

four-wire equivalent circuit A circuit using the same pair of conductors to give "transmit" and "receive" channels by means of different carrier frequencies for the two channels.

fox message A diagnostic test message that uses all the letters (and that sometimes includes numerals): "THE QUICK BROWN FOX JUMPED OVER A LAZY DOG'S BACK 1234567890." (In French, "VOYEZ LE BRICK GEANT QUE J'EXAMINE PRES DU WHARF." Often run continuously during system testing and fault isolation.

FPLA Field Programmable Logic Array.

fractional T1 A communications service which subscribers can obtain a portion of a T1 channel's bandwidth, such as 2, 4, 6, 8 or 10 64 Kbps channels.

frame 1. Same as transmission block. 2. The sequence of bits and bytes in a transmission block. 3. The overhead bits and bytes which surround the information bits in a transmission block. 4. In T1/DS1 signals, the collection of 8 bits from each of the 24 channels plus one frame bit for a total of 193 bits.

frame alignment pattern In a T1 signal, a logical sequence of six bit values carried in the framing bits of the 1st, 3rd, 5th, 7th, 9th, and 11th frames of a D4 framed circuit, or the 1st, 5th, 9th, 13th, 17th, and 21st frames of an ESF circuit. Also called the "frame synchronization sequence." The frame alignment pattern is the pattern 101010. The European Frame Alignment Pattern contains 7 bits—0011011.

frame bits Digital bits used in the frame which are added to the basic data rate to identify the start of a collection of individual channel data.

Frame Check Sequence (FCS) Usually a 16-bit field used for error detection in bit-oriented communications protocols.

frame errors In a T1 environment, errors in the 12 bit, D4 frame word. An error is counted when the 12-bit frame word received does not conform to the standard 12-bit frame word pattern.

frame level In packet switching, Level 2 of the CCITT X.25 Recommendation which defined the link access procedure for reliable data exchange over a link between a DTE and a DCE.

frame mnemonics A set of acronyms used to identify the frame types as defined by the appropriate bits in the control field.

frame relay A high-speed packet switching technology which achieves up to 10 times the speed of conventional X.25 packet switching networks using the same hardware.

frame synchronization sequence A logical sequence of six bit values carried in the framing bits of the 1st, 3rd, 5th, 7th, 9th, and 11th frames of a D4 framed circuit, or the 1st, 5th, 9th, 13th, 17th, and 21st frames of an ESF circuit. The frame synchronization pattern is the pattern 101010. The European frame synchronization pattern contains 7 bits—0011011. Also called the frame alignment pattern.

framed bert A Bit Error Rate Test in which the D4 Framing or Extended Superframe Format line discipline is imposed on the test pattern.

framing 1. The process of establishing a reference so that time slots or elements within the frame can be identified. 2. Process of inserting control bits to identify channels. Used in TDM signals such as the formatted version of T1.

framing bits In multiplexed digital transmissions, the leading bit of each frame, not included in any octet channel of user data; also known as "F-Bits." On any T1 circuit, framing bits occupy 8Kbps bandwidth. In D4 framing, the framing bits of successive frames carry the Frame Synchronization Pattern and the Signaling Synchronization Pattern. In ESF, the framing bits of sequential frames carry four signaling patterns: Frame Synchronization; Signaling Synchronization; CRC-6; and Link Data Channel (LDC).

free list A list structure holding all the units of store which are not currently in use. It may be operated as a kind of queue so that store units which become free are added to the tail, while new requests for store are met from the head of the queue, but other regimes are possible.

free running mode This mode of using a virtual terminal allows two associated systems or users to have access to the data structures simultaneously. The possibility that conflict between them may cause difficulties must be taken account of by the users. This problem is avoided in alternate mode.

Freefone A British Telecom communications service

which bills the cost of a public switched telephone call to the destination party. Similar to AT&T's WATS.

freespace Transmission of information using light-wave signals directly through the air, rather than through a conductor.

free-wheeling An old approach to terminal and communications system design in which terminals were able to transmit an operator initiative, without regard to whether the host or switching system was able to receive and handle all the data sent.

freeze frame A television transmission technique in which pictures are captured, transmitted, and displayed every few seconds at a rate considerably slower than full-motion television, thus lowering the required transmission bandwidth and storage. Commonly used in video-based teleconferencing systems.

frequency The rate at which a signal alternates. Typically, the number of complete cycles per second, normally expressed in Hertz (Hz).

frequency, carrier The frequency of the unmodulated carrier.

frequency band An arbitrarily defined bandwidth of the electromagnetic spectrum for which the boundary frequencies are stated.

Frequency Division Multiple Access (FDMA) A technique for sharing a multipoint or broadcast channel. This technique allocates different frequencies to different users.

Frequency-Division Multiplexer (FDM) A device that divides the available transmission frequency range into narrower bands, each of which is used for a separate channel.

Frequency Division Multiplexing (FDM) A multiplexing technique that partitions the composite bandwidth into channels, assigning a specific range of frequencies to each channel.

Time Channel 1 Channel 2 Channel 3 Channel 4 Channel 5 Channel 6 Frequency

Channel allocation

Frequency Division Multiplexing (FDM)

frequency hopping A spread spectrum technique by which the information is hopped between several communications channels.

Frequency Modulation (FM) One of three basic ways to add information to a sine wave signal. The frequency of the sine wave, or carrier, is modified in accordance with the information to be transmitted.

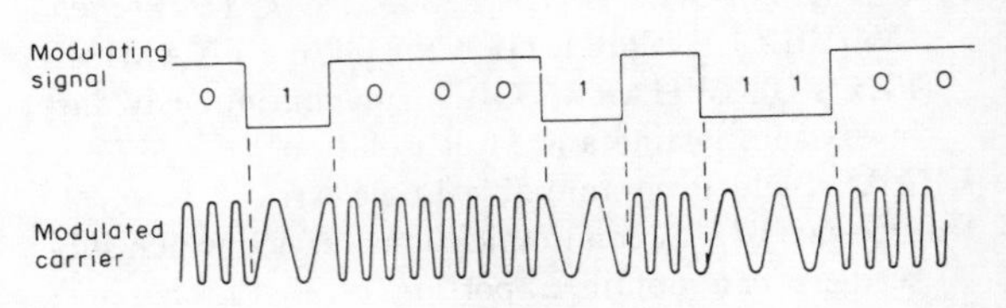

Frequency Modulation (FM)

frequency offset Analog line frequency changes which is one of the impairments encountered on a communications line.

frequency range The lowest to the highest frequency that can be transmitted over a band.

frequency response The variation in relative strength (measured in decibels) between frequencies.

Frequency Shift Keyed Modulation (FSK) A modulation technique where frequency shifts occur due to binary digital level changes. The carrier shifts between two predetermined frequencies (commonly 1200 Hz and 2200 Hz). Typically when the binary modulating signal is positive, the lower of the two frequencies is generated. The higher of the two frequencies is generated when the modulating signal is negative.

Frequency Shift Keying (FSK) Frequency shift key (FSK) modulation varies the number of waves per

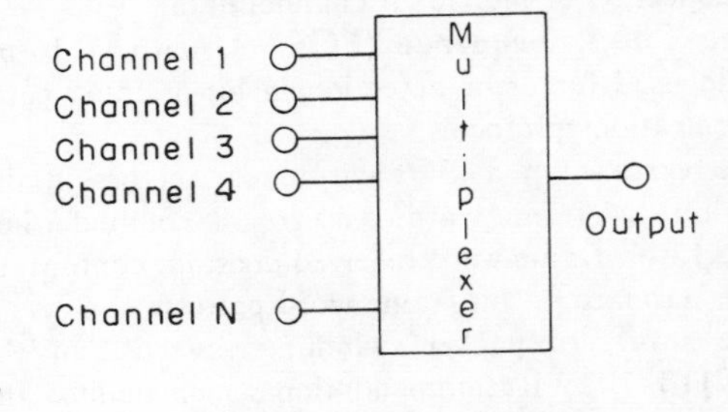

Multiplexer

unit of time using different tones to represent a binary "1" and a binary "0." It is used with low speed, asynchronous transmission. It cannot transmit more than one bit per baud.

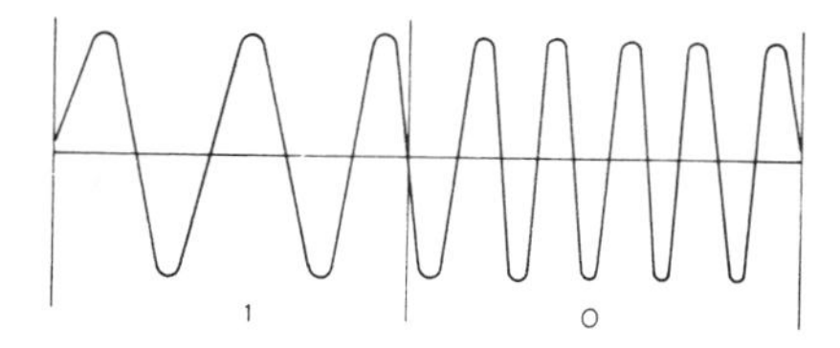

Frequency Shift Keying (FSK)

frequency translation The amount of movement of a single frequency signal at the input end of the transmission line to a different frequency when received at the output end.

frequency translator See headend.

frequency-derived channel Any of the channels obtained from multiplexing a channel by frequency division.

FRL Facility Restriction Level.

Front End Processor (FEP) A specialized computer processor generally used in conjunction with a larger mainframe computer that interfaces the computer to communications facilities and remote users. The FEP performs data communications functions which serve to preprocess and offload the attached computer(s) of network processing functions. In an IBM network, an FEP is called a communications controller.

FRU Field Replaceable Unit.

FS File Separator.

FS bits (framing bit signaling) In D4 or ESF framing, framing bits that carry the signaling synchronization sequence.

FSA startup That part of system initialization when the FSA is loaded into the functional subsystem address space and begins initializing itself.

FSI Functional Subsystem Interface.

FSI Connect The FSI communication service which establishes communication between IBM's JES2 and the FSA or functional subsystem.

FSI Disconnect The FSI communication service which severs the communication between IBM's JES2 and the FSA or functional subsystem.

FSI services A collection of services available to users (JES2) of the FSI. These services compromise communication services, data set services, and control services.

FSK Frequency-Shift Keying.

FTAM File Transfer, Access, and Management.

FT-bits (framing bits tracking) In D4 or ESF framing, framing bits that carry the frame synchronization sequence.

FTCMS File Transfer Customer Monitor System.

FTS Federal Telecommunications System.

full-duplex (FD, FDX) Synonym for duplex.

full-duplex transmission (FDX) Simultaneous two-way independent transmission in both directions. Compare with half-duplex transmission. Also used to describe terminals in the echoplex mode.

full-duplex channel A channel capable of transmitting data in both directions at the same time.

full-motion video Television transmission by which images are captured, transmitted and displayed at a repetition rate fast enough that a human observer perceives full motion. Compare with freeze frame.

fully connected network A network topology in which each node is directly connected by branches to all other nodes, forming a mesh structure. Normally impracticable as the number of nodes in the network increases.

function A standard, pre-packaged set of coded instructions for carrying out a computer operation.

Function Management Data (FMD) In IBM's SNA, an RU category used for end-user data exchanged between logical units (LUs) and for requests and responses exchanged between network services components of LUs, PUs, and SSCPs.

Function Management Header (FMH) In IBM's SNA, one or more headers, optionally present in the leading request units (RUs) of an RU chain, that allow one half-session in an LU–LU session to: (1) select a destination at the session partner and control the way in which the end user data it sends is handled at the destination, (2) change the destination or the characteristics of the data during the session, and (3) transmit between session partners status or user information about the destination (for example, a program or device). *Note*: FM headers can be used on LU–LU session types 0,1,4, and 6.

function management (FM) profile In IBM's

SNA, a specification of various data flow control protocols (such as RU chains and data flow control requests) and FMD options (such as use of FM headers compression, and alternate codes) supported for a particular session. Each function management profile is identified by a number.

Functional Subsystem (FSS) An address space uniquely identified as performing a specific function related to the JES. For IBM's JES2, an example of an FSS is the program Print Services Facility that operates the 3800 Model 3 and 3820 printers.

Functional Subsystem Application (FSA) The functional application program managed by the functional subsystem.

Functional Subsystem Interface (FSI) The interface through which IBM's JES2 or JES3 communicate with the functional subsystem.

functional subsystem startup That process part of system initialization when the functional subsystem address space is created.

fuse A device used for the protection against excessive currents. Consists of a short length of fusible metal wire or strip which melts when the current through it exceeds the rated amount for a definite time. Placed in series with the circuit it is to protect.

fusible links Short lengths (about 25 feet) of fine gauge wire pairs inside metallic sheath cable that melt to interrupt an electrical circuit, preventing overheating in building wiring and equipment.

FX Foreign Exchange.

G

G Giga.

G.703 The CCITT Standard (current version, 1984) for the physical and logical characteristics of transmissions over digital circuits. G.703 includes specifications for both the North American (1.544 Mbps) and European (2.048 Mbps) T1 circuits, as well as for circuits of larger bandwidth in the North American and European digital hierarchies. *Note*: In common usage, the term G.703 usually refers only to the standard for 2.048 Mbps circuits.

G.703 64K Those transmission facilities at the 64 Kbps rate that use the CCITT recommended physical/electrical interface specified under CCITT publication G.703.

G.703 2.048 Mbps Those transmission facilities at the 2.048 Mbps rate that use the CCITT recommended physical/electrical interface specified under CCITT publication G.703.

G.821 The CCITT Recommendation that specifies performance criteria for digital circuits in the ISDN.

gain The degree the amplitude is increased. The amplification realized when a signal passes through an amplifier, repeater, or antenna. Normally measured in decibels. Opposite of loss, or attenuation (negative gain).

gain hits A cause of errors in telephone line data transmission, usually when the signal surges more than 3 dB and lasts for more than 4 milliseconds. Bell standards call for eight or fewer gain hits per 15-minute period.

gain/slope The measurement of frequency levels over a specific frequency bandwidth on an analog circuit. Typically measured at frequencies of 404, 1004, and 2804 Hz.

garbage An informal term used to refer to corrupted data.

gas tube surge protector Surge limiting device similar in operation to the carbon block surge protector except that it has specially configured electrode with a more precise narrow gap and a sealed gas composition. The gas tube results in a more accurate and precise operating voltage range and extended service life under conditions of repeated operation.

gateway A combination of hardware and software used to interconnect otherwise incompatible networks, network nodes, subnets, or other network devices. Two examples are PADs and protocol converters. Gateways operate at the 4th through 7th layers of the OSI model.

gateway access protocol The protocol used between a host system and a system that is a DTE on a PSDN, to provide the X.25 gateway access facility to a user on the host.

gateway control functions In IBM's SNA, functions performed by a gateway SSCP in conjunction with the gateway NCP to assign alias network address pairs for LU–LU sessions, assign virtual routes for the LU–LU sessions in adjacent networks, and translate network names within BIND RUs.

gateway facilities A device that connects two systems, especially if the systems use different protocols. For example, a gateway is needed to connect two independent local networks or to connect a local network to a long-haul network.

gateway host In IBM's SNA, a host node that contains a gateway SSCP.

gateway NCP In IBM's SNA, an NCP that performs address translation to allow cross-network session traffic. The gateway NCP connects two or more independent SNA networks.

gateway node In IBM's SNA, synonym for gateway NCP.

gateway SSCP In IBM's SNA, an SSCP that is capable of cross-network session initiation, termination, takedown, and session outage notification. A gateway SSCP is in session with the gateway NCP; it provides network name translation and assists the gateway NCP in setting up alias network addresses for cross-network sessions.

gaussian noise Undesirable, random, low-level background electrical energy introduced into a transmission, consisting of frequency components such that, over a period of time, a continuous frequency spectrum within the band limits of the system is observed. Also called white noise, ambient noise, and hiss.

Gbytes (gigabytes) 1,024,000,000 bytes (1000 Mbytes).

GCOS General Comprehensive Operating Supervisor.

GCOS 6 Operating system that executes in a Honeywell Level 6 processor and supports multiple interrupt-driven activities.

GDSU Global Digital Service Unit.

GE Remote Terminal Supervisor (GRTS) Originally developed by General Electric (GE) when that firm was manufacturing computers. It now references a software system resident in a Honeywell DATANET 6600 or 335 Front-End Network Processor that controls communications. Now more commonly known as remote terminal supervisor; however, still abbreviated as GRTS.

gender mender A pair of connectors of the same gender (male/plug or female/socket) connected back to back which permits a cable end and device having the same connector genders to connect.

General Comprehensive Operating Supervisor (GCOS) The Honeywell operating system for that vendor's Series 60 and 6000 mainframe computers.

general poll Used in BSC for abbreviated addressing on lines that include cluster controllers. The cluster controller responds to a general poll and indicates if any of its connected devices has data to send.

General Switched Telephone Network (GSTN) Same as the public switched telephone network.

general topology subnetwork A non-broadcast subnetwork.

Generalized Path Information Unit Trace (GPT) In IBM's SNA, a record of the flow of path information units (PIU's) exchanged between the network control program and its attached resources. PIU trace records consist of up to 44 bytes of transmission header (TH), request/response header (RH), and request/response unit (RU) data.

general-purpose cable Cable specifically used to connect computers or telephones inside a building. This type of cable is not for use in risers (vertical shafts) and plenums unless enclosed in non-combustible tubing.

generic A term with various definitions. It may refer to the fundamental version of a widely distributed program package, or to such a package and accompanying hardware. Generic also may refer to the latest version of a program package.

generic bind In IBM's SNA, a synonym for a session activation request.

generic unbind In IBM's SNA, a synonym for a session deactivation request.

geosynchronous orbit The orbit where communications satellites will remain stationary over the same earth location. The geosynchronous orbit altitude is approximately 23,300 miles above the equator.

GFE Government Furnished Equipment.

GFI Group Format Identifier.

GHz (gigahertz) A frequency unit equal to 1E+9 hertz.

giga (G) A prefix for one billion (1E+9) times a specific unit.

gigabyte One billion bytes.

gigahertz A term used to state 1,000,000,000 (giga) cycles per second (hertz).

GIGO An acronym for Garbage In, Garbage Out.

global name That part of the entity name which identifies the node in the network.

global network addressing domain In Digital Equipment Corporation Network Architecture (DECnet), the addressing domain consisting of all NSAP addresses in the OSI environment.

GOS Grade of Service.

GOSIP Government OSI Profile.

Government Data Network (GDC) A packet switching network being installed by Racal Data Group Ltd in the United Kingdom. The company is to sell packet switching network services to the UK Government.

GPT Generalized Path information unit Trace (IBM's SNA).

grade of service (GOS) 1. In telephone transmission, a measure of users' options of the quality of speech heard during a telephone conversation. GOS is measured on a scale of 1 (unacceptable) to 5 (excellent). 2. When used for access measurement, GOS refers to the probability of a call being blocked or delayed, expressed as a percentage.

graded index fiber A fiber lightguide with a graduated refractive index higher at the center of the core than at the cladding. Because light travels faster where the refractive index is lower, it is speeded along the longer, outer paths and is slowed over the shorter inner paths, thus equalizing the time needed to travel the length of the fiber lightguide. This is a way of reducing dispersion.

graded index profile Any refractive index profile that varies with radius in the core. Distinguished from a step index profile.

grandfathered installation Any existing system, serving existing customers where existing intra-building network cable and network terminating wire are currently being used to provide both network service and intra-system or inside wire-like service is considered to be grandfathered. Such use will continue to be allowed for the inservice life of the associated equipment.

graphics station A collection of hardware and software which operates a high-resolution video display to present information to the user in the form of both text and graphic images.

groom-and-fill A process by which channels (DS0s) from several DS1s can be segregated for transmission to different remote locations and channels from separate DS1s are combined to share a DS1 for transmission to a remote location.

Grosch's Law The amount of computational power obtained is proportional to the square of the price paid. First stated by Herbert Grosch.

ground An electrical common conductor that at some point connects to the earth.

ground start Signaling method which detects a circuit is grounded at the other end.

ground station An assemblage of communications equipment to include a signal generator, transmitter, receiver, and antenna which receives signals to/from a communications satellite (earth station).

grounding The act of connecting a circuit or metal parts to earth or other pieces of equipment that are connected to earth.

Group 1 The CCITT standard for four-to-six minute analog facsimile.

Group 2 The CCITT standard for two-to-three minute analog facsimile.

Group 3 The CCITT standard for digital facsimile transmission in less than one minute per page.

Group 4 The CCITT standard for digital facsimile transmission in less than 10 seconds per page.

group access A method of granting rights to several users so they may access computer ports or files on a LAN.

group address Address of two or more stations in a group.

group addressing In transmission, the use of an address which is common to two or more stations.

group channel A unit on telephone carrier (multiplex) systems. A full group is a channel equivalent to 12 voice grade channels (48 kHz). A half-group has the equivalent bandwidth of six voice grade channels (24 kHz). Group channels can be used for high speed data communication (wideband).

group delay If a complex signal wih a narrow bandwidth is sent down a transmission path, at the receiving end the envelope of the signal will appear to have suffered a delay, called the group or envelope delay.

Group Format Identifier (GFI) In X.25 packet switched networks, the first 4 bits in a packet header. Contains the Q bit, D bit, and modulus value.

Group Separation (GS) A control character used to separate and qualify data logically. Normally delimits data item called a group.

GRTS GE Remote Terminal Supervisor.

GS Group Separator.

GSTN General Switched Telephone Network.

guard band The unused bandwidth separating channels to prevent crosstalk in a frequency division multiplexing (FDM) system.

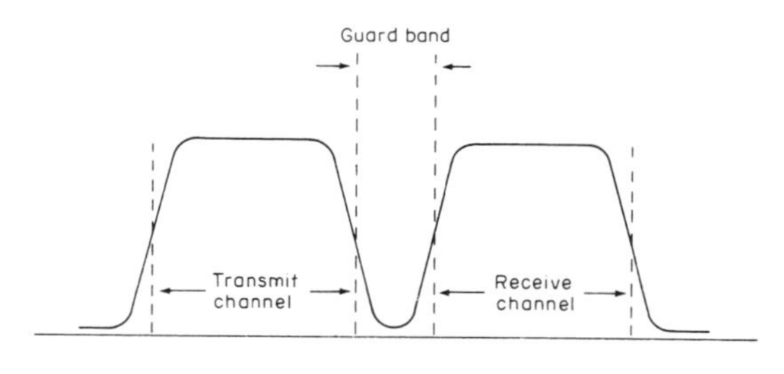

guard band

guard frequency A single frequency carrier tone used to indicate a line is prepared to send data. The frequencies between subchannels in FDM systems used to guard against adjacent channel interferences. Also called guard band.

Guardian Tandem Computer Company Operating System.

H

H.261 A CCITT standard for video compression called "Codes for Audiovisual Services at NX384 Kilobits Per Second." This standard establishes a common algorithm for converting analog video signals to digital signals operating at or above 384 Kbps.

H channel The ISDN packet switched channel on a Basic Rate Interface (BRI) which is designed to carry user information streams at varying rates, depending on type: H11, 1536 Kbps; H0, 384 Kbps; and H12, 1920 Kbps.

H0 channel In ISDN, a digital channel of 384 Kbps, available in primary rate service to the end user for high-speed data. An North American ISDN primary rate circuit can carry up to four H0 channels; a European ISDN primary rate circuit can carry up to five H0 channels.

H1 channel In North American ISDN, digital channel of 1.536 Mbps, available in primary rate service to the end user for highspeed data, image, or video. An H1 channel occupies the entire available bandwidth of the primary rate channel (1.544 Mbps aggregate—8 Kbps framing and 1.536 Mbps data).

half-duplex In data communication, pertaining to an alternate, one way at a time, independent transmission.

half-duplex channel A channel capable of transmitting data in either direction but only in one direction at any time.

half-duplex (HD or HDX) circuit 1. CCITT definition: A circuit designed for duplex operation, but which, on account of the nature of the terminal equipments, can be operated alternately only. 2. Definition in common usage (the normal meaning in computer literature): A circuit designed for transmission in either direction but not both directions simultaneously.

half-duplex operation The use of a circuit only in one direction at a time. The alternative word "simplex" is best avoided because it is used in two different senses.

half-duplex terminal operation A terminal whose transmitted data is printed locally.

half-modular cable A cable that has a modular plug on one end and spade lugs on the other end.

half-session In IBM's SNA, a component that provides FMD services, data flow control, and transmission control for one of the sessions of a network addressable unit (NAU).

halon A fire-suppression gas used to protect electrical rooms and systems.

Hamming code A forward error correcting code named for its inventor that corrects for as little as a single bit received in error.

Hamming distance Between two binary words (of the same length), the number of the corresponding bit positions in which the two words have different bit values. Also known as signal distance.

handler A software package designed to run in a processor module in a master network processor.

handset Another name for any ordinary telephone. As distinguished from a headset. Also used to refer to the part of the telephone containing the mouthpiece and receiver.

handset switch A switch frequently used in local-battery telephone sets. When pressed the switch connects the transmitter in the talking circuit. When released, the transmitter is placed out of the circuit, thereby conserving the battery when the transmitter is not used.

handshake, handshaking A preliminary procedure, usually part of a communications protocol, to establish a connection.

hard copy A printed copy of machine output in readable form.

hard disk drive A high-capacity magnetic storage device which allows a user to save, read, or erase data.

hardware Equipment (as opposed to a computer program or a method of use), such as mechanical, electrical, magnetic, or electronic devices. Compare with firmware and software.

hardware interface Physical hardware used to connect electrically two devices to one another.

hardwired A communications link of telephone lines or local cable, that permanently connects two devices.

hardwired FEP Non-programmable FEP, or FRONT END. Also line adapter.

hardwired logic A procedure in which the decision-making elements cannot be altered by external means but only by changing the internal connnections.

HA1-receiving weighting A telephone class of noise weighting used in a noise measuring set to measure noise across the HA1 receiver of a 302-type or similar subset.

harmonic A frequency equal to a whole-number multiple of a fundamental frequency present in the resultant frequencies because of a sinusoidal stimulus.

harmonic distortion The resultant presence of harmonic frequencies due to non-linear characteristics of a transmission line. Harmonic distortion occurs when attenuation of the receive level varies with the amplitude of the signal. For example, a 1 V signal may be attenuated by one-half, while a 5 V signal many be attenuated by two-thirds. In this way, the signal may be flattened at the peaks, which is equivalent to adding low amplitude harmonics. This distortion is measured according to the relative degree of second and third order harmonics (two and three times the standard frequency, respectively). The result is distorted data transmission.

HARNET Hong Kong Academic & Research NETwork.

hashing Automatic conversion of input information (such as a network address) into a table location (offset).

HASP Houston Automatic Spooling Program.

HASP/MLI Houston Automatic Spooling Program/Multi-Leaving Interface.

HASP workstation A remote batch terminal capable of interfacing with the Houston Automatic Spooling Program by means of a special multileaving version of BSC designed for this application. Often a System/360 Model 20 computer.

HCTDS High Capacity Terrestrial Digital Service.

HD Half-Duplex.

HDB3 High-Density Binary 3.

HDLC High-Level Data Link Control.

HDFI Host Digital Facilities Interface.

HDX Half Duplex Transmission.

headend unit In LAN technology, an item of hardware on a single or dual cable broadband network using split frequency bands to provide multiple services. The headend allows devices on a network to send and receive on a single cable.

header The control information added to the beginning of a message—either a transmission block or a packet—for control, synchronization, routing and sequencing of a transmitted data block or packet.

headset Any configuration of earpiece and speaker which fits over the head to allow hands-free telephone use.

HEANET Higher Education Authority NETwork.

hearing The perception of sound by the brain.

heat coil An electrical protection device used to prevent equipment from overheating as a result of foreign voltages on a conductor that do not operate voltage-limiting devices. It typically consists of a coil of fine wire around a brass tube that encloses a pin soldered with a low-melting alloy. When abnormal currents occur, the coil heats the brass to soften the solder allowing the spring-loaded pin to move against a ground plate directing currents to ground.

HEPNET High Energy Physics NETwork.

hertz (Hz) A measure of frequency or bandwidth equal to one cycle per second. Named after the experimenter Heinrich Hertz.

heterodyning The mixing of frequencies which produces the original signals as well as signals representing their sum and differences in the frequency domain. The use of a mixer and filters permits frequency division multiplexing.

heuristic Pertaining to a procedure which uses trial and error or random searching and therefore cannot be certain of its results. In contrast an algorithm is expected always to arrive at a correct or optimal result.

heuristic routing A routing method proposed by Baran in which delay data produced by normal data routing packets coming in on different links from a given source node are used to guide outgoing packets as to the best link for getting to that node.

HEX Hexadecimal.

hexadecimal A digital system that has sixteen states, 0 through 9 followed by A through F. Any 8-bit byte can be represented by two hexadecimal digits. The decimal and hexadecimal notation of values 0 through 9 is identical. The equivalent notation of the remaining basic hexadecimal values is as follows:

HEXADECIMAL	DECIMAL
A	10
B	11
C	12
D	13
E	14
F	15

The basic hexadecimal values are 0 through F.

HF High Frequency.

hierarchic data base model Scheme of logical representation of entities and relationships between entities of a data base by use of tree or hierarchic structures.

hierarchical network A communications network consisting of one host processing center and one or more satellite processors.

hierarchical network structure Network structure in which functions are broken down into layers, each having a specific role. (OSI Reference Model.)

hierarchical office class The functional ranking of a telephone company network switching center based upon its transmission requirements and its hierarchical relationship to other switching centers.

hierarchical switching In LAN technology, similar to star switching. The switching is done in stages.

hierarchically distributed processing The distribution of functions over components connected in a tree-structured hierarchy. Also called vertical distribution.

high frequency (HF) Portion of the electromagnetic spectrum typically used in short-wave radio applications. Frequency approximately in the 3 to 30 MHz range.

high order The port(s) of communications controllers, multiplexers and other devices connected to the end of a rotary or hunt group.

high pass Frequency level, above which a filter will allow all frequencies to pass.

high performance option Same as D1 conditioning.

High-Capacity Satellite Digital Service (HCSDS) See Skynet 1.5.

high-capacity service A term which generally refers to tariffed, digital data transmission service equal to, or in excess of T1 data rates (1.544 Mbps).

High Capacity Terrestrial Digital Service (HCTDS) AT&T's original name for its T1 service. Replaced in 1983 by Accunet T1.5 Service.

High-Density Binary 3 (HDB3) A technique for maintaining one's density by substituting a specific pattern for a string of three consecutive zeros. Similar in concept to B8ZS, HDB3 is used primarily in Europe.

High-Density Bipolar (HDB) A modified bipolar code which avoids the long absence of pulses and thus eases timing recovery. There are many versions.

High-Level Data Link Control (HDLC) The international standard communications protocol (similar to SDLC) developed by ISO.

High-Level Data Link Control Packet Assembler/Disassembler (HPAD) Supports Synchronous Data Link Control (SDLC) devices using two-way alternate transmission (logically half-duplex).

High-Level Data Link Control (HDLC) Procedure A set of protocols defined by ISO for carrying data over a link with error and flow control. Versions of HDLC are also being developed for multipoint lines. In spite of its name HDLC is not a high-level protocol.

High-Level Data Link Control (HDLC) Station A process located at one end of the link which sends and receives HDLC frames in accordance with the HDLC procedures.

High-Level Language Application Program Interface (HLLAPI) An IBM standard which defines and simplifies access to mainframe data.

high-level protocol A protocol to allow network users to carry out functions at a higher level than merely transporting streams or blocks of data.

HIM Host Interface Module.

HIT Any random deviation during transmission which causes data to be received in error.

hit on the line General term used to describe errors caused by external interferences such as impulse noise caused by lightning or man-made interference.

HLF High Level Function.

HLLAPI High-Level Language Application Program Interface.

HLSC High-Level Service Circuit.

HMINET I, II Hahn-Meitner Institut NETwork.

HN Host Network.

HNS Hospitality Network Service.

holding time The length of time that a communications channel is in use for each transmission.

Holidex 2000 The advanced reservation system used by more than 1700 U.S. Holiday Inn Hotels.

home directory In a local area network, the directory to which a user's first network drive is mapped when the user logs in to a file server.

home loop An operation involving only those input and output units associated with the local terminal.

hookswitch The device on which the telephone receiver hangs or on which a telephone headset hangs or rests when not in use. The weight of the receiver or handset operates a switch which opens the telephone circuit, leaving only the bell connected to the line.

Horizontal Redundancy Check (HRC) A validity method to check blocks in which redundant information is included with the information to be checked.

horizontal wiring subsystem That part of a premises distribution system installed on one floor that includes the cabling and distribution components connecting the backbone subsystem and equipment wiring subsystem to the information outlet via cross-connects, components of the administration subsystem.

Hospitality Network Service (HNS) An AT&T flat-rate, discounted long distance direct dial service. Under HNS calls to any domestic location at any time of the day are billed at a flat rate per minute.

host Any computer to which remote terminals are attached and in which the application program resides.

host computer The central computer (or one of a collection of computers) in a data communications system which provides the primary data processing functions such as computation, data base access, or special programs or programming languages. Often shortened to host.

host LU In IBM's SNA, a logical unit located in a host processor, for example, a VTAM application program.

host master key In IBM's SNA, deprecated term for master cryptography key. The cryptography key is used to encipher operational keys that will be used at the host processor.

host processor 1. A processor that controls all or part of a user application network. 2. In a network, the processing unit in which the data communication access method resides. 3. In an SNA network, the processing unit that contains a system services control point (SSCP).

hot key switch A keystroke sequence enabling a fast change from the session of one program to another.

hot potato routing Packet routing which sends a packet out from a node as soon as possible even though this may mean a poor choice of an outgoing line when the preferred choices are not available.

hotline The facility used to report problems for correction.

house cables Conductors within a building used to connect communications equipment to outside lines.

Houston Automatic Spooling Program (HASP) A mainframe job control program developed by IBM and widely used in the late 1960s and 1970s. HASP was developed to facilitate the flow of remote jobs into and out of mainframes. It required a special multileaving version of the BSC protocol which took its name from the program.

HPIB Hewlett Packard Interface Bus.

HPO High Performance Option.

HRC Horizontal Redundancy Check.

HSM Host Switching Module.

HTR Hard-To-Reach.

hub 1. In LAN technology the center or a star topol-

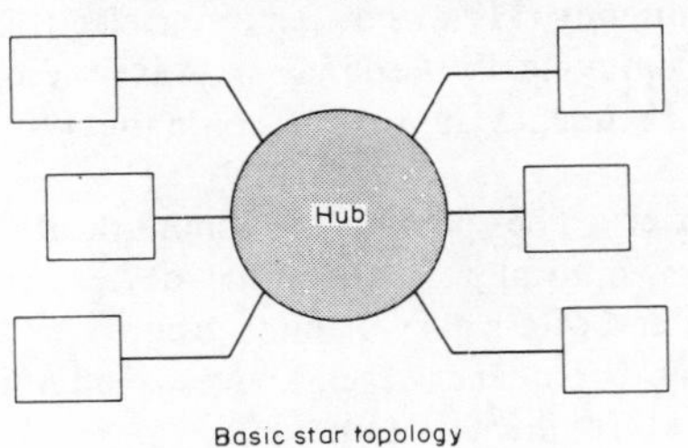

Basic star topology

hub

ogy network or cabling system. 2. A multinode network topology that consists of a central multiplexer node with multiple nodes feeding into and through the central node. The outlying nodes do not normally interconnect with one another.

hub (DDS) An office in a digital data system that combines the T1 data streams from a number of local offices into signals suitable for transmission.

Huffman coding A compression technique based upon the statistical probability of occurrence of characters in the character set to be compressed. The resulting code assigned to each character is a maximum efficient code and is often referred to as an optimum code.

hundred call seconds See CCS.

hunt group The telephone numbers assigned to a rotary such that a person dialing a single listed number is automatically connected by telephone company switching equipment to an available line in the group.

hybrid 1. In transmission systems, a passive electrical component used for bridging between legs of a circuit, such as interfacing a backbone multidrop circuit to individual drop circuits. 2. Combination of two or more technologies. 3. Hybrid Telephone Circuit, which interfaces two-wire local loop.

hybrid coil A two- to four-wire converter used to connect subscriber telephone lines to a common carrier transmission facility. The hybrid coil suppresses echoes caused by the impedance mismatch of the two- to four-wire conversion.

hybrid interface structure An interface that has a mixture of labeled and positioned channels.

Hyperaccess A communications software program from Hilgraeve, Inc., of Monroe, MI, designed to operate on IBM PC and compatible computers. The program is noted for its powerful script language, informative tutorial and reliable and efficient file transfer capability.

Hz See hertz.

I

I.441 "ISDN User-Network Interface Data Link Layer Specification." The CCITT recommendation in the I-Series that describes LAP-D. Identical to Recommendation Q.921.

I.451 "ISDN User-Network Interface Layer 3 Specification." the CCITT recommendation in the I-Series that describes the Network layer of the ISDN end user interface. Identical to Recommendation Q.931.

IATA International Air Transport Association.

Ibertex The videotex service marketed by Spain's national carrier Telefonica.

IBM International Business Machines.

IBM 3270 The term "3270" originally referred to an IBM product—a dumb terminal that was interactive with its host. The term now refers to a whole group of hardware, manufactured by IBM as well as other vendors that emulate or imitate the functions of the 3270. The group includes:

- IBM'S 3270 family of display stations.
- 3270 emulators produced by plug and system compatible vendor's.
- Proprietary terminals produced by the other major mainframe, manufacturers (e.g. Unisys, NCR, Control Data, and Honeywell). These terminals use data communications equipment configurations similar to the 3270's, but are not compatible with the 3270. Proprietary terminals are compatible only with their own hosts.
- Multifunction terminal controllers that allow terminals to do 3270 applications, such as data entry, word processing and electronic mail.

IBM Cabling System (IBMCS) A cabling system originally introduced by IBM in 1984 which is designed to support most types of cabling requirements to include the vendor's token-ring network. The cabling system consists of specified cables, faceplates, distribution panels, and a special connector.

IBM TSO Timesharing System (IBM Network).

IBM Type 1 (indoor) cable A cable used to connect terminal work stations and other equipment to computers. This cable is for indoor use and consists of two twisted pairs, 22 AWG solid conductors in a foil braid shield. Type 1 cable is also available in plenum.

IBM Type 1 (outdoor) cable A cable used for instllation in aerial or in underground conduits. The cable consists of two twisted pairs, 22 AWG solid conductors in a corrugated metallic shield.

IBM Type 2 cable A cable which provides the same interconnections as a Type 1 indoor cable as well as supporting PBX requirements. The cable consists of two twisted pairs, 22 AWG solid conductors in a foil, braid shield, and four voice-grade twisted pair 22 AWG for telephone use. Type 2 is also available in plenum.

IBM Type 3 cable A PVC media cable which consists of four pair 24 AWG wire and which is used for telephone cable. Type 3 is also available in plenum.

IBM Type 5 cable A cable which consists of two optical fiber conductors. This cable can be used for indoor installations or aerial installations, or placed in an underground conduit.

IBM Type 6 cable This cable is designed for patching applications within a wiring closet or from a faceplate in the back of a work station or computer terminal. Type 6 cable consists of two twisted pairs, 26 AWG stranded conductors which provides flexiblity.

IBM Type 8 cable This cable is designed for use under carpets and is useful for an open office environment. Type 8 cable consists of two parallel pairs of 26 AWG solid conductors arranged flat.

IBM Type 9 Cable This cable is considered as an economy version of Type 1 plenum and has a maximum transmission distance approximately two-thirds of that of Type 1 cable. Type 9 cable consists of two twisted pairs of 26 AWG stranded conductors and is only available in plenum.

IBM VNET Virtual NETwork.

IBMCS IBM Cabling System.

IBS International Business Services.

IC Integrated Circuit.

ICA 1. Integrated Communications Adapter. 2. International Communications Association.

ICSU Intelligent Channel Service Unit.

ICV Initial Chaining Value (IBM's SNA).

ID Insulation Displacing.

IDA Integrated Digital Network.

IDCMA Independent Data Communications Manufacturers Association.

IDF Intermediate Distribution Frame.

idle character A transmitted character which indicates a "no information" status. It does not manifest itself as part of the received message at the destination.

idle condition A non-busy state for equipment or circuits, indicating that they are not in operation and/or ready for use.

IDN Integrated Digital Network.

IDI Initial Domain Identifier.

IDP Initial Domain Part.

IDT Integrated Digital Terminal.

IEC Interexchange Carrier.

IEEE Institute of Electrical and Electronic Engineers.

IEEE 488 An IEEE standard parallel interface often used to connect test instruments to computers.

IEEE 802.2 In LAN technology, a data link layer standard used with IEEE 802.3, IEEE 802.4, and IEEE 802.5.

IEEE 802.3 A specification published by the Institute of Electrical and Electronic Engineers which defines a physical cabling standard for local area networks as well as the means of transmitting data and controlling access to the cable. The physical layer standard uses the CSMA/CD access method on a bus topology LAN. Similar to Ethernet.

IEEE 802.4 In LAN technology, a physical layer standard that uses the token-passing access method on a bus topology LAN. Nearly identical to MAP.

IEEE 802.5 In LAN technology, a physical layer standard that uses the token-passing access method on a ring topology LAN.

IEEE Project 802 In LAN technology, an IEEE team that developed the IEEE 802 family of LAN standards.

IFIPS International Federation of Information Processing Societies.

IM Interface Module.

IMINT IMagery INTelligence.

IMLT Integrated Mechanized Loop Testing.

immediate-request mode In IBM's SNA, an operational mode in which the sender stops sending request units (RUs) on a given flow (normal or expedited) after sending a definite-response request chain on that flow until that chain has been responded to.

immediate-response mode In IBM's SNA, an operational mode in which the receiver responds to request units (RUs) on a given normal flow in the order it receives them, that is, in a first-in, first-out sequence.

IMP Interface message processor.

impact printer A printer in which printing is the result of mechanical impact.

impedance (z) The total opposition (resistance, inductance, and capacitance) to the flow of current in an electrical circuit.

impulse A high-amplitude, short-duration, pulse. A rapid change in current flow or intensity of a magnetic field.

impulse hits Cause of errors in telephone line data transmission. AT&T suggests no more than 15 impulse hits per 15 minute period (spikes).

impulse noise A type of interference on communication lines characterized by high amplitude and short duration. This type of interference may be caused by lightning, electrical sparking action, or by the make–break or switching devices, etc.

IMS Information Management System.

IMS/VS Information Management System/Virtual Storage.

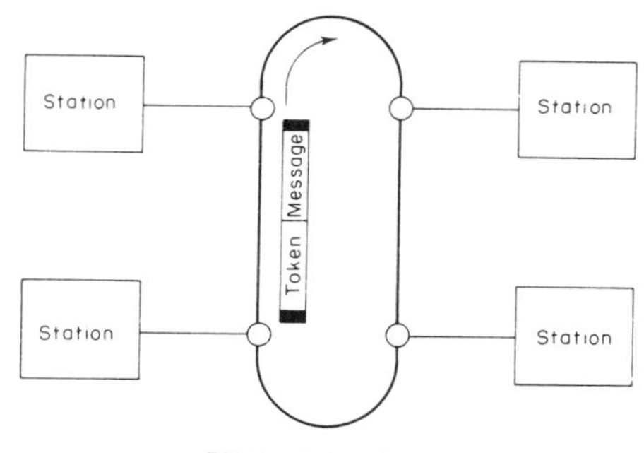

IEEE 802.5

INACTIVE In IBM's VTAM, pertaining to a major or minor node that has not been activated or for which the VARY INACT command has been issued.

in-band signaling Call control signaling transmitted within the bandwidth of the call it controls. Also called channel-associated signaling. In T1 transmission, in-band signaling is performed by bit-robbing.

Independent Company (ICO) A generic acronym for any provider of local telephone service in the U.S. that is not part of a regional Bell operating company.

Indepedent Telephone Company (ITC) A non-Bell Telephone operating company that furnishes telecommunications service within a geographic area. There are approximately 1400 independent telephone companies across the U.S.

index matching material A material, often a liquid or cement, whose refractive index is nearly equal to the core index. Used to reduce Fresnel reflections from a fiber end face.

indexed file A file whose records are organized to be accessed sequentially in key sequence or directly by key value.

indirect activation In IBM's VTAM, the activation of a lower-level resource of the resource hierarchy as a result of SCOPE or ISTATUS specifications related to an activation command naming a higher-level resource.

indirect deactivation In IBM's VTAM, the deactivation of a lower-level resource of the resource hierarchy as a result of a deactivation command naming a higher-level resource.

induction The process by which an electrically charged object induces an electrical current in a nearby conductor.

induction coil A transformer in which the common connection between the windings does not affect the mutual inductance between the coils.

inductive reactance The opposition caused by inductance.

inflection Variation in the pitch or loudness which a speaker uses for emphasis or special meaning.

INFNET Istituto Nazionale Fisica Nucleare.

Info-2 An AT&T automatic number identification facility (an ISDN service feature) which forwards the calling party's telephone number to the receiving party (ISDN customer). Info-2 is available with AT&T's Megacom 800 service.

Infomaster An on-line database and retrieval service marketed by Western Union which provides subscribers access to over 300 data bases located on computer systems around the world.

Infonet The public data network operated by Computer Sciences Corporation of El Segundo, CA.

Infoplex An electronic mail service marketed by CompuServe to corporations.

information 1. The meaning assigned to data by people. 2. The organizational content of a signal.

information bit A data bit, as opposed to an overhead bit.

Information Management Data Base In IBM's NPDA, a system management tool that helps collect, organize, and keep track of problems and their resolutions.

Information Management System A major IBM mainframe software system responsible for controlling extensive batch and transaction processing activities.

Information Management Systems/Virtual Storage (IMS/VS) An IBM host operating system, oriented toward batch procesing, and communications-based transaction processing.

information outlet A connecting device designed for a fixed location (usually a wall in an office) on which horizontal-wiring-subsystem cable pairs terminate and which receives an inserted plug. It is an administration point located between the horizontal wiring subsystem and work location wiring subsystem. Although such devices are also referred to as "jacks," the term "information outlets" encompasses the integration of voice, data, and other communication services that can be supported via a premises distribution system.

information path The route by which information is transferred.

Information Separator (IS) A category of control characters used to separate and qualify data logically. Its specific meaning has to be defined for each application. Samples of IS characters are: IS1(US), IS2(RS), IS3(GS), IS4(FS).

information service Commercial dial-up service that provides many and various services to its subscribers, to include large on-line databases, travel booking, banking, bulletin board services, special interest groups, electronic mail and other features.

Information Technology Management (ITM) A term which is used to encompass telecommunications and data management. As telecommunications technology and computer technology advance and merge, the distinctions between the two blur. ITM is used to describe the merging of these technologies.

Infostream A registered trademark of Infotron Systems for equipment that provides data PBX functionality.

infrared Portion of the electromagnetic spectrum used for optical fiber transmission as well as for short-haul, open-air transmission. Infrared transmission wavelengths are approximately 0.7 micrometres or longer.

infrared light Light with a frequency just below that of visible light which moves through many types of lightguides with little loss of power.

inhibited In IBM's VTAM, pertaining to an LU that has indicated to its SSCP that it is not ready to establish LU–LU sessions. An initiate request for a session with an inhibited LU will be rejected by the SSCP. The LU can separately indicate whether this applies to its ability to act as a primary logical unit (PLU) or as a secondary logical unit (SLU).

Initial Chaining Value (ICV) In IBM's SNA, an eight-byte pseudo-random number used to verify that both ends of a session with cryptography have the same session cryptography key. The initial chaining value is also used as input to the Data Encryption Standard (DES) algorithm to encipher or decipher data in a session with cryptography. Synonymous with session seed.

Initial Domain Identifier (IDI) In Digital Network Corporation Network Architecture (DECnet), the part of an NSAP address which identifies the authority responsible for the assignment of the DSP.

Initial Domain Part (IDP) In Digital Network Corporation Network Architecture (DECnet), the part of an NSAP address assigned by the first level addressing authority.

Initial Program Load (IPL) The set of programs loaded into a computer's active memory areas at startup.

initialization Concerns specific preliminary steps required before execution of iterative cycles to establish or determine efficient start procedures. Usually a single, non-repetitive operation after a cycle has begun and/or until a full cycle is again begun.

initialization parameter An installation-specified parameter that controls the initialization and ultimate operation of JES2.

initialization statement An installation-specified statement that controls the initialization and ultimate operation of JES2.

initialize The process of setting hardware, such as a protocol analyzer, to a known operating state.

initiate In IBM's SNA, a network services request, sent from an LU to an SSCP, requesting that an LU–LU session be established.

inline exit routine In IBM's VTAM, a SYNAD or LERAD exit routine.

INN In IBM's SNA, a deprecated term for "Intermediate Routing Node" (IRN).

in-plant system A system whose parts, including remote terminals, are all situated in one building or localized area. The term is also used for communication systems that span several buildings and sometimes covering a large distance, but in which no common carrier facilities are used.

input 1. The data to be processed. 2. The state or sequence of states occurring on a specified input channel. 3. The device or collective set of devices. 4. A channel for impressing a state on a device or logic element. 5. The process of transferring data from an external storage to an internal storage.

input devices Devices which allow the users to "talk" to the computer system. For example, a keyboard on a CRT terminal.

input service processing In JES2, the process of performing the following for each job: reading the input data, building the system input data set, and building control table entries.

input/output (I/O) The process of moving information between the central processing unit and peripheral devices. May refer to the particular peripheral hardware used.

input/output device A peripheral or communications device.

inquiry A request or interrogation regarding the stored information at the computer system.

inquiry/response An exchange of messages and responses with one exchange usually requesting information (inquiry) and the response that provides information.

insertion loss Signal power loss due to connecting communications equipment units with dissimilar impedance values.

Inside Wiring (IW) Customer premises wiring. A simple twisted pair wire which might be used for a service which goes from one floor in the building to another.

installation exit routine In IBM's VTAM, a user-written exit routine that can perform functions related to initiation and termination of sessions and is run as part of VTAM rather than as part of an application program. Examples are the accounting, authorization, logon-interpret, and virtual route selection exit routines.

Institute of Electrical and Electronic Engineers (IEEE) An international professional society that issues its own standards and is a member of ANSI and ISO. It created IEEE Project 802.

insulation Material that provides a high resistance path for current flow between wires, preventing the current from flowing from one wire to another.

integrated circuits (IC) A complete complex electronic circuit that is capable of performing all the functions of a conventional circuit containing numerous transistors, diodes, capacitors, and or resistors.

Integrated Communication Adapter (ICA) In IBM's SNA, an integrated adapter that allows connection of multiple telecommunication lines to a processing unit.

Integrated Digital Access (IDA) British Telecom's ISDN-type network based on System X digital exchanges that does not support the CCITT I.420 interface for basic rate ISDN service. IDA provides one 80 Kbps and one 16 Kbps signalling control channel while ISDN offers two 64 Kbps and one 16 Kbps signaling control channel.

integrated modem A modem designed into an information product, such as a terminal or computer, rather than used externally to it and connected to it by means of a cable.

Integrated Network Management Services (INMS) A service of MCI Communications Corporation which enables subscribers to monitor and reconfigure their networks.

integrated optical circuit The optical equivalent of a microelectronic circuit, it acts on the light in a lightwave system to carry out communications functions. Generates, detects, switches and transmits light.

Integrated Service Unit (ISU) A single device that combines the functions of both a CSU and a DSU.

Integrated Services Digital Network (ISDN) A CCITT standard, currently under development, that will cover a wide range of data communications issues but primarily the total integration of voice and data. Access channels include basic (144 Kbps) and primary rate (1.544 and 2.048 Mbps). The ISDN is described in the CCITT's I-series of recommendations (1984 Red Book). Already having major effects on exchange and multiplexer design.

Integrated Services Line Unit (ISLU) The AT&T 5ESS switch line unit which provides ISDN standard access to the 5ESS switch.

integrated system A computer system which extends the concept of integrated software to all end users. Integrated software is groups of programs that perform different tasks but which share data. A major advantage results from the different applications all updating the one common data base, keeping it current.

Integrated Voice/Data Terminal (IVDT) A family of devices that feature a terminal keyboard/display and voice telephone. May work with a specific vendor's data PBX.

integrity of data The status of information after being processed by software.

Intel A semiconductor (chip) manufacturer, one of the sponsors of Ethernet.

Intelli-flex A communications service marketed by Cable & Wireless Communications which offers subscribers fractional T1 capabilities, such as 4, 6, 8, 10 or more 64 Kbps channels.

intelligence, intelligent A term for equipment (or a system or network) which has a built-in processing power (often furnished by a microprocessor) that allows it to perform sophisticated tasks in accordance with its firmware.

Intelligent Channel Service Unit (ICSU) A combination Channel Service Unit (CSU) and Data Service Unit (DSU) marketed by Timeplex of Woodcliff Lake, NJ, which permits the user to view diagnostic Extended Superframe (ESF) data.

intelligent communications network A network which usually permits alternate routing and other complex control functions.

intelligent port selector Same as data PABX.

Intelligent Synchronous (I-SYNC) A General DataComm, Inc., proprietary feature found on some GDC modems. With I-SYNC a special "y" cable is used to provide simultaneous connection to both the asynchronous and synchronous boards on a PC. Once the asynchronous auto dialer has established a connection to the remote end modem, GDC communications software automatically switches the modem into synchronous mode.

intelligent TDM Same as concentrator.

intelligent terminal A terminal with some level of programmability. Contains processor and memory. Used for pre-transmissions or post-transmission communications processing as well as for off-loading of data processing.

Intelligent Time Division Multiplexer (ITDM) A device which assigns time slots on demand rather than fixed subchannel basis.

Intellipath A trademark and service of NYNEX Corporation. Intellipath is a digital Centrex service.

Intelsat INTErnationaL SATellite service.

interactive Pertaining to a time-dependent (real-time) data communications operation where a user typically enters data and awaits a response message from the destination prior to continuing. Contrast with batch processing.

Interactive System Productivity Facility (ISPF) An IBM program product that serves as a full screen editor and dialogue manager. Used for writing application programs, it provides a means of generating standard screen panels and interactive dialogues between the application programmer and terminal user.

Interactive Terminal Interface (ITI) In packet switched networks, a PAD that supports network access by asynchronous terminals.

interactive testing Testing performed through user and system communication. The system acknowledges and reacts to requests entered from the keyboard or to conditions detected on the data lines.

interbuilding cable The communications cable that is part of the campus subsystem and runs between buildings. There are four methods of installing interbuilding cable: in-conduit (in underground conduit); direct-buried (in trenches); aerial (on poles); and in-tunnel (in steam tunnels).

interbuilding cable entrance The point at which campus subsystem cabling enters a building.

intercept The process of rerouting a call from an invalid or out-of-service number to an alternative number, or to an operator position.

interchange circuit In any interface, a circuit with an associated pin assignment on the interface connector that is assigned a data, timing, testing, or control function.

intercom An option on many business phone systems which allows you to use an abbreviated dialing sequence to reach another telephone connected to the same communications system.

interconnect 1. (*verb*) To associate the electrical and functional aspects of two different services, often provided by different suppliers. 2. (*noun*) Any supplier of non-telephone-company alternative equipment.

interconnect company A provider of communications terminal equipment for connection of telephone lines.

interconnect industry The industry to manufacture, market, and service equipment and services that connects to telephone lines.

interconnected networks In IBM's SNA, networks connected by gateways.

interconnection See SNA network interconnection (IBM's SNA).

Interexchange Carrier (IEC) Any carrier registered with the Federal Communications Commission (FCC) that is authorized to carry customer transmissions between LATA's interstate or, if approved by a state public utility commission (PUC), intrastate.

interface 1. A shared boundary defined by common physical interconnection characteristics, signal characteristics, and meanings of interchanged signals. 2. The equipment which provides this shared boundary.

Interface-CCITT The present mandatory method in Europe (and a possible international standard) for the interface requirements between data terminal equipment and data circuit-terminating equipment (modems). The CCITT V.24 modem interface standard resembles the American EIA RS-232 standard.

Interface-EIA RS-232 A standardized method adopted by the Electronic Industries Association, in order to ensure uniformity of interface between data communication equipment and data circuit-terminating equipment. This standard interface has been generally accepted by a majority of the manufacturers of data transmission (modems) and data terminal equipment in the U.S. The latest version of the RS-232 standard is version D.

interface computer Part of a packet switching network which mediates between network subscriber and high-level or trunk network. It can be regarded as containing a local area switching terminal processors.

Interface Message Processor (IMP) A computer used to interface terminals and other computers to a network. Commonly used in packet switching networks to interface incompatible devices with the network.

interface processor Communications processor specialized for handling the function of interfacing computers and terminals to a network.

interference The addition of unwanted signals through a communications facility.

interLATA A call between local access and transport areas.

interleave The process of putting bits or characters alternately into time slots in a TDM (time division multiplexer).

interleaving The process of alternately transmitting characters, blocks, ..., messages.

interlock code In packet switching, a numerical value indicating Closed User Group (CUG) membership. The interlock code is sent in a Call Request Packet.

Intermediate Block Character (ITB) A transmission control character that terminates an intermediate block. A Block Check Character (BCC) usually follows. Use of ITBs allows for error checking of smaller transmission blocks.

Intermediate Distribution Frame (IDF) An equipment unit used to connect communications equipment by use of connection blocks.

intermediate node A node that is capable of routing path information units to another subarea and that contains neither the origin NAU, nor the destination NAU, nor any associated boundary function for these NAUs.

intermediate routing function In IBM's SNA, a path control capability in a subarea node that receives and routes path information units (PIUs) that neither originate in nor are destined for network addressable units (NAUs) in the subarea node.

Intermediate Routing Node (IRN) In IBM's SNA, a subarea node with intermediate routing function. A subarea node may be a boundary node, an intermediate routing node, both, or neither, depending on how it is used in the network.

intermediate SSCP In IBM's SNA, an SSCP along a session initiation path that owns neither of the LUs involved in a cross-network LU–LU session.

intermediate system A system which relays data between other systems.

intermodulation distortion An analog line impairment where two frequencies create an erroneous frequency which in turn distorts the data signal.

internal clock Clock signals generated by the modem.

International Access Code The code that must prefix the country code, city code, and local number of a directly dialed international telephone call. If international dialing is available, dial:

International Access Code + Country Code + City Code + Local Number

International Access Codes vary based upon the country you are located in and are listed in the following table.

Australia	0011
Austria—Linz	00
—Vienna	900
Bahrain	0
Belgium	00
Brazil	00
Colombia	90
Costa Rica	00
Cyprus	00
Czechoslovakia	00
Denmark	009
El Salvador	0
Finland	990
France	19*
French Antilles	19
Germany, Fed. Rep. of	00
Greece	00
Guam	001
Guatemala	00

Hong Kong	106
Hungary	00
Iran	00
Iraq	00
Ireland, Rep. of	16
Israel	00
Italy	00
Ivory Coast	00
Japan	001
Korea	001
Kuwait	00
Lebanon	00
Libya	00
Liechtenstein	00
Luxembourg	00
Malaysia (Kuala Lumpur & Penang only)	00
Monaco	19
Morocco	00*
Netherlands	09*
Netherlånds Antilles	00
New Zealand	00
Nicaragua	00
Norway	095
Panama	00
Philippines	00
Portugal (Lisbon only)	097
Qatar	0
Saudi Arabia	00
Senegal	12
Honduras	00
Singapore	005
South Africa/Namibia	09
Spain	07*
Sweden	009
Switzerland	00
Taiwan	002
Thailand	001
Tunisia	00
Turkey	99
United Arab Emirates	00
United Kingdom	010
United States	011
Vatican City	00
Venezuela	00

*Await dial tone

International Alphabet (IA) An internationally defined code for communication. Baudot code is IA2 and ASCII code is IA5.

International Business Machines Corporation (IBM) A very large computer company known more recently for its Personal Computer (IBM PC) and 3270 Information Display System products.

International Business Service (IBS) An all digital private line transmission facility offered by Comsat World Systems.

International Communications Association The trade association representing the largest users of communication and telecommunication products and services.

International Federation for Information Processing (IFIP) A federation of professional and technical societies concerned with information processing. One society is admitted from each participating nation. IFIP has established a number of technical committees and these have formed working groups. The technical committee for data communication is TC 6 and was formed in 1971. Among its working groups is WG 6.1 which is also known as the International Network Working Group. Working Group WG 6.3 deals with human–computer communications.

International Federation of Information Processing Societies (IFIPS) An organization comprising of trade and professional organizations related to the computer industry.

International Kilostream A private leased circuit facility from British Telecom International which provides point-to-point digital communications between the UK and other countries.

International Network Working Group (INWG) INWG was formed in 1972 to be a forum for discussion of network standards and protocols. In 1973 it was adopted as working group 1 of IFIP TC 6 with the title "International Packet Switching for Computer Sharing." It has about 100 members organized into four study groups. This working group distributes *INWG Notes* to exchange ideas on protocols, on interworking and on network research generally.

International Record Carrier (IRC) A common carrier originally dedicated to carrying telecommunication traffic from various gateway cities in the U.S. to foreign locations. Now, any common carrier who performs these services, whether on a dedicated basis or along with providing domestic services.

International Record Carrier Selection (IRC Select) An optional X.25 facility that allows the caller to select the gateway service for an international call.

International Standard An ISO standard document that has been approved in final balloting.

International Standards Organization (ISO) An international and voluntary standards organization, closely aligned with CCITT. Its OSI model is widely quoted, and its OSI communications protocols are widely accepted. Membership includes standards organizations from pariticipating nations (ANSI is the U.S. representative). ISO standards are listed under ISO.

International Telecommunciations Union (ITU) The telecommunications agency of the United Nations, established to provide standardized communications procedures and practices, including frequency allocation and radio regulations, on a worldwide basis. The International Telecommunications Union regulates telecommunications at the international level. It was established in 1863 and served as one of the founding agencies in the United Nations. Membership now stands at over 150 nations.

Internet Protocol A protocol used in gateways to connect networks at OSI network level 3 and above.

Internetwork Two or more networks of the same or different hardware types connected by means of special bridge hardware and software.

Internetwork Protocol Exchange (IPX) A subset of the Xerox Network Services (XNS) protocol used by Novell Corporation in its Netware software.

internetwork router In LAN technology, a device used for communications between subnetworks. Only messages for the corrected subnetwork are transmitted by this device. Internetwork routers function at the network layer of the OSI model.

internetworking Communication between two or more different networks.

internodal delay The delay of customer data that results from the time it takes to process a bypass channel through a multiplexer node.

interoffice trunk A direct circuit between telephone company central offices.

interpacket gap The time between the conclusion of transmission or reception of one packet and the transmission or reception of the next packet.

interpret table In IBM's VTAM, an installation-defined correlation list that translates an argument into a string of eight characters. Interpret tables can be used to translate logon data into the name of an application program for which the logon is intended.

interrupt 1. The initation, by hardware, of a routine intended to respond to an external (device-originated) or internal (software-originated) event that is either unrelated or asynchronous with the executing program. 2. A break in the normal flow of a system or routine that allows the flow to be resumed from that point at a later time. An interrupt is usually caused by a signal from an external source.

interrupt driven An interactive system whose operations and transmissions are determined by user interrupts (e.g., requests, transmission).

interstate Any connection made between two states.

inter-symbol interference An (analogue) waveform which carries binary data is generally transmitted as a number of separate signal elements. When received, the signal elements may be influenced by neighboring elements. This is called inter-symbol interference.

intertoll trunk A trunk between toll offices in different telephone exchanges.

interval timer A timer can be set by a program to produce an interrupt after a specified interval.

Interview A trademark used by Atlantic Research Corporation of Alexandria, VA, for a series of protocol analyzers.

Interview 7000 Series A series of high-speed, high-performance protocol analyzers from Atlantic Research Corporation.

interworking unit Another name for a gateway.

intrabuilding network cable Cable (formerly referred to as house or riser cable) that extends to the outside plant distribution from the building entrance point to equipment rooms, cross-connection points, or other distribution points within the building.

intraLATA A call within a local access and transport area.

intraoffice trunk A trunk connection within a central office.

intrastate Any connection made that remains within the boundaries of a single state.

intrasystem wiring Wire which is used to connect system components, e.g., a PBX to its stations.

INTUG International Telecommunications User's Group.

inventory control Logistics functions that include management, cataloging, requirements determination, procurement, inspection, storage, distribution, overhaul and disposal of material.

in-WATS The use of a WATS service permits toll-free calling (by means of an "800" number) into a subscriber's facility, from various extended geographical areas. Billing to the subscriber is based on zones and time blocks rather than metered usage.

INWG International Network Working Group.

I/O Input/Output.

I–O channel An equipment forming part of the input–output system of a computer. Under the control of I–O commands the "channel" transfers blocks of data between the main store and peripherals.

IOMI Input Output Microprocessor Interface.

IOP Input Output Processor.

IP 1. Internal Protocol. 2. Internet Protocol.

IPARS International Passenger Airline Reservation System (IBM).

IPL Initial Progam Load.

IPN Instant Private Network.

IPX Internetwork Protocol Exchange.

IRC International Record Carrier.

IRC Select International Record Carrier Selection.

Irma The name used by Digital Communications Associates of Alpharetta, GA, for a series of communications products.

Irmalan A family of hardware and software products marketed by Digital Communications Associates designed to provide personal computers on a local area network with access to mainframe applications.

IRN Intermediate Routing Node (IBM's SNA).

IS Information Separator.

isarithmic control The control of flow in a packet switching in such a way that the number of packets in transit is held constant.

ISDN Integrated Services Digital Network.

ISDN access line A digital transmission facility between a customer's premises and a local exchange which provides a mechanism to transmit data via ISDN. The ISDN interfaces are illustrated in the diagram.

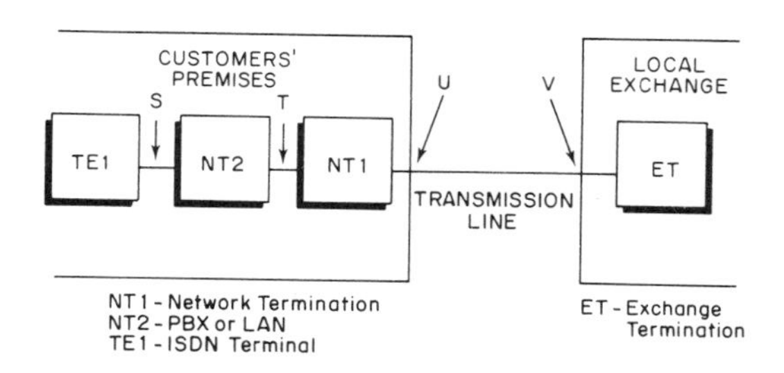

ISDN access line

ISDN user part The higher layer protocol in Common Channel Signalling System No. 7 that deals with end user signalling on the ISDN.

I-Series Recommendations CCITT recommendations on standards for ISDN services, ISDN networks, user–network interfaces, and internetwork and maintenance principles.

I-Series Standards The collected series of standards for the ISDN and its parts, in the CCITT 1984 Red Book.

ISO International Standards Organization.

ISO 646 7-bit character set for information interchange.

ISO 2022 Code extension technique for use with 7-bit code.

ISO 2110 25-pin DTE/DCE interface connector with pin assignments (RS232/V.24).

ISO 2593 Connector pin allocations for high-speed data terminal equipment (DTE).

ISO 3309 High-level data link procedures; frame structure (HDLC).

ISO 4902 HDLC unbalanced classes of procedures.

ISO 4903 15-pin DTE/DCE interface connector and pin assignments.

ISO–OSI International Organization for Standardization–Open Systems Interconnect.

isochronous Pertaining to a form of data transmission in which individual characters are only separated by a whole number of bit-length intervals.

isochronous distortion Caused by clock jitter in synchronous modems. The jitter can be early (positive jitter), in which the bits are too short, or the jitter can be late (negative jitter), in which the bits are too long. The isochronous distortion is the percentage of the sum of the absolute peak values of the two jitters compared to the ideal pulse width.

Isochronous Distortion = [(positive jitter + negative jitter)/total ideal pulse width] × 100%

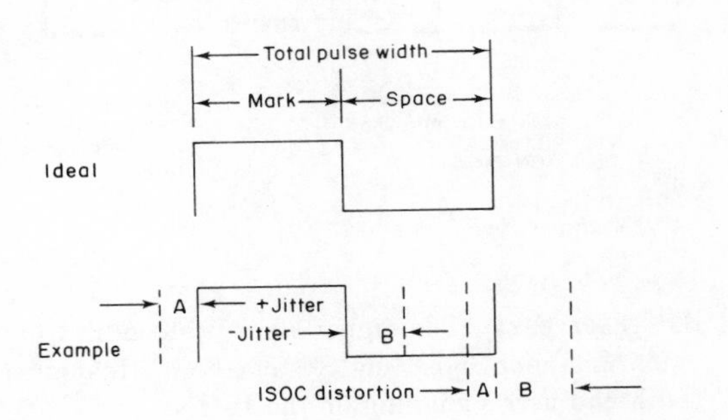

isochronous distortion

isolated adaptive routing A method of routing in which the decisions are made solely on the basis of information available in each node.

ISPF Interactive System Productivity Facitily (IBM's SNA).

ISSN Integrated Special Services Network.

ISTF Integrated Services Test Facility.

ISU Integrated Service Unit.

ISUP ISDN User Part.

I-SYNC Intelligent Synchronous (General DataComm, Inc.).

IT Intelligent Terminal.

ITD Intermediate Block Character.

ITDM Intelligent Time Division Multiplexer.

item In IBM's CCP, any of the components, such as communication controllers, lines, cluster controllers, and terminals, that comprise an IBM 3710 Network Controller configuration.

ITI Interactive Terminal Interface.

ITS Invitation To Send.

ITU International Telecommunication Union.

IUP ISDN User Part.

IW Inside Wiring.

J

jabber control In local area networking, the ability of a station to automatically interrupt transmission to inhibit an abnormally long data stream.

jabbering In LAN technology, the continuous sending of random data (garbage). Normally used to describe the action of a station (whose circuitry or logic has failed) that locks up the network with its incessant transmission.

jack A device used for terminating the permanent wiring of a circuit to which access is obtained by inserting a plug.

jam In local area networking, a collision enforcement technique that ensures detection of a collision by all stations.

jamming Intentional interference of open-air radio transmission to prevent communications.

JANET Joint Academic NETwork.

JCL Job Control Language.

JES Job Entry System.

JES2 A functional extension of the HASP II program that receives jobs into the system and processes all output data produced by the job.

jitter Slight movement of a transmitted signal in time or phase that can cause errors or loss of synchronization. An error condition on a digital circuit caused by short-term deviations of signal pulses from phase (i.e. the drifting of the centers of pulses from the center of the specified bit time slots).

job One or more related programs. A set of data including programs, files, and instructions to a computer.

job class Any one of a number of job categories that can be defined. With the classification of jobs and direction of initiator/terminators to initiate specific classes of jobs, it is possible to control the mixture of jobs that are performed concurrently.

job deck A sequence of job control and data cards used by an operating system to activate a job.

Job Entry Subsystem (JES) A system facility for spooling, job queueing, and managing the scheduler work area.

Job Entry System (JES) An IBM mainframe control protocol and procedure designed to handle the execution of specific jobs and tasks input to the computer.

Job Output Element (JOE) Information that describes a unit of work for the HASP or JES2 output processor and represents that unit of work for queueing purposes.

Job Queue Element (JQE) A control block that represents an element of work for the system (job) and is moved from queue to queue as that work moves through each successive stage of JES2 processing.

Job Separator Page Data Area (JSPA) A data area that contains job-level information for a data set. This information is used to generate job header, job trailer or data set header pages. The JSPA can be used by an installation-defined JES2 exit routine to duplicate the information currently in the JES2 separator page exit routine.

job separator pages Those pages of printed output that delimit jobs.

joule A unit of energy expended when a force of one newton (a unit of measurement) moves the point of application one meter in the direction of the force (1 watt = 1 joule/second).

jumbo group A FDM carrier system multiplexing level containing 3600 voice frequency (VF) or telephone channels (6 master groups). Also called hyper group.

jump A change in the normal sequence of obeying instructions in a computer. A conditional jump is a jump which happens only if a certain criterion is met. Also known as control jump or control transfer.

jumper A cable (wire) used to establish a circuit, usually for testing or diagnostics.

junctor Part of a circuit switching exchange. The switching matrix brings to the junctor the two lines which are to be connected. In the junctor there is the common equipment needed in the circuit during the call.

JUNET Japanese Unix NETwork.

K

K Abbreviation for kilo, which means a thousand. 1. Used as such in unit symbols (e.g., Kbps). 2. Used as a measurement of computer memory capacity. However, 1K in these terms means 1024 bits of memory. Thus, a 4K memory chip really carries 4096 bits, not 4000.

K band Portion of the electromagnetic spectrum in the 10 GHz to 12 GHz range increasingly used for satellite communications.

Ka band A portion of the electromagnetic spectrum; frequencies approximately in the 18 GHz to 30 GHz range.

KAK Key-Auto-Key.

Katakana A character set of symbols used in one of the two common Japanese phonetic alphabets.

KAU Keystation Adapter Unit.

Kbit 1024 bits (128 bytes).

Kbps Thousands of bits per second (bps).

Kbytes (kilobytes) 1,024 bytes.

KDS Keyboard Display Station.

KDU Keyboard Display Unit.

Kermit Asynchronous file transfer protocol originally developed at Columbia University.

Kermit 370 A version of the Kermit protocol which operates on all IBM 370 series mainframes to include the 43XX, 308X, 309X, and 937X models.

key In encryption systems, the digital code used with a standard combination algorithm to render an encrypted data stream unique.

key generator A device used to generate keys used in encryption systems for communication security.

key management The process of generating, distributing, monitoring, and destroying keys used in the security system. Usually, there are levels of keys. A root key (or base key) protects the generation or transport of another key. It is changed relatively rarely (yearly) and is transported and loaded manually for security and accountability reasons. A session key (sometimes called a data encrypting key) is used to encrypt your data. It is changed relatively frequently (weekly or daily) and, since it is protected by root keys, can be transported securely over the network and changed electronically without visits to remote sites.

key pad 12 pushbuttons or "keys" on the face of a touch- tone telephone set, used by a human to dial a circuit to another telephone or sometimes to control certain computer functions.

key set A telephone set where each telephone line is accessed by pushing a button or key directly on the set.

key system A telephone system in which each telephone line is accessed by a push-button telephone for incoming or outgoing calls. Incoming calls are routed by an operator to the requested extension. In a pure key system, an outgoing line is similarly accessed by pushing a button.

Key Telephone System (KTS) Limited PBX type equipment typically with up to 12 trunk lines and up to 40 extensions.

keyboard A manual coding device in which pressing keys is the means of generating the desired code elements.

keyboard lockout An interlock feature used with teleprinters that prevent sending from the keyboard while the tap transmitter or another station is sending on the same circuit. This feature is used to avoid breaking up a transmission by simultaneous sending.

keyboard perforator A perforator provided with a bank of keys, the manual depression of any one of which will cause the code of the corresponding character or function to be punched in a tape.

keyboard send/receive (KSR) A combination teleprinter transmitter and receiver with transmission capability from the keyboard only.

key-encrypting key In IBM's SNA, a key used in sessions with cryptography to encipher and decipher other keys.

keying Modulation of a carrier, usually by frequency or phase, to encode data. Also the interruption of a dc circuit for the purpose of signaling information.

KG Key Generator.

kilo (K) 1. A prefix for one thousand times a specific unit. 2. A measure of computer memory size equal to 1024 units.

kilobit per second (Kbps) One thousand bits per second. A measurement of the rate of data transmission.

kilohertz A term used to state 1000 (kilo) cycles per second (hertz).

kilopacket (kpkt) One thousand packets.

Kjoebenhauns Telefon Aktieselskab (KTAS) A regional Danish telephone company.

Knowledge Index An information service which provides on-line access to approximately 50 of Dialog's most popular data bases at night and on weekends. For additional information telephone 1-800-334-2564.

kpkt Kilopacket.

KSR Keyboard Send/Receive.

KTAS Kjoebenhauns Telefon Aktieselskab.

KTS Key Telephone System.

Ku band Portion of the electromagnetic spectrum being used increasingly for satellite communications. Frequencies approximately in the 14/12 GHz range.

L

L band A portion of the electromagnetic spectrum commonly used in satellite and microwave applications with frequencies approximately in the 1 GHz region.

LADC Local Area Data Channel.

LADT Local Area Data Transport.

LAN Local Area Network.

LAN segment A portion of a local area network (LAN) separated from other portions of the LAN by one or more bridges.

language A set of terms or symbols used according to very precise rules to write instructions, or programs, for computers.

LAP Link Access Procedure.

LAPB Link Access Procedure-Balanced.

LAPD Link Access Procedure D-channel.

LAPM Link Access Procedure M.

Large Message Performance Enhancement Outbound (LMPEO) In IBM's VTAM, a facility in which VTAM reformats function management (FM) data that exceeds the maximum request unit (RU) size (as specified in the BIND) into a chain or partial chain of RUs.

Large-Scale Integration (LSI) A term used to describe a multifunction semiconductor device, such as a microprocessor, with a high density (up to 1000 circuits) of electronic circuitry contained on a single silicon chip. See the following table for comparison of circuit density ranges.

SCALE	CIRCUIT RANGE
Small (SSI)	2 to 10 circuits
Medium (MSI)	10 to 100 circuits
Large (LSI)	100 to 1,000 circuits
Very large (VLSI)	1,000 to 10 000 circuits
Ultra large (ULSI)	Over 10 000 circuits

laser (light amplification by stimulated emission of radiation) A device that transmits a very narrow and coherent (separate waves in phase with one another rather than jumbled as in natural light) single-frequency beam of electromagnetic energy in the visible light spectrum.

LASS Local Area Signaling Services.

Last In, First Out (LIFO) A method used to process the latest item to a queue first, without regard to the age of the oldest item.

last number dialed A PBX or telephone set feature which allows a user to quickly redial the previously dialed number.

Last-In-Chain (LIC) In IBM's VTAM, a request unit (RU) whose request header (RH) end chain indicator is on and whose RH begin chain indicator is off.

LATA Local Access and Transport Area.

latency In local area networking, the time (measured in bits at the transmission rate) for a signal to propagate around or throughout the network.

layer 1. One of the divisions of the OSI model (see following table). 2. One of the division of SNA and other communications protocols.

7-LAYER OSI MODEL

LAYER	*DESCRIPTION*
7. Application	Provides interface with network users.
8. Presentation	Performs format and code conversion.
5. Session	Manages connections for application programs.
4. Transport	Ensures error-free, end-to-end delivery.
3. Network	Handles internetwork addressing and routing.
2. Data Link	Performs local addressing and error detection.
1. Physical	Includes physical signaling and interfaces.

layered protocol A protocol designed to obtain services from, and deliver services to other protocols in the manner described by the ISO layered model.

lays The twists in twisted pair cable. Two single wires are twisted together to form a pair. By varying the length of the twists, or lays, the potential for signal interference between pairs is reduced.

LCB Line Control Block.

LCD 1. Liquid Crystal Display. 2. Line Current Disconnect.

LCN Logical Channel Number.

LDC Link Data Channel.

LDM Limited Distance Modem.

LDSU Local Digital Service Unit.

leakage The current flowing through insulation. The symbol for leakage is the capital letter *G*.

learning bridge In a local area network, a bridge which adaptively creates its own tables of network topology by performing an analysis of the traffic it processes.

leased line A telephone line reserved for the exclusive use of a leasing customer without interexchange switching arrangements. A leased line may be point-to-point or multipoint.

least-cost routing The process of determining the optimum route or service to use for communications based upon existing tariffs.

LEC Local Exchange Carrier.

LED Light Emitting Diode.

LEN Low Entry Networking.

LEQ Line Equalizer.

lerad exit routine In IBM's VTAM, a synchronous EXLST exit routine that is entered automatically when a logic error is detected.

Letters Shift (LTRS) 1. A physical shift in a terminal using Baudot code that enables the printing of alphabetic characters. 2. The character that causes the shift.

level 1. Magnitude, as in signal level or power level. 2. Used as a synonym for layer.

Level Of Repair (LOR) The locations and facilities where items are to be repaired. Typical levels are operator, field technician, bench, and factory.

level 1 router In Digital Equipment Corporation Network Architecture (DECnet), a router which performs routing within a single area. Messages for destinations in other areas are routed to the nearest Level 2 router.

level 2 router In Digital Equipment Corporation Network Architecture (DECnet), a router which acts as a Level 1 router within its own area, but in addition routes messages between areas.

LF 1. Line Feed. 2. Low Frequency.

liaison A virtual connection (like a virtual circuit) which can be set up between two transport stations. A concept used in the end-to-end transport protocol defined by International Network Working Group (INWG).

LIC Last-In-Chain.

LIFO Last In, First Out.

Light Emitting Diode (LED) A semiconductor which emits incoherent light for a p-n junction (when biased with an electrical current in the forward direction). Light may exit from the junction strip edge or from its surface (depending on the device's structure).

lightguide Optical waveguide.

lightguide cable An optical fiber, multiple fiber, or fiber bundle which includes a cable jacket and strength members, fabricated to meet optical, mechanical, and environmental specifications.

Lightnet A fiber optic communications carrier based in Rockville, Md.

lightwave Electromagnetic wavelengths of approximately 0.8 to 1.6 micrometres in the region of visible light.

LIM Line Interface Module.

Limited Distance Modem (LDM) A signal converter which conditions and boosts a digital signal so that it may be transmitted much further than a standard RS-232 signal.

line 1. A communication medium connecting two or more points. 2. A physical path which provides direct communications among some number of stations.

Line Access Procedure (LAP) In packet switched networks, superseded by LAPB.

line adapter A device used for switching between switched (DDD) backup and dedicated lines. This adapter compensates for differences in receive and transmit power levels.

line analysis The process of measuring telecommunication circuit (line) parameters and analyzing the condition and quality of the circuit.

line buildout An attenuator which stimulates cable loss with a frequency rolloff. The amount of attenuation is specified in units of dB.

line conditioning Conditioning is a procedure that is used to make the levels of transmission impairments fall within specified limits. Line conditioning can be used on a leased line circuit to improve transmission quality. The level of conditioning required is a function of both speed and distance. Several different types are offered which result in a higher transmission rate and/or a reduction in data errors. Conditioning cannot ensure a "clean" line, but that distortion will fall within the limits prescribed.

line control In IBM's SNA, the scheme of operating procedures and control signals by which a data link is controlled; for example, sychronous data link control (SDLC). Synonym for data link control protocol.

Line Control Block (LCB) An area of main storage containing control data for operations on a line. The LCB can be divided into several groups of fields; most of these groups can be identified as generalized control blocks.

Line Control Unit (LCU) A communications controller.

line discipline Archaic term for communications protocol, e.g., the sequence of operations involving the actual transmission and reception of data.

line driver A signal converter which conditions the digital signal transmitted by an RS-232 interface to ensure reliable transmission beyond the 50-foot RS-232 limit and often up to several miles. It is a baseband transmission device.

Line Equalizer (LEQ) A fixed equalizer used to improve transmission and the effectiveness of a modem. LEQ offsets the sloping high frequency rolloff of a metallic local balanced pair cable, which is a function of cable length. LEQ may improve transmission on long cable runs which can experience greater losses and rolloff.

Line Feed (LF) A control character used to move to the next line on a printer or display terminal.

line folding The procedure which is necessary when a text message has a line longer than the maximum allowed by a printer. The excess characters are printed on the next line by generating a local new line signal. The appearance of the message is marred, but its sense is preserved.

line group In IBM's SNA, one or more communication lines of the same type.

line hits An incident of electrical interference causing unwanted signals to be introduced onto a transmission circuit. There are four different types of line hits (or momentary electrical disturbances) on the line caused by atmospheric conditions, telephone company switching equipment, and radio/microwave transmission.

line monitor A device capable of passively intercepting a data transmission and, based on a prior knowledge of the protocol in use and other factors, providing its operator with a display and analysis of the transmission.

line of sight transmission A characteristic of some open-air transmission technologies where the area between a transmitter and a receiver must be unobstructed. Examples of line of sight transmission include microwave, laser, and infrared.

line printer A device that prints all the characters of a line as a unit.

Line Processing Unit (LPU) A card in the Telenet Processor (TP) from which lines (links) to the network or users emanate. LPUs can contain four or eight lines and can run at low, medium, or high speed.

line protocol A set of rules used to organize and control the flow of information between two or more stations connected by a common transmission facility.

Line Protocol Handler (LPH) A communications program that processes messages, interrupts, and timeouts, and handles protocol acknowledgements, error recovery and other communications functions.

Line Quality Analysis (LQA) Diagnostic software that operates on a network management system in conjunction with modems to provide the measurement and display of analog line parameters.

line segment scrambling A method of video signal scrambling in which segments of lines are moved to other lines, totally obscuring the picture.

line sharing A form of X.21 switched line sharing in which many clients have access to a line, but only one client has access to any single call.

line shuffle scrambling A method of video signal scrambling in which lines are randomly interchanged within the image.

line speed The transmission rate of signals over a circuit, usually expressed in bits per second.

line splitter A device which splits a single line among a cluster of terminals. A modem sharing unit.

line switching Switching in which a current path is set up between the incoming and outgoing lines. Contrast with message switching in which no such physical path is established.

line test set Analog test equipment that measures characteristics of a circuit: level, frequency, noise, signal-to-noise ratio, and equalization.

Line Turnaround (LTA) On a two-wire circuit, the time required for one end to stop transmitting and then start receiving from the other end.

Linear Predictive Coding (LPC) A voice digitization technique in which speech parameters to include pitch, voice, and unvoiced sound parameters are used to develop an algorithm that can be encoded at a low data rate, typically 2400 or 4800 bps.

linearity The property of a transmission medium or of an item of equipment that allows it to carry signals without introducing distortion.

linelock Modem sharing software developed by Crystal Point, Inc., of Kirkland, WA, for Ungermann–Bass, Inc., and IBM Network BIOS compatible local area networks.

line-mode data A type of data that is formatted on a physical page by a printer only as a single line.

link 1. Communications circuit or transmission path connecting two points. 2. In IBM's SNA, the combination of the link connection and the link stations joining network nodes; for example: (1) a System/370 channel and its associated protocols, (2) a serial-by-bit connection under the control of Synchronous Data Link Control (SDLC). Synonymous with data link. *Note*: A link connection is the physical medium of transmission; for example, a telephone wire or a microwave beam. A link includes the physical medium of transmission, the protocol, and associated communication devices and programming. It is both logical and physical.

Link Access Procedure (LAP) The data link-level protocol specified in the CCITT X.25 interface standard, Supplemented by LAPB (LAP-Balanced) and LAPD.

Link Access Procedure, D-channel (LAPD) The Layer 2 protocol for the ISDN D-channel. Specified in CCITT Recommendation Q.921. LAPD is a framed, bit-oriented protocol similar to the LAP and LAPB protocols specified for X.25 circuits.

Link Access Procedure-Balanced (LAPB) In X.25 packet-switched networks, a link initialization procedure which establishes and maintains communications between the DTE and DCE. LAPB involves the T1 timer and N2 count parameters. All public data networks (PDNs) now support LAPB.

link communication The physical means of connecting one location to another for the purpose of transmitting and receiving information.

link connection In IBM's SNA, the physical equipment providing two-way communication between one link station and one or more other link stations; for example, a communication line and data circuit terminating equipment (DCE).

Link Control Procedure (LCP) A standard procedure by which data is transferred over any communications link to ensure order and accuracy. Same as link control protocol.

link control protocol A set of rules and procedures for conducting data communications across a transmission link.

Link Data Channel (LDC) A channel of 4 Kbps carried in the framing bits of the 4th, 8th, 12th, 16th, 20th, and 24th frames of a T1 channel using the Extended Superframe format. AT&T has specified that the Link Data Channel on its circuits will carry circuit performance information in a proprietary protocol. CCITT Recommendation G.703 makes no provision for the contents of the LDC.

link exchange A technique for improving network topology by trying the effect of substituting one link for another.

link layer In the OSI model, the layer between the physical and networks layers.

link level 2 test Same as link test.

link loopback A diagnostic technique in which a signal transmitted at the aggregate bit rate of a digital link (e.g. T1) is returned to the sending device in the opposite direction on the same link. Link loopback tests can be performed by some BERT testers, and by some channel service units equipped for diagnostic testing.

Link Problem Determination Aids (LPDA) In IBM's SNA, a series of testing procedures initiated by the NCP that provide modem status, attached device status, and the overall quality of a communications link.

Link State PDU (LSP) A protocol data unit (PDU) used by the routing algorithm to exchange information about each system's neighbors.

link station 1. In IBM's SNA, the combination of hardware and software that allows a node to attach to and provide control for a link. 2. In IBM's ACF/VTAM, a named resource within a subarea node representing another subarea node directly attached by a cross-subarea link. In the resource hierarchy, the link station is subordinate to the cross-subarea link. *Note*: An SDLC link station is defined in an NCP subarea node with a PU PUTYPE=4 macro in the NCP definition. A channel link station is defined dynamically by ACF/VTAM when a channel-attached NCP is activated.

Link Telecommunications A subsidiary of BellSouth Enterprises, the holding company for all unregulated BellSouth companies which provides paging and telephone answering service in Australia.

link test In IBM's SNA, a test in which one link station returns data received from another link station without changing the data in order to test the operation of the link. *Note*: Three tests can be made; they differ in the resources that are dedicated during the test. A link test, level 0 requires a dedicated subarea node, link, and secondary link station. A link test, level 1 requires a dedicated link and secondary link station. A link test, level 2 requires only the dedicated link station.

link-attached Pertaining to describe devices that are connected to a communications link or telecommunications circuit. Compare with channel-attached.

link-attached communications controller In IBM's SNA, an IBM communications controller that is attached to another communication controller by means of a link.

link/design A support service of Timeplex, Inc., of Woodcliff Lake, NJ, to both existing and prospective customers that can be used to reconfigure and expand an existing Timeplex network or to create a new network from scratch.

list A data structure in which each item of data can contain pointers to other items. Any data structure can be represented in this way, which allows the structure to be independent of the storage of the items.

LIU Line Interface Unit.

LLC Logical Link Control.

LLF Low Layer Functions.

LMOS Loop Management Operations System.

LMPEO Large Message Performance Enhancement Outbound.

LNI Local Network Interface.

load coils Coils used to reduce the attenuation distortion by making the attenuation nearly constant across the frequency band of 300 to 3000 Hz.

load module In ISO, a program unit that is suitable for loading into main storage for execution. Itis usually the output of a linkage editor.

load sharing A multiple-computer system that shares the load during peak hours. During non-peak periods or standard operation, one computer can handle the entire load with the others acting as fallback units.

loaded line A telephone line equipped with coils (called load coils or loading coils) which minimize voice frequency amplitude distortion by restoring the response at the higher frequencies within the voice bandwidth. To be used only with analog voice. Digital data cannot be used on these lines without experiencing severe attenuation of the data signal.

loading Adding inductance by the use of load coils to a transmission line to minimize amplitude distortion.

loading coil An induction device used in telephone local loops, generally exceeding 18 000 feet in length, that compensates for the wire capacitance and serves to boost voice-grade frequencies.

LOC Local Operating Company.

local In IBM's SNA: 1. Synonymous with channel-attached. 2. Pertaining to a device that is attached to a controlling unit by cables, rather than by a communication line.

Local Access and Transport Area (LATA) One of 161 U.S. geographical subdivisions used to define local (as opposed to long distance) telephone service.

local address In IBM's SNA, an address used in a peripheral node in place of a network address and transformed to or from a network address by the boundary function in a subarea node.

local analog loopback An analog loopback test that forms the loop at the line side (analog output) of the local modem.

Local Area Data Channel (LAD) Same as Bell 43401 circuit.

Local Area Data Transport (LADT) A method by which AT&T customers can send and receive digital data over existing wires between their premises and a telephone company office. Dial-up LADT lets customers use their lines for occasional data services. Direct Access LADT transmits simultaneous voice and data traffic on the same lines.

Local Area Network (LAN) A data communications network confined to a limited geographic area (up to 6 miles radius or about 10 kilometers) with moderate to high data rates (100 Kbps to 50 Mbps). The area served may consist of a single building, a cluster of buildings, or a campus-type arrangement. It is owned by its user, includes some type of switching technology, and does not use common carrier circuits—although it may have gateways or bridges to other public or private networks. Because it uses physical media (wires or coaxial cables) owned by the operator and does not normally cross public roads, it is not regulated by a body such as the FCC.

Local Area Signaling Services (LASS) Intra-LATA voice and data services using the AT&T 1A ESS switch and data services using the 1A ESS switch and signal transfer point (STP).

local attachment In IBM environments, the connection of a peripheral device or control unit directly to a host channel.

local bridge A bridge is a combination of hardware and software used to connect two local area networks (LANs).

local channel A cable pair within the cable that goes from the building complex to the telecommunications carrier's office. The local channel can be a major source of trouble for data services.

local channel loopback A channel loopback test that forms the loop at the input (channel side) to the local multiplexer.

local composite loopback A composite loopback that forms the loop at the output (composite side) of the local multiplexer.

local dataset A signal converter which conditions the digital signal transmitted by a RS-232 interface to ensure reliable transmission over a dc continuous metallic circuit without interfering with adjacent pairs in the same telephone cable. Normally conforms with Bell Publication 43401. Also called baseband modem, limited distance modem, local modem, or short-haul modem.

local digital loopback A digital loopback test that forms the loop at the DTE side (digital input) of the local modem.

local echo The ability of a device to echo, or send a copy of the incoming data stream back to the sending device.

local echoplex A method of checking data transmission accuracy whereby data characters are returned to the sending terminal's display screen for comparison with the original transmitted data. Some host computers can provide echoplex to the terminal. However, local echoplex is especially desirable when operating over links where long time delays are encountered, e.g., satellite communications.

local exchange company/local operating company The local telephone company responsible for service within LATAs.

local exchange, local central office The exchange or central office in which the subscriber's lines terminate.

local line A term equivalent to local loop.

local loop A channel connecting the subscriber's equipment to the line terminating equipment in the central office, usually a metallic circuit, either two-wire or four-wire.

local 3270 major node See local non-SNA major node (IBM's SNA).

local name That part of the entity name which identifies an entity within a node.

Local Network Interface (LNI) A device that attaches to a transceiver on a LAN cable and branches out to multiple ports. It both transmits and receives signals.

local non-SNA major node In IBM's ACF/VTAM, a major node whose minor nodes are channel-attached non-SNA terminals.

local operating company The telephone company supplying local services to a customer.

local service area The entire area within which a customer may call at the local rates without incurring toll charges.

Local Session Identification (LSID) In IBM's SNA, a field in a FID3 transmission header that contains an indication of the type of session (SSCP–PU, SSCP–LU, or LU–LU) and the local address of the peripheral logical unit (LU) or physical unit (PU).

local SNA major node In IBM's ACF/VTAM, a major node whose minor nodes are channel-attached peripheral nodes.

local-attached In IBM's SNA, pertaining to a device that is attached to a controlling unit by cables, rather than by a telecommunication line. Synonym for channel-attached.

location sharing A software-defined network (SDN)

service feature offered by AT&T which permits a user to share access circuits to another customer's SDN location.

lock-up An unwanted state of a system from which it cannot escape, such as a 'deadly embrace' in the claiming of common resources.

logfile An exact duplicate of a program session recorded in a file.

logging The act of recording something for future reference such as error events or transactions. Also called log, tape, logrec.

logic 1. In computer programming, the procedure used to perform a task or solve a problem. 2. In computer hardware, the circuits that carry out logical operations and do arithmetic.

logic channel, logical connection See virtual circuit.

logic error In IBM's ACF/VTAM, an error condition that results from an invalid request. A program logic error.

logical channel number In packet switched networks, a number assigned when a virtual call is placed. Up to 4095 independent logical channels may exist on a single link.

logical connection A call following a physical connection to a packet network, which establishes communication with another device in the network. A logical call continues until the user initiates a disconnect request.

logical group, logical group number In packet switched networks, logical channels are divided into one of 16 logical groups.

Logical Link Control (LLC) A data link protocol based on HDLC developed for local area networks. The protocol was developed by the IEEE 802 committee and is common to all of its LAN standards for data link-level transmission control.

logical record A collection of items independent of their physical environment. Portions of the same logical record may be located on other physical records.

Logical Unit (LU) 1. On an SNA network, a type of network-addressable unit that represents end users (application programs or operators at a device) to the network; a collection of programs that provide the interface through which end users access network resources and than manage information transmission between end users. 2. The combination of programming and hardware of a teleprocessing subsystem that comprises a terminal.

logical unit 6.2 A logical unit used to implement program-to-program communications.

logical unit services In IBM's SNA, capabilities in a logical unit to: (1) receive requests from an end user and, in turn, issue requests to the system services control point (SSCP) in order to perform the requested functions, typically for session initiation; (2) receive requests from the SSCP, for example to activate LU–LU sessions via Bind Session requests; and (3) provide session presentation and other services for LU–LU sessions.

logoff 1. The procedure by which a user ends a terminal session. 2. In IBM's ACF/VTAM, an unformatted session-termination request.

logon 1. The process of establishing communications with a computer, including identification of the user and verification (by use of a password) of the user's identity. 2. In IBM's VTAM, a request that a terminal be connected to a VTAM application program.

logon data In IBM's ACF/VTAM: 1. The user data portion of a field-formatted or unformatted session-initiation request. 2. The entire logon sequence or message from an LU. Synonymous with logon message.

logon message Synonym for logon data (IBM's ACF/VTAM).

logon mode In IBM's ACF/VTAM, a subset of session parameters specified in a logon mode table for communication with a logical unit.

logon mode table In IBM's ACT/VTAM, a set of entries for one or more logon modes. Each logon mode is identified by a logon mode name.

logon-interpret routine In IBM's ACF/VTAM, an installation exit routine, associated with an interpret table entry, that translates logon information. It may also verify the logon.

long distance access code A code used to gain access to a long distance network.

long distance call A call placed by a subscriber in one area code to another subscriber in a different area code.

long distance company/carrier A telephone company that provides services between LATAs and, in some cases, international service.

long distance service information As indicated in the following table, five communications carriers in the United States have toll-free telephone numbers that subscribers and potential users can call for information about their services.

AT&T	1-800-222-0300
ITT	1-800-526-3000
Metromedia Long Distance	1-800-292-1052
U.S. Sprint	1-800-521-4949
Western Union	1-800-562-0240

long line A communication line of a long distance.

long-haul Long distance telephone circuits that cross out of the local exchange. Generally applied to any inter-LATA circuits.

longitudinal parity check See Longitudinal Redundancy Check.

Longitudinal Redundancy Check (LRC) An error detection method in which the Block Check Character (BCC) consists of bits calculated on the basis of odd or even parity for all the characters in the transmission block. The first bit of the LRC is set to produce an odd (or even) number of first bits that are set to 1; the second through eighth bits are set similarly. Also called horizontal parity check.

loop 1. Instructions in a program that cause a computer to repeat an operation until a task is completed or until a predetermined condition is achieved. 2. Local transmission that connects your telephones to the nearest Central Office. Also known as subscriber loop.

loop adapter In IBM's SNA, a feature of the 4331 processor that supports the attachment of a variety of SNA and non-SNA devices. To ACF/VTAM, these devices appear as channel-attached devices.

loop checking A method of checking the accuracy of transmission of data in which the received data are returned to the sending end for comparison with the original data, which are stored there for this purpose.

loop current A teletypewriter to line interface and operating technique without modems.

Loop Management Operation System (LMOS) An AT&T system which automates the entry of trouble tickets into the telephone company's repair service bureau and provides a mechanism to monitor the status of repairs.

loop network A central network topology that includes a continuous circuit connecting all nodes in which messages are routed around the loop to and through a central controller.

loop start Method of signaling an off-hook condition between an analog telephone set and switch. Picking up the receiver closes a loop, allowing dc current to flow.

loopback Type of diagnostic test in which the transmitted signal is returned to the sending device after passing through all, or a portion of, a data communications link or network. A loopback test permits the comparison of a returned signal with the transmitted signal. In a digital transmission system, the electrical connection of a CSU's receive circuit back to its transmit circuit. With this connection made, all signals received by the CSU are retransmitted back to the sending unit which allows one-ended testing. Loopbacks may be made by sending an up-loop code, and broken with a down-loop code.

loopback test Type of diagnostic test in which the transmitted signal is returned to the sending device after passing through all, or a portion of, a data communications link or network. This allows a technician (or a built-in diagnostic circuit) to compare the returned signal with the transmitted signal. This comparison provides the basis for evaluating the operational status of the equipment and the transmission paths through which the signal travelled.

loosely coupled Pertaining to processors that are connected by means of channel-to-channel adapters (IBM).

LOR Level of Repair.

LOS Loss Of Signal.

loss Power decrease in a circuit or portion of a circuit, normally expressed in decibels. Opposite of gain.

Loss Of Signal (LOS) An error condition on a T1 circuit defined by AT&T as the absence of a DS1 signal on the circuit for more than 150 milliseconds. A Loss-Of-Signal condition triggers a Blue Alarm in the receiving hardware.

loss of synchronization An error condition on a framed T1 circuit, detected when two or more consecutive framing bits are received in error.

loudness of sound The intensity or strength of the sensation sound causes in the human ear.

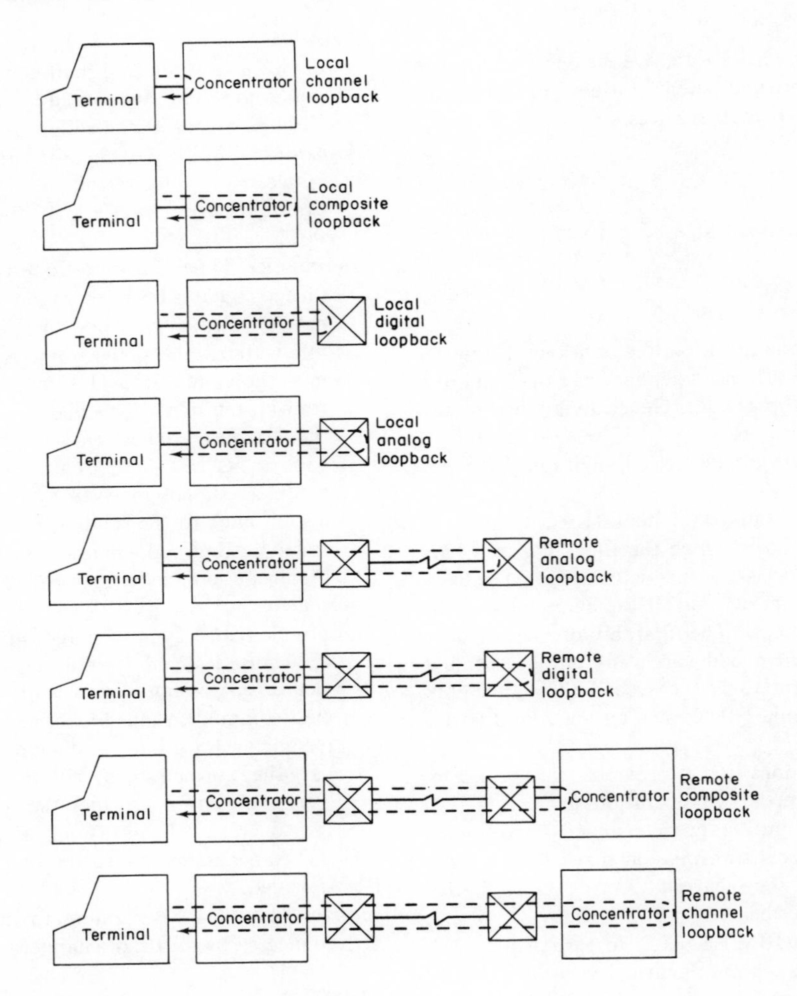

loopback test

Low Entry Networking (LEN) A peer-oriented extension to IBM's SNA first implemented on the System/36. LEN allows networks to be much more easily built and managed due to dynamic route selection, data base exchange, and other features in the extension.

Low Frequency (LF) A portion of the electromagnetic spectrum with frequencies of approximately 30 to 300 kHz.

low pass A specific frequency level, below which a filter will allow all frequencies to pass. Opposite of high pass.

low-speed line A narrowband line that is usually used for telegraph transmissions and teletypewriter transmissions.

Low-Speed Line Processing Unit (LSLPU) A card of the Telenet Processor (TP) computer. It interfaces asynchronous hosts and terminals to the network using start/stop protocol.

LPC 1. Linear Predictive Coding. 2. Longitudinal Parity Check.

LPDA Link Problem Determination Aids (IBM's SNA).

LPH Line Protocol Handler.

LQA Line Quality Analysis.

LRC Longitudinal Redundancy Check.

LSD Line Signal Detect.
LSI Large Scale Integration; Large Scale Integrated (circuit).
LSID Local Session IDentification (IBM's SNA).
LSLPU Low-Speed Line Processing Unit.
LSP Link State PDU.
LTA Line TurnAround.
LTRS LeTteR Shift (teletypewriters).
LTP Logical Test Port.
LTS 1. Loop Testing System. 2. Line Test Set.
LU 1. Line Unit. 2. Logical Unit.
LU 6.2 Logical Unit 6.2.
LU connection test In IBM's SNA, a diagnostic aid that permits a terminal operator to check whether the path between a system services control point (SSCP) and logical unit (LU) is operational.
LU type In IBM's SNA, deprecated term for LU–LU session type.
LU–LU session In IBM's SNA, a session between two logical units in an SNA network. It provides communication between two end users, or between an end user and an LU services component.
LU–LU session type In IBM's SNA, the classification of an LU–LU session in terms of the specific subset of SNA protocols and options supported by the logical units (LUs) for that session, namely: (1) The mandatory and optional values allowed in the session activation request. (2) The usage of data stream controls, FM headers, RU parameters, and sense codes. (3) Presentation services protocols such as those associated with FM header usage. LU–LU session types 0,1,2,3,4,6, and 7 are defined. *Note*: At session activation, one LU–LU half-session selects the session type and includes or excludes optional protocols of the session type by sending the session activation request, and the other half-session concurs with the selection by sending a positive response or rejects the selection by sending a negative response. In LU–LU session types 4 and 6, the half-sessions may negotiate the optional parameters to be used. For the other session types, the primary half-session selects the optional protocols without negotiating with the secondary half-session.

M

M Mega. Designation for one million.

M24 A feature offered on AT&T T1 circuits that allows the customer to specify separate destinations for individual channels on a T1 circuit that terminates at a DACS-equipped Central Office.

M44 A multiplexing service offered by common carriers that uses ADPCM to compress voice channels and combines up to 44 such channels with signaling onto a single DS1. To access this service, the DS1 format must adhere to M44 specifications as outlined in AT&T Technical Publication 54070.

m Milli. Designation for one thousand.

M bit The More Data mark in an X.25 packet which allows the DTE or the DCE to indicate a sequence of more than one packet.

mA Milliampere.

MAC Media Access Control.

Mac3270 A terminal-emulation and file-transfer product offered by Simware, Inc., of Ottawa, Canada. Also a trademark of Simware, Inc.

MacAPPC Apple Computer Corporation's implementation of IBM's Advanced Program-to-Program Communications (APPC) protocol.

Macintosh A series of Apple Computer Corporation microcomputers that represent the first wide-scale use of a window-based display system using icons and a mouse for the selection of specific operations.

MacIRMA A 3270 emulation board for the Macintosh SE manufactured by Digital Communications Associates (DCA) of Alpharetta, GA.

macro instruction In IBM's SNA, 1. An instruction in a source language that is to be replaced by a defined sequence of instructions in the same source language. The macro instruction may also specify values for paramenters in the instructions that are to replace it. 2. In assembler programming, an assembler language statement that causes the assembler to process a predefined set of statements called a macro definition. The statements normally produced from the macro definition replace the macro instruction in the program.

Macstar end customer management system A trademark of AT&T for a system which performs customer station rearrangements. The end customer can rearrange and/or move analog, Centrex, and all digital lines as well as perform automatic route selection with Macstar.

Magnetic Ink Character Recognition (MICR) A method of character recognition in which printed characters containing particles of magnetic material are read by a scanner and converted into a computer-readable digital format.

magnetic medium Any data-storage medium, including disks, diskettes, and tapes.

magnetic stripe A stripe of magnetic material, similar to a piece of magnetic tape, usually affixed to a credit card, badge or other portable item, on which data is recorded and from which data can be read.

mail server A computer system and associated software which performs the functions analogous to that provided by a post office box accessible by a number of persons. Users may send or forward electronic mail messages to anyone served by the system.

main (PBX or Centrex) Switch into which other PBXs are homed.

Main Distribution Frame (MDF) An equipment unit is used to connect communications equipment to lines, by use of connecting blocks.

main network address In IBM's SNA, the logical unit (LU), network address, within ACF/VTAM that is used for SSCP-to-LU sessions and for certain LU-to-LU sessions.

mainframe, mainframe computer A large-scale computer (such as those made by IBM, Unisys, Control Data, and others) normally supplied complete with peripherals and software by a single large vendor, often with a closed architecture. Also called host or CPU.

mainline program In IBM's ACF/VTAM, that part of the application program that issues OPEN and CLOSE macro instructions.

Maintain System History Program (MSHP) In IBM's SNA, a program that facilitates the process of installing and servicing a VSE system. For NPDA, MSHP is required for installation in a VSE system.

maintenance An activity intended to eliminate faults or to keep hardware or programs in satisfactory working condition.

Maintenance and Operator Subsystem (MOSS) In IBM's SNA, a subsystem of the 3725 Communication Controller that contains a processor and operates independently of the rest of the controller. It loads and supervises the 3725, runs problem determination procedures, and assists in maintaining both hardware and software.

maintenance mode In DDCMP, the mode of operation used by Maintenance Operations Protocol (MOP).

Maintenance Operations Protocol In DDCMP, a management protocol used for low-level communications with a system which is not fully operational or which is being tested.

maintenance services In IBM's SNA, network services performed between a host SSCP and remote physical units (PUs) that test links and collect and record error information. Related facilities include configuration services, management services, and session services.

major alarm Any alarm condition that causes loss of two or more channels of customer data.

major node In IBM's ACF/VTAM, a set of minor nodes that can be activated and deactivated as a group.

male-to-female connector A connector typically used to connect a DTE interface to a modem or multiplexer, to connect a multiplexer's network port to its trunk modem, or to connect a modem to a digital patch device.

male-to-male connector A connector typically used to connect a modem to a digital patch panel or to connect a DCE interface to a digital patch panel.

MAN 1. MANual. 2. Metropolitan Area Network.

Management Domain (MD) The set of Message Handling System (MHS) entities managed by an organization that includes at least one Message Transfer Agent (MTA) (X.400 specific).

Management Event Notification protocol (MEN) An Application layer management protocol used in DNA Phase V for communication between an event dispatcher and an event sink.

Management Information Control and Exchange (MICE) An Application layer management protocol used in DNA Phase V.

management services In IBM's SNA, management services are network services performed between a host SSCP and remote physical units (PUs) that include the request and retrieval or network statistics.

Manchester encoding A binary signaling mechanism that combines data and clock pulses. Each bit period in Manchester encoding is divided into two complementary halves: a negative-to-positive voltage transition in the middle of the bit period designates a binary "1" while a positive-to-negative transition represents a binary "0."

mandatory cryptographic session In IBM's SNA, a synonym for required cryptographic session.

manual answering In data communications, manual answering is performed by a person who hears the telephone ring, lifts the receiver, causes the called modem to send an answer tone to the calling modem and places the called modem into its data transmission mode of operation.

manual calling In data communications, calling performed by a person who dials a number, waits for an answer, then places the calling modem into its data transmission mode of operation.

Manufacturing Automation Protocol (MAP) In LAN technology, a token-passing bus designed for factory environments by General Motors. The IEEE 802.4 standard is nearly identical to MAP.

MAP Manufacturing Automation Protocol.

mapping 1. The establishment of one-to-one correspondence between two sets of data. 2. The translation of one group of data into a new set of data.

marine telephone Marine telephones operate on assigned radiotelephone frequencies, much as a radio broadcast does. Marine telephones can be used to contact other marine telephones or to reach land-based telephones through an operator.

mark 1. In single-current telegraph communications, a mark represents the closed, current-flowing condition. 2. In data communications, a mark represents a binary 1, the steady-state, no-traffic state for asynchronous transmission. 3. The idle condition. 4. In the context of the virtual terminal, a mark is a signal inserted into an output data stream by the virtual terminal, to acknowledge that an attention or interrupt input signal has been received.

mark-hold The normal no-traffic line condition whereby a steady mark is transmitted.

Markov constraint A constraint on the routing method according to which the future route of a packet is independent of its past history, such as its source or its route so far. This constraint is implied by directory routing.

mark-to-space transition The transition, or switching, from a marking impulse to a spacing impulse.

maser Microwave Amplification by Simulated Emission of Radiation. A device that generates signals in the microwave range, with low-noise characteristics.

mask Pattern of bits (1s or 0s) specified by the user that can be used with the trap mode of a communications test set.

masking A method of transforming one set of data into another while blocking or excluding some data from this process on the basis of code patterns or position.

master clock The source of timing signals, or the signals themselves, which all network stations use for synchronization.

master cryptography key In IBM's SNA, a cryptographic key used to encipher operational keys that will be used at a node.

master group In Frequency Division Multiplexing (FDM), an assembly of 10 supergroups occupying adjacent bands in the transmission spectrum for purposes of simultaneous modulation and demodulation.

master modem In a multipoint system, the modem that transmits constantly in the outbound direction. Usually the modem at the central site. In a multitier system, the term represents a remote master, or a master modem that is not located at the central site.

master station 1. In multipoint circuits, the unit which controls/polls the nodes. 2. In point-to-point circuits, the unit which controls the slave station. 3. In LAN technology, the unit on a token-passing ring that allows recovery from error conditions, such as lost, busy, or duplicate tokens. A monitor station.

mathematical model A mathematical description or approximation of some real event.

matrix In switch technology, that portion of the switch architecture where any input leads and any output leads meet.

matrix switch A device that allows any input to be cross-connected to any output.

MAU Multistation Access Unit.

maximum SSCP rerouting count In IBM's SNA, the maximum number of times a session initiation request will be rerouted to intermediate SSCPs before the request reaches the destination SSCP. This count is used to prevent endless rerouting of session initiation requests.

M-bit 1 000 000 (data rate) or 1 048 576 (data storage) bits.

Mbps Millions of bits per second (bps).

Mbytes (megabytes) 1 000 000 (data rate) or 1 048 576 (data storage) bytes (1000 Kbytes).

MCI800 A WATS service offered by MCI Communications Corporation.

MCP Multi-location Calling Plan.

MCVF Multi-Channel Voice Frequency.

MD 1. Management Domain. 2. Multiple Dissemination.

MDF Main Distributing Frame.

MDS Multiple Dataset System.

Mean Recovery Time (MRT) The normal repair time over a given period. Is sometimes used as a measure of assessing equipment reliability.

Mean Time Between Failures (MTBF) A figure of merit for electronic equipment or systems that indicates the average duration of periods of fault-free operation. Used in conjunction with Mean Time To Repair (MTTR) to derive availability figures.

Mean Time to Failure (MTF) The average length of time for which the system, or a component of the system, works without fault.

Mean Time to Repair (MTTR) A figure of merit for electronic equipment or systems that indicates the average time required to fix the equipment or system. Used in conjunction with Mean Time Between Failures (MTBF) to derive availability figures.

measured local service Telephone service where a change is made in accordance with measured usage (measured units).

Media Access Control (MAC) A local network control protocol that governs station access to a shared transmission medium. Examples are token passing and Carrier Sense, Multiple Access (CSMA).

media conversion Transformation of electrical signals used to transmit information to or from human usable form, e.g., words or numbers on a printed page, characters on a digital display.

medium Any material substance by which signals are conducted from point to point. Includes wire, coaxial cable, fiber optics, water, air, or free space.

medium band A term used for voice telephone transmissions and data transmissions linked to visual display terminals and similar devices. A medium band is also called a voice-band.

Medium-Scale Integration (MSI) A term used to describe a multifunction semiconductor device with a medium density (up to 100 circuits) of electronic circuitry contained on a single silicon chip.

mega (M) A prefix for one million times (10*E6) a specific unit.

megabit One million binary digits, or bits.

megabyte (Mbyte or M) 1 048 576 bytes; equal to 1024 Kbytes.

megahertz (MHz) A unit of analog frequency equal to 1 000 000 hertz.

MEMO An electronic mail system developed by Volvo Data and marketed by Verimation of Northvale, NJ. MEMO runs directly under VTAM and can be operated under MVS or DOS/VE. The MEMO system can be accessed from any IBM 3270 or compatible display terminal.

memory The part of a computer system or terminal where information is stored.

memory manager Software which controls dynamic requests for memory or returns unused memory to a memory pool.

MEN protocol Management Event Notification protocol.

menu An organized collection of fixed captions (field headers) and fields to accept variable data associated with each caption.

Mercury Communications A wholly owned subsidiary of Cable & Wireless which operates communications facilities within the United Kingdom.

Meridian Mail A voice processing system marketed by Northern Telecom, Inc.

Meridian Norstar A sophisticated but easy to use phone system manufactured by Northern Telecom which is equipped with a visual display.

Merlin A trademark used by British Telecom for a series of modems.

mesh A multi-node network topology that consists of more than three multiplexer nodes each interconnected with another node such that more than one aggregate path exists for each channel circuit.

message 1. A complete transmission; used as a synonym for packet, but a message is often made up of several packets. 2. In IBM's VTAM, the amount of FM data transferred to VTAM by the application program with one SEND request.

message address The information contained in the message header that indicates the destination of the message.

message format Rules for the placement of message elements such as the header, text, and closing.

Message Handling Service (MHS) A proprietary architecture developed by Action Technologies of Emeryville, CA, which uses a different addressing scheme and file format than the CCITT X.400 Recommendation. MHS electronic messaging software is supported by Novell in that firm's Netware LAN network software.

Message Handling System (MHS) The standard defined by the CCITT X.400 Recommendation and by ISO as the Message-Oriented Text Interchange Standard (MOTIS).

message header That portion of a message which contains the control, routing, and identification information for the message.

message numbering The identification of each message in a communications system by the assignment of a sequential number. This numbering is frequently used to facilitate message tracking and accounting.

message reference block An area of storage allocated to a message until it is processed.

message routing The selection of a path or a channel for sending a message.

message switch A device used to receive a message, store it until the proper outgoing line is available, and then retransmit it.

message switching A data communications technique in which a complete message is stored and then forwarded to one or more destinations when the required destination(s) are free to receive traffic. Because of the high transit time through the network, message switching is not used much for computer data but rather for administrative messages.

message switching network A public communications network over which subscribers send primarily textual messages to one another. TWX and Telex are examples of message switching networks.

Message Telecommunications Service (MTS) The official designation for long distance dialed (switched), tariffed telephone service.

message text That portion of a message which contains the data or information content of the message.

Message Transfer Agent (MTA) The functional component that with other MTAs constitutes the Message Transfer System (MTS). The MTA provides Message Transfer Service by: (1) interacting with originating User Agents (UAs); (2) relaying messages to other MTAs based on recipient designations; and (3) interacting with recipient UA(s) (X.400 specific).

Message Transfer Part (MTP) Collectively, the lower layers (layers 2 and 3) of CCITT Signaling System No. 7.

Message Transfer Protocol (P1) The protocol that defines the relaying of messages between Message Transfer Agents (MTAs) (X.400 specific).

Message Transfer Service The set of optional service elements provided by the Message Transfer System (MTS) (X.400 specific). *Note*: Message Transfer Service is not referred to by the acronym MTS. This acronym is used only for Message Transfer System.

Message Transfer System (MTS) The collection of Message Transfer Agents (MTAs) that provide the Message Transfer Service elements (X.400 specific).

message unit In IBM's SNA, a message unit is that portion of data within a message that is passed to, and processed by a particular network layer (e.g., path information unit (PIU) or request/response unit (RU)).

Message-Oriented Signaling A common channel signaling protocol for AT&T's Digital Multiplexed Interface. Based on the CCITT's Common Channel Signaling System No. 7.

Metal Oxide Semiconductor (MOS) Technology describing a transistor composed of a semiconductor layer including "source" and "drain" regions separated by a channel. Above the channel is a thin layer of oxide and over that a metal electrode called a gate. A voltage applied to this gate controls the current between the source and drain regions or, in another format, stops a flow between the two areas.

metered usage A type of telecommunications service offering in which the amount of usage of the service is monitored and billed accordingly. The opposite of a full period service in which usage is not a factor in billing.

metric prefixes A series of terms and their associated abbreviations used in the metric system to indicate multiples or portions of quantities which can be expressed as positive or negative powers of 10.

Prefix	*Symbol*	*Multiple*
nano-	n	0.000 000 001
micro-	μ	0.000 001
milli-	m	0.001
centi-	c	0.01
deci-	d	0.1
deka-	da	10
hecto-	h	100
kilo-	k	1 000
mega-	M	1 000 000
giga-	G	1 000 000 000

Metropolitan Area Network (MAN) A communications network that spans geographical areas whose size is between that of a local area network and a wide area or long distance network. Other characteristics are a high data rate, moderate delay and moderate error rate.

MFC Modular Feature Construction.

MFJ Modified Final Judgement.

MG Master Group.

MHD Moving Head Disk.

MHP Message-Handling Processor.

MHS Message-Handling System.

MHz Megahertz.

MIC Middle-In-Chain (IBM's SNA).

MICE Protocol Management Information Control and Exchange Protocol.

Micom, Micom Systems, Inc A supplier of data communications equipment.

MICR Magnetic Ink Character Recognition.

micro (μ) A prefix for one millionth of a specific unit.

micro programming The process of building a sequence of instructions into read only memory to carry out functions that would otherwise be directed by stored program instructions at a much lower speed.

microbend loss The leakage of light caused by very tiny, sharp curves in a lightguide that may result from imperfections where the glass fiber meets the sheathing that covers it.

microcode A set of software instructions which execute a macro instruction.

Microcom Networking Protocol (MNP) An error-correction and data compression protocol developed by the Microcom Corporation of Norwood, MA. The Microcom Networking Protocol (MNP) communications protocol supports interactive and file-transfer applications, divided into six classes, or performance levels. MNP performance levels include the following classes:

Class 1—the lowest performance level, uses an asynchronous byte-oriented half-duplex method of exchanging data. The protocol efficiency of a Class 1 implementation is about 70 percent; (a 2400-bps modem using MNP Class 1 will have a 1690-bps throughput).

Class 2—uses asynchronous byte-oriented full-duplex data exchange. The protocol efficiency of a Class 2 modem is about 84 percent (a 2400-bps modem will realize 2000-bps throughput).

Class 3—uses synchronous bit-oriented full-duplex data exchange. This approach is more efficient than the asynchronous byte-oriented approach, which takes 10 bits to represent 8 data bits because of the "start" and "stop" framing bits. The synchronous data format eliminates the need for start and stop bits. Users still send data asynchronously to a Class 3 modem, but the modems communicate with each other synchronously. The protocol efficiency of a Class 3 implementation is about 108 percent (a 2400-bps modem will actually run at a 2600-bps throughput).

Class 4—adds two techniques, Adaptive Packet Assembly and Data Phase Optimization. In the former technique, if the data channel is relatively error-free, MNP assembles larger data packets to increase throughput. If the data channel is introducing many errors, then MNP assembles smaller data packets for transmission. Although smaller data packets increase protocol overhead, they concurrently decrease the throughput penalty of data retransmissions—more data is successfully transmitted on the first try. Data Phase Optimization is a technique for eliminating some of the administrative information in the data packets, which further reduces protocol overhead. The protocol efficiency of a Class 4 implementation is about 120 percent (a 2400-bps modem will effectively yield a throughput of 2900 bps).

Class 5—adds data compression, which uses a real-time adaptive algorithm to compress data. The real-time capabilities of the algorithm allow the data compression to operate on interactive terminal data as well as file-transfer data. The adaptive nature of the algorithm refers its ability to continuously analyze user data and adjust the compression parameters to maximize data throughput. The effectiveness of data compression algorithms depends on the data pattern being processed. Most data patterns will benefit from data compression, with performance advantages typically ranging from 1.3 to 1.0 and 2.0 to 1.0, although some files may be compressed at an even higher ratio. Based on a 1.6 to 1 compression ratio, Microcom gives Class 5 MNP a 200 percent protocol efficiency, or 4800-bps throughput in a 2400-bps modem installation.

Class 6—adds 9600-bps V.29 modulation, universal link negotiation and statistical duplexing to MNP Class 5 features. Universal link negotiation allows two unlike MNP Class 6 modems to find the highest operating speed (between 300 and 9600 bps) at which both can operate. The modems begin to talk at a common lower speed, and automatically "negotiate" the use of progressively higher speeds. Statistical duplexing is a technique for simulating full-duplex service over half-duplex high-speed carriers. Once the modem link has been established using full-duplex V.22 modulation, user data streams move via the carrier's faster half-duplex mode. However, the modems monitor the data streams, and allocate each modem's use of the line to best approximate a full-duplex exchange. Microcom claims that a 9600-bps V.29 modem using MNP Class 6 (and Class 5 data compression) can achieve 19.2 Kbps throughput over dial circuits.

microcomputer 1. A desktop (or knee-top) computer. A personal computer. 2. A microprocessor system.

Micro-Fone II A data terminal which is a microprocessor-controlled telephone unit for both data and voice communications. It communicates with host computers either through a packet network or through the dial telephone network. (Micro-Fone II is designed primarily for transaction [credit card] communication).

microprocessor An electronic circuit on the surface of a small silicon chip which can be programmed to perform a wide variety of functions within the computer system or terminal. The Intel 8088, 80286, and 80386, Motorola 68000, and Zilog Z80 are examples of popular microprocessors.

microprocessor chip A single device cut from a wafer of silicon used in microcircuitry.

microsecond One millionth of a second.

Microsoft Disk Operating System (MS-DOS) A microcomputer operating system developed by Microsoft Corporation for the IBM PC and compatible personal computers. Also known as PC-DOS by IBM.

microwave 1. An electromagnetic wave between 1 centimeter (10*10 Hz) and 100 centimeters (10*8 Hz) in length. 2. Those frequencies in the super-high frequency (SHF) band above 890 megahertz (MHz) used for data and voice communications.

microwave communications A line-of-sight communications system which transmits on microwave frequencies.

Microwave Pulse Generator (MPG) A device that generates electrical pulses at microwave frequencies.

microwave relay system A microwave relay system consists of towers spaced as much as thirty miles apart. Using microwave signals, transmissions can criss-cross the country. Microwave transmissions travel in straight lines, so the towers must be in "line-of-sight" of each other. The curvature of the earth, mountains, and other obstacles can block transmission between land-based microwave towers. Repeaters at each tower amplify and retransmit the signals until they reach their destination.

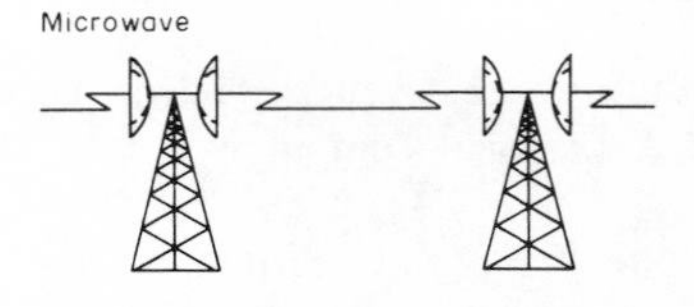

microwave relay system

microwave transmission Transmission of voice or data messages by data communications systems in which ultra-high frequency waveforms are used with line-of-site transmission between antennas, and with repeaters every 20–30 miles.

MIDAS Multiple Indexed Data Access System.

Middle-In-Chain (MIC) In IBM's VTAM, a request unit (RU) whose request header (RH) begin chain indicator and RH end chain indicator are both off.

midicomputer Computer that has a 24- or 32-bit word size. Sometimes called a super mini.

midsplit A method of allocating the available bandwidth in a single-cable broadband system. Transmissions from the headend to users are in the 168 to 300 MHz frequency range; transmissions from the users to the headend are in the 5 to 116 MHz sequence range.

MIF Minimum Internetworking Functionality.

military standards (United States) Two major U.S. military standards affecting the communications industry are MIL-STD-188C and MIL-STD-188-114. MIL-STD-188C defines the technical design of U.S. military communications systems such as permissible signal deterioration, modulation schemes, and hardware operation. MIL-STD-188-114 defines the electrical characteristics of military data communications systems.

Military Strategic/Tactical and Relay System (MILSTAR) A planned replacement for the US Air Force's defense satellite communications system (DSCS) satellites. It will use a 44-GHz uplink and a 20-GHz downlink.

milli (m) A prefix for one thousandth of a specific unit.

milliampere (mA) A measurement unit of electric current.

Millions of Instructions Per Second (MIPS) One measure of processing power.

millisecond One thousandth of one second.

milliwatt One thousandth of one watt.

MILNET MILitary NETwork.

MILSTAR MILitary Strategic/Tactical And Relay system.

MIL-188C A shielded, military standard interface, equivalent to RS-232C with the exception that the data and clocks are inverted and signal levels are +6, −6 volts.

Mind A network design tool of Contel Business Networks of Great Neck, NY.

minibased Any device containing and operating with a minicomputer.

minicall In packet-switched networks, the process of sending a datagram.

minicomputer A small-scale or medium-scale computer (such as those made by DEC, Data General, Hewlett-Packard, and others) usually operated with interactive, dumb terminals and often having an open architecture. Also called mini, for short.

Mini-Manufacturing Automation Protocol (MINI-MAP) A version of the Manufacturing Automation Protocol (MAP) consisting of only physical, link, and application layers that is intended for use in lower cost process-control networks. Under Mini-MAP a device with a token can request a response from an addressed device. However, unlike a standard MAP protocol, the addressed mini-MAP device does not have to wait for the token to respond.

MINI-MAP MINI-Manufacturing Automation Protocol.

Minimum Internetworking Functionality (MIF) A general principle within the ISO that calls for minimum local area network station complexity when interconnecting with resources outside the local area network.

Minimum Point Of Penetration (MPOP) A convenient point within a building or on a multibuilding facility where the communications carrier may choose to terminate its entrance cable. Beyond the MPOP all cable and wire responsibility falls on the customer.

minimum weight routing A routing scheme which minimizes the sum of the weights of the links employed route. These "weights" could be link delays, cost or error rate—anything which adds together in transit and should be minimized.

Miniset 270 A trademark of Siemens for that firm's quality telephones designed for home and business use.

Minitel A terminal developed in France for videotex usage.

Minitel 1 The first Minitel terminal which had an alphamosaic screen display of 25 rows by 40 characters.

Minitel 1B The basic Minitel terminal since 1986. This terminal is a dual-standard device, which supports both the Teletel standard and the 80-column ASCII standard.

minor node In IBM's VTAM, a uniquely defined resource within a major node.

MIPS Millions of Instructions Per Second.

MIS Management Information System.

MIU Multistation Interface Unit.

MJU Multipoint Junction Unit.

MLCP Multiline Communications Processor.

MLT Mechanized Loop Testing.

MMSU Modular Metallic Service Unit.

MNCS Multipoint Network Control System.

mnemonic code Instructions for the computer written in a form that is easy for the programmer to remember. A program written in mnemonics must be converted to machine code prior to execution.

MNP Microcom Networking Protocol.

mobile phone Any telephone which can be operated without a physical line.

mobile telephone service Telephone service for moving vehicles that use both the telephone network and radio.

Mobile Telephone Switching Office (MTSO) An office which controls individual call switching for all traffic emanating from or terminating on cellular radios within a certain area.

mode The path a light ray follows through a fiber.

model A representation of a system that is frequently constructed from the mathematical expressions which characterize the behavior of the system components.

modem A contraction of the term MOdulator–DEModulator. A modem is an electronic device used to convert digital signals to analog form for transmission over the telephone network. Since the telephone network was designed for analog voice transmission, it is not possible to transmit digital information from a terminal or a computer in its

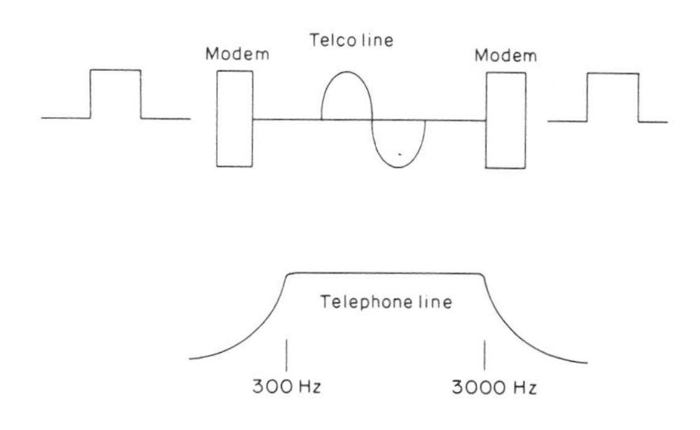

modem

binary form. Since the telephone network has a bandwidth of approximately 3000 Hz, modems using the telephone network must condition signals to fit within this band. Also known as a data set.

modem, multiport A device that combines a multiplexer and a modem allowing two or more DTEs to be connected to the same line. Also called split stream modem.

modem, quick turnaround A modem with minimal turnaround time when the line is used in a half-duplex mode. Also called Quick Poll (QP), Fast Poll (FP).

modem, short haul Description of both line drivers and limited distance modems.

modem, wideband A modem designed to operate at speeds greater than those used with high speed modems, such as 19.2 or 56 Kbps. Wideband modems will not operate over voice-grade circuits but require a wideband circuit.

Modem-7 Communications software program supporting the public-domain, X-modem, error-correcting file transfer protocol. This version of the X-modem has multifile transfer capability.

modem operation characteristics See p. 143.

modem connect The name used in DNA for that class of communications links governed by industry standards for modem connection.

modem eliminator A device used to connect a local terminal and a computer port in lieu of the pair of modems that they would expect to connect to. Allows DTE-to-DTE data and control signal connections otherwise not easily achieved by standard cables or connectors. Modified cables (crossover cables) or connectors (adapters) can also perform this function.

modem pooling A feature of a PABX and other communications products that permits subscribers to be automatically or manually connected to a group of shared or "pooled" modems.

modem sharing unit A device that splits a signal among a cluster of terminals and allows them to share one modem.

modem substitution switch An external option that allows you to reroute your data through a "hot" spare (a modem that is already powered up) in the event the original modem fails.

modes Discrete optical waves that can propagate in optical waveguides. Whereas in a single-mode fiber, only one mode, the fundamental mode, can propagate. There are several hundred modes in a multimode fiber which differ in field pattern and propagation velocity (multimode dispersion). The upper limit to the number of modes is determined by the core diameter and numerical aperture of the waveguide.

Modified Chemical Vapor Deposition An AT&T Bell Laboratories-patented process that uses high temperatures to speed the manufacture of large quantities of fiber lightguide. The glass is made by allowing hot vapors to form a coating inside a tube of heated silica, which is later drawn into fiber. Temperatures reach 4000 degrees F. (The melting point of steel is 2800 degrees F.)

Modified Final Judgement (MFJ) The 1982 Federal Court ruling that determined the rules governing the divestiture of the Bell Operating Companies from AT&T and other antitrust and deregulation issues. Presided over by Judge Harold Greene, as was the AT&T Antitrust settlement which the MFJ modified. Judge Greene continues his involvement in enforcing and interpreting the provisions of this settlement.

Modular Feature Construction (MFC) The unique capability on the AT&T 5ESS switch that permits operating companies to actually create new features by varying the way in which existing features work and interwork.

modular jack cable A cable designed to connect modems or telephones to modular telephone jacks, or to connect telephones to modems.

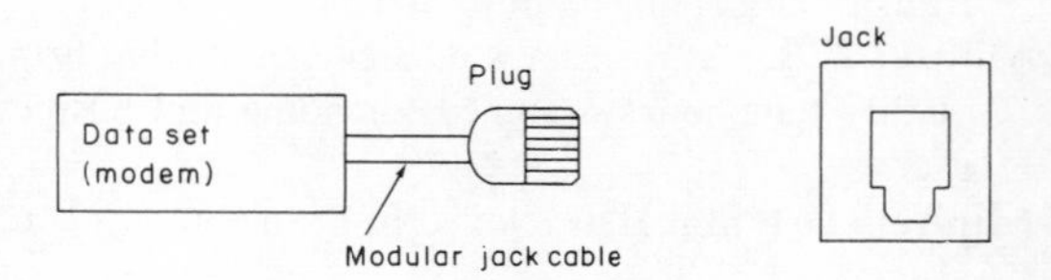

modular jack cable

modulation The process by which a carrier is varied to represent an information-carrying signal. See amplitude modulation, frequency modulation and phase modulation.

modulation frequency The frequency of the modulating wave.

modulation with a fixed reference A type of modulation in which the choice of the significant con-

modem operation characteristics

Modem type	Data rate	Transmission technique	Modulation technique	Transmission mode	Line use
Bell System					
103A,E	300	asynchronous	FSK	Half, Full	Switched
103F	300	asynchronous	FSK	Half, Full	Leased
201B	2400	synchronous	PSK	Half, Full	Leased
201C	2400	synchronous	PSK	Half, Full	Switched
202C	1200	asynchronous	FSK	Half	Switched
202S	1200	asynchronous	FSK	Half	Switched
202D/R	1800	asynchronous	FSK	Half, Full	Leased
202T	1800	asynchronous	FSK	Half, Full	Leased
208A	4800	synchronous	PSK	Half, Full	Leased
208B	4800	synchronous	PSK	Half	Switched
209A	9600	synchronous	QAM	Full	Leased
212	0–300	asynchronous	FSK	Half, Full	Switched
	1200	asynchronous/ synchronous	PSK	Half, Full	Switched
CCIT					
V.21	300	asynchronous	FSK	Half, Full	Switched
V.22	600	asynchronous	PSK	Half, Full	Switched/ Leased
	1200	asynchronous/ synchronous	PSK	Half, Full	Switched/ Leased
V.22 bis	2400	asynchronous	QAM	Half, Full	Switched
V.23	600	asynchronous/ synchronous	FSK	Half, Full	Switched
	1200	asynchronous synchronous	FSK	Half, Full	Switched
V.26	2400	synchronous	PSK	Half, Full	Leased
	1200	synchronous	PSK	Half	Switched
V.26 bis	2400	synchronous	PSK	Half	Switched
V.26 ter	2400	synchronous	PSK	Half, Full	Switched
V.27	4800	synchronous	PSK		
V.29	9600	synchronous	QAM	Half, Full	Leased
V.32	9600	synchronous	QAM	Half, Full	Switched
V.33	14400	synchronous	TCM	Full	Leased

dition for any signal element is based on a fixed reference.

modulator An electronic circuit that combines digital data to be transmitted with a carrier signal. The combination of the digital data and the carrier form a composite signal that is suitable for transmission over data communications lines.

module 1. (In hardware) short for card module. 2. (In software) a program unit or subdivision that performs one or more functions.

modulo A term used to express the maximum number of states for a counter. This term is used to describe several packet-switched network parameters, such as packet number (usually set to modulo 8—counted from 0 to 7). When the maximum count is exceeded, the counter is reset to 0.

modulo 2 addition A method of adding binary digits which gives: 0+0 = 0; 0+1 = 1; 1+0 = 1; 1+1 = 0 (binary addition without carries). Other names for the operation are "not equivalent" and "exclusive OR".
modulo-n A quantity such as of messages or frames, that can be counted. (Modulo 8, Modulo 128).
Molniya The name for a series of Soviet communications satellites used to form a domestic communications network.
Molniya 1-1 The first Soviet operational communications satellite that was launched on 23 April, 1965.
monitor A function or device that involves the observation of activity on a data communication line without interference.
monitor station In LAN technology on ring networks, the unit responsible for removing damaged packets and for making sure that the ring is intact.
monitored bulletin board An information system or electronic mail service bulletin board that remembers a user's last access and reminds the user when new mail has been sent to that board.
monitoring A testing function in which a protocol analyzer displays, records, or gathers statistics on the data transmitted over a circuit without interrupting the circuit or transmitting test data.
MOP Maintenance Operations Protocol.
MOS 1. Metal Oxide Semiconductor. 2. A common channel signaling protocol for AT&T's Digital Multiplexed Interface, based on the CCITT's Common Channel Signaling System No. 7.
MOSFET MOS Field-Effect Transistor.
MOSS Maintenance and Operator SubSystem (IBM's VTAM).
MOTIS Message-Oriented Text Interchange Standard.
mouse Electronic device that controls movement of a cursor on a video display terminal or monitor, when the user rolls the device along a flat surface by hand.
move A change in the physical location of facilities or equipment provided by a common carrier, either within the same customer or authorized user premises or to different premises, when made at the request of the customer without discontinuance of service.
MP Modem Port.
MPCC Multi-Protocol Communication Controller.
MPG Microwave Pulse Generator.
MPL Multi-schedule Private Line.
MPOP Minimum Point of Penetration.
MRT Mean Recovery Time.
MS DOS MicroSoft Disk Operating System.
MS NET A networking software package by Microsoft, for use with the IBM Personal Computer.
MSHP Maintain System History Program (IBM's VTAM).
MSI 1. Medium-Scale Integrated (circuit). 2. Medium-Scale Integration.
MSLPU Medium-Speed Line Processing Unit.
MSNF Multi-System Networking Facility.
MSU Multipoint Signaling Unit.
MT Measured Time.
MTA Message Transfer Agent.
MTBF Mean Time Between Failures.
MTF Mean Time to Failure.
MTL Master Test Line.
MTP 1. Message Transfer Part. 2. Message Transfer Point.
MTS 1. Message Telecommunications Services. 2. Message Telephone Service. 3. Message Transfer System.
MTSO Mobile Telephone Switching Office.
MTTR Mean Time To Repair.
MU-LAW The North American standard algorithm for the PCM encoding of PAM samples for digital voice communications.
multiaccess A system which can be independently accessed by several users simultaneously.
multiaccess spool configuration Two to seven systems sharing the JES2 input, job, and output queues through the use of shared DASD.
multiaggregate Where more than one aggregate or transmission facility is interfaced to a single multiplexer. This term does not imply multinode.
multicast bit In LAN technology, a bit in the Ethernet addressing structure used to indicate a broadcast message (a message to be sent to all stations).
multicast message In a LAN, a message intended for a subset of the stations on a network as opposed to an individual station or all stations.
multichannel The use of a common channel to derive two or more channels through frequency- or time-division multiplexing.
Multicom 3270 A trademark of MultiTech Systems

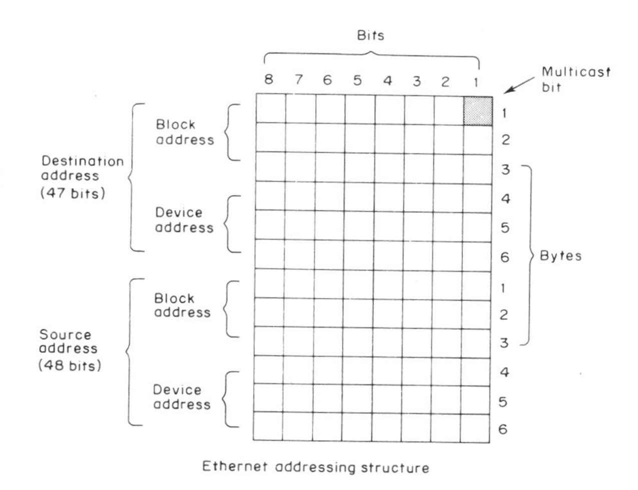

Ethernet addressing structure

multicast bit

of New Brighton, MN, as well as a terminal emulation system consisting of a coaxial card designed for insertion into a PC and 3278/79 terminal emulation and file transfer software.

multidomain network In IBM's SNA, a network consisting of two or more host-based system services control points, usually multi-mainframes.

multidrop A telephone line configuration in which a single transmission facility is shared by several end stations.

multidrop line Line or circuit interconnecting several stations. Also called a multipoint line.

Multiframe An ordered, functional sequence of frames on a multiplexed digital circuit. "Multiframe" is the CCITT standard term for a "Superframe." D4 framing uses a 12-frame Multiframe; ESF uses a 24-frame Multiframe; the European format uses a 16-frame Multiframe.

multileaving A characteristic of some character-oriented protocols in which more than one job's set of data can be transmitted, intermixed, at the same time.

Multiline Communications Processor (MLCP) A programmable interface between a central processor and one or more communications devices. The MLCP can be programmed to service specific communications devices.

Multi-line IDA A 2.048 Mbps primary rate ISDN service marketed by British Telecom in the U.K. This service is based upon British Telecom's proprietary signaling system for private to public exchange, and not on the CCITT Q.931 international standard.

Multi-location Calling Plan (MCP) An AT&T bulk-rated multi-location discount service that provides separate bills to each participating location instead of billing a single customer of record.

multimode A fiber-optic waveguide capable of propagating light signals of two or more frequencies or phases.

multi-mode fiber The relatively large core of this lightguide allows light pulses to zigzag along many different paths. It is ideal for light sources larger than lasers, such as LEDs.

multimode optical fiber A fiber that will allow more than one bound mode to propagate. May be either graded index or step index fiber.

multi-node Pertaining to a network with three or more nodes where networking functionality is employed, such as channel pass-through and alternate routing.

multiple access A network connecting many devices or terminals, all of which use the same channel for broadcasting messages.

multiple connection port A dedicated packet network Central Office access port that supports multiple, simultaneous, virtual connections. Stations connected to the network through such ports must use the X.25 protocol.

multiple domain network In IBM's SNA, a network with more than one System Services Control Point (SSCP).

multiple gateways In IBM's SNA, more than one gateway serving to connect the same two SNA networks for cross-network sessions.

Multiple Indexed Data Access System (MIDAS) A Prime database.

multiple routing The process of sending a message to more than one recipient, usually when all destinations are specified in the header of the message.

multiple speed selection A modem feature which permits continued transmission, even while the quality of the phone line is degrading, by falling back to a lower transmission speed to reduce error rate until line quality is restored.

Multiple Virtual Storage (MVS) An IBM operating system whose full name is the Operating System/Virtual Storage (OS/VS) with Multiple Virtual Storage/System Product for the System/370. It controls the execution of programs.

Multiple Virtual Storage for Extended Architecture (MVS/XA) An IBM program product whose full name is the Operating System/Virtual Storage (OS/VS) with Multiple Virtual Storage/System Product for Extended Architecture. Extended architecture allows 31-bit storage addressing. MVS/XA is a software operating system controlling the execution of programs.

multiplex To interleave or simultaneously transmit two or more messages on a single channel.

multiplex systems Used to increase the call carrying capacities of the different types of carrier systems. Through the use of bundled-coax, microwave, radio systems, and fiber optics. It is possible to stack 100 000 individual voice/data channels on one system.

multiplexed channel A communications channel capable of servicing a number of devices or users at one time.

multiplexer (MUX) A data communications device that receives signals over several low-speed circuits and combines them in an orderly fashion on a single high-speed circuit. A T1 multiplexer (North American) receives data from 24 circuits at 64 Kbps (or from some number of lower-speed circuits at an equivalent combined bit rate) and combines them according to D4 or ESF framing onto a single circuit at 1.544 Mbps.

multiplexer channel A channel designed to operate with a number of devices simultaneously. May perform its multiplexing function by interleaving either bits or bytes.

multiplexing The process of combining two or more digital signals of low bandwidth into a single signal of higher bandwidth, where the aggregate (combined) bandwidth of the low-speed signals is less than or equal to the bandwidth of the higher-speed signal.

multiplexing, frequency division Multiplexing by splitting the bandwidth of the transmission facility into some number of lower speed channels or subbands.

multiplexing, statistical Multiplexing by providing bandwidth on the transmission facility only for those data streams that have data available for transmission. No bandwidth is wasted on terminals not sending data.

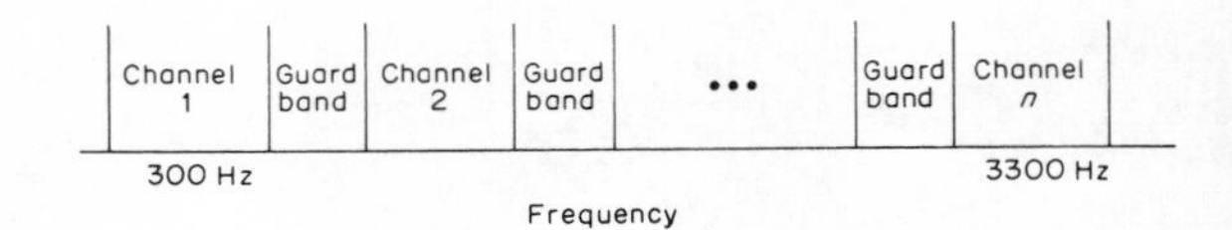

multiplexing, frequency division

multiplexing, time division Multiplexing by assigning each data stream its own time slot during which it transmits data over the transmission facility.

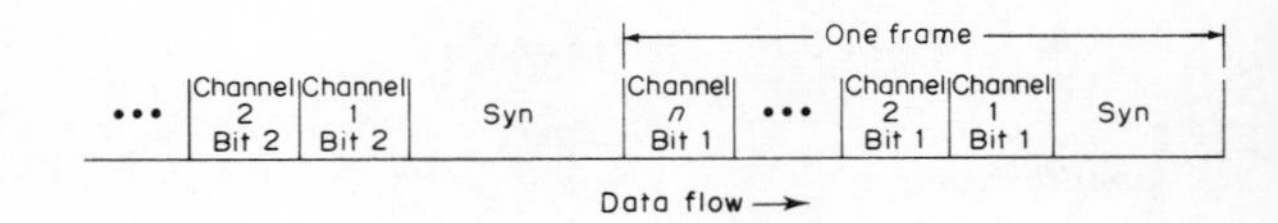

multiplexing, time division

multipoint Pertaining to a data channel which connects terminals at more than two points. The term multidrop is sometimes used.

multipoint line, multipoint connection A single communications line or circuit interconnecting several stations supporting terminals in several different locations. Use of this type of line usually requires some kind of polling mechanism, with each terminal having a unique address. Also called a multidrop line.

multipoint network A multipoint network is a communications line in which three or more stations are connected. This results in saved lines and saved

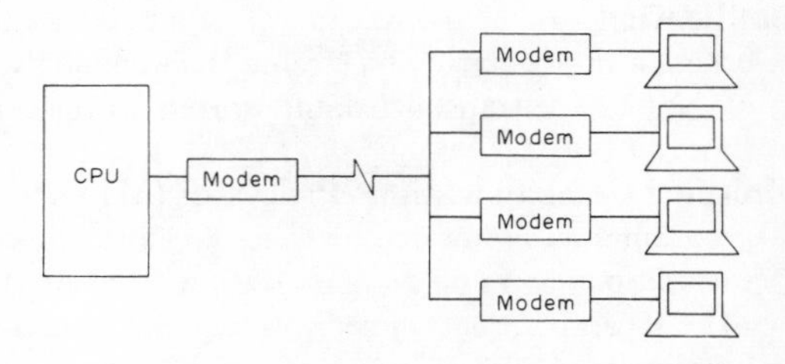

multipoint network

computer ports. A multipoint network should be connected by a leased line, as opposed to a switched line. There is a master/slave relationship between the modems that organize the communication network. Terminals must possess some intelligence so they can be addressed and polled.

Multipoint Junction Unit (MJU) Used to provide multipoint Digital Service.

Multipoint Signaling Unit (MSU) A device used in conjunction with a Digital Data test equipment to isolate and test various segments of a digital service multipoint circuit.

multiport Describes a network configuration in which several transmission facilities connect several end stations to a master station.

multiprocessing Strictly, this term refers to the simultaneous application of more than one process in a multi-CPU computer system to the execution of a single "user job" which is only processed if the job can be effectively defined in terms of a number of independently executable components. The term is more often used to denote multiprogramming operation of multi-CPU computer systems.

multiprogramming A method of operation of a computer system whereby a number of independent "user jobs" are processed together. Rather than allow each job to run to completion in turn, the computer switches between them so as to improve the utilization of the system hardware components.

Multi-schedule Private Line (MPL) AT&T's tariff for a voice-grade leased line.

multiserver network A single network which has two or more file servers. On a multiserver network, users may access files from any server to which they are attached.

Multistation Access Unit (MAU) A data concentrator/distributor used in local area networks that permits the attachment of up to eight devices. By connecting MAUs to one another, a larger network can be formed.

Multisystem Networking Facility (MSNF) An optional software feature that can be added to certain IBM telecommunications access methods which permits more than one host entity running ACF/TCAM or ACF/VTAM to jointly control an ACF/NCP network.

multitasking The concurrent execution of two or more tasks. Can also be the concurrent execution of a single program that is used by many tasks.

multiterminal controller A terminal controller having more than one terminal device connected to it for subsequent access to the communications line.

multithread application program In IBM's VTAM, an application program that processes requests for more than one session concurrently.

multithreading Concurrent processing of more than one message (or similar service-request) by an application program.

multiuser Pertaining to operating systems that allow many users at separate terminals to share a system's processing power, and possibly to also share data and peripherals.

multiway comparator An electronic circuit that can simultaneously compare one combination of bits with several other combinations of bits and determine if one combination is equal to any of the others.

MUX MUltipleXer.

MVS Multiple Virtual Storage.

MVS/XA Multiple Virtual Storage for eXtended Architecture operating system.

M/W MicroWave.

N

N connector A threaded connector for coax. N is named after Paul Neill.

N2 count In X.25 packet switched networks, the count for allowable number of retransmissions.

NACK Negative ACKnowledgement.

NAK Negative AcKnowledgement.

NAM Network Access Method.

name translation In IBM's SNA network interconnection, converting logical unit names, log-on mode table names, and class of service names used in one network into equivalent names to be used in another network. This function can be provided through NCCF and invoked by a gateway SSCP when necessary.

nameserver In Digital Equipment Corporation Network Architecture (DECnet), a system with at least one active clearinghouse.

namespace A tree of directories, starting at a root directory.

nano (n) A prefix for one billionth of a specific unit.

nanosecond One billionth of a second.

NAPLPS North American Presentation Level Protocol Syntax.

narrowband channel Sub-voice-grade channel with a speed ranging from 100 to 200 bits per second.

NARUC National Association of Regulatory Utility Commissioners.

National Association of Regulatory Utility Commissioners (NARUC) The trade association for state utility commissioners, representing their interests and concerns before Congress, the FCC, and federal courts.

National Bureau of Standards (NBS) A U.S. government agency that produces Federal Information Processing Standards (FIPS) for all U.S. government agencies, except the DoD. Membership includes other U.S. government agencies and network users.

National Cable Television association Trade organization for cable television carriers.

National Electrical Code (NEC) A nationally recognized safety standard for the design, construction, and maintenance of electrical circuits. The NEC, sponsored by the National Fire Protection Association, generally covers electrical power wiring in buildings.

National Exchange Carrier Association Trade association for interexchange carriers mandated by the FCC upon the divestiture of AT&T.

national facilities In packet switched networks, non-standard facilities selected for a given (national) network, which may or may not be found on other networks.

National Science Foundation Network (NSFNET) A network operated by the US National Science Foundation which ties together several supercomputer centers and which is designed to support academia.

National Telecommunications and Information Administration (NTIA) An agency of the US Department of Commerce concerned with telecommunication standards.

native mode The use of a protocol on the type of network for which that protocol was developed.

native network In IBM's SNA, the network that is attached to a gateway NCP and in which that NCP's resources reside.

NAU Network Addressable Unit.

NBS National Bureau of Standards.

NC 1. Network Control (IBM's SNA). 2. Network Concept.

NCC Network Control Center.

NCCF Network Communications Control Facility.

NCCS Network Control Center System.

NCL Network Control Language.

NCP Network Control Program.

NCP major node In ACF/VTAM, a set of minor nodes representing resources, such as lines and peripheral nodes, controlled by a network control program.

NCP/EP Definition Facility (NDF) In IBM's SNA, a program that is part of System Support

Programs (SSP) and is used to generate a partitioned emulation programming (PEP) load module or a load module for a Network Control Program (NCP) or for an Emulation Program (EP).

NCR–DNA NCR Corp.–Distributed Network Architecture.

NCS National Communications Systems.

NCTA National Cable Television Association.

NCTE Network Channel Terminating Equipment.

NDF NCP/EP Definition Facility.

NDT Net Data Throughput.

Near Instantaneous Companding (NIC) The real time process of quantizing an analog signal into digital data.

Near-End-Crosstalk (NEXT) Undesired energy transferred from one circuit to an adjoining circuit. Occurs at the end of the transmission link where the signal source is located.

near-end noise A noise measurement performed by the Master Test Line unit.

NEC National Electrical Code.

NECA National Exchange Carrier Association.

negative acknowledgement (NAK; NACK) 1. In BSC communications protocol, a control character used to indicate that the previous transmission block was in error and the receiver is ready to accept retransmission of the erroneous transmission block 2. In multipoint systems, the not-ready reply to a poll.

negative polling limit For a start–stop or BSC terminal in IBM's SNA, the maximum number of consecutive negative responses to polling that the communication controller accepts before suspending polling operations.

negative response In IBM's SNA, a response indicating that a request did not arrive successfully or was not processed successfully by the receiver.

negotiable bind In IBM's SNA, a capability that allows two logical unit half-sessions to negotiate the parameters of a session when the session is being activated.

negotiation When a connection is established between the virtual terminals in different systems, an initial dialogue is needed to establish the parameters that each will use during the transaction. This is known as the negotiation phase.

neighbor A system reachable by traversal of a single subnetwork.

nest Control or Expander Mainframe of a LINK Family Facilities Management System. Each Mainframe accepts various plug-in modules.

NET Used as an abbreviated form of network. Within the Telemail system, NET is used to indicate a device connected to a Telenet network.

NET 1000 Advanced Information System (ATT-IS).

Net Partner A trademark and network management system of AT&T. Developed by AT&T Bell Laboratories, it allows Integrated Services Digital Network (ISDN) and Centrex customers to work directly with most of the key operations systems controlled by their local exchange carrier for testing, monitoring and maintaining voice and data networks.

NETBIOS NETwork Basic Input–Output System.

Netline An X.25 Data Terminal Equipment (DTE) link from a host or concentrator into the network.

Net/Master A network management system developed by Cincom Systems, Inc., of Cincinnati, OH which rivals IBM's Netview network management system.

NetStream A British Telecom service that represents a range of local area network products. Under NetStream, users can interconnect PABX's, computers, kilostream packet services and other transmission facilities.

Netview In IBM's System Network Architecture, a mainframe product which provides an unified approach to SNA and non-SNA, IBM and non-IBM network management.

Netview Status Monitor (STATMON) A Netview subsystem which provides the network operator with a summary status of network resources.

Netview/PC An adaptation of the Netview approach for use in personal computers.

NetWare LAN networking products produced by Novell, Inc.

NetWare Operating System LAN network operating system produced by Novell, Inc.

NetWare Shell Novell LAN software which is loaded into the memory of each work station and which "shells" DOS to enable the work station to communicate with the LAN file server. The NetWare shell intercepts workstation requests before they reach DOS and reroutes network requests to a file server. The shell allows different types of workstations to use Novell NetWare.

network 1. A series of points connected by communications channels. 2. The switched telephone network is the network of telephone lines normally used for dialed phone calls 3. A private network is a network of communications channels confined to the use of one customer. 4. In IBM's SNA, an interconnected group of nodes; a user application network in data processing.

Network II.5 An analysis tool marketed by CACI of La Jolla, CA, which accepts computer or communications system descriptions and provides measures of hardware utilization, software execution, and conflicts.

network address 1. In packet switching, a unique identifier for every device (data terminal, host computer, switch or concentrator) that identifies the device for connection through the network. The address consists of a 12- or 14-digit number made up of the Data Network Identifier Code (DNIC), the area code, the server (TP or host computer), and the individual port subaddress. 2. In IBM's SNA, an address consisting of subarea and element fields that identifies a link, a link station, or a network addressable unit. Subarea nodes use network addresses; peripheral nodes use local addresses. The boundary function in the subarea node to which a peripheral node is attached transforms local addresses to network addresses and vice versa.

network address translation In IBM's SNA network interconnection, conversion of the network address assigned to a logical unit in one network into an address in an adjacent network. This function is provided by the gateway NCP that joins the two networks.

network architecture A computer vendor's plan for configuring networks by interfacing hardware and software products, e.g., SNA.

Network Basic Input–Output System (NETBIOS) An interface used by application programs in an IBM Personal Computer to access networks and network resources.

Network Channel Terminating Equipment (NCTE) Devices required at the customer premises for interfacing to a telephone circuit. The general name for equipment that provides line transmission termination and layer-1 maintenance and multiplexing, terminating a two-wire U interface in ISDN.

Network Communications Control Facility (NCCF) A host-based IBM program through which users can monitor and control network operations.

network configuration tables In IBM's SNA, the tables through which the system services control point (SSCP) interprets the network configuration.

Network Control (NC) In IBM's SNA, an RU category used for requests and responses exchanged between physical units (PUs) for such purposes as activating and deactivating explicit and virtual routes and sending load modules to adjacent peripheral nodes.

Network Control Center (NCC) Any centralized network diagnostic station used for the control and monitoring of a telecommunications network.

network control mode In IBM's SNA, the functions of a network control program that enable it to direct a communication controller to perform activities such as polling, device addressing, dialing, and answering.

Network Control Program (NCP) The software provided by IBM to run in its 3705, 3720, 3725, or 3745 communication controllers, responsible for handling links, controlling terminals, etc.

network control program generation In IBM's SNA, the process, performed in a host system, of assembling and link-editing a macro instruction program to produce a network control program.

network controller An IBM concentrator/protocol converter used with SDLC links. By converting protocols, which manage the way data is sent and received, the IBM 3710 Network Controller allows the use of non-SNA devices with an SNA host processor.

network facilities In packet switched networks, standard facilities are divided into (1) essential facilities (found on all networks) and (2) additional facilities (selected for a given network but which may or may not be selected for other networks).

Network File System (NFS) A distributed file system protocol developed by Sun Microsystems and adopted by other vendors which allows computers on a network to use the files and peripherals of other computers on the network as if they were local.

Network Identifier (Network ID) In IBM's SNA, the network name defined to NCPs and hosts to indicate the name of the network in which they reside. It is unique across all communicating SNA networks.

Network Information Services Company Ltd (NIS) A company, jointly owned by McDonnell Douglas and Marubeni, a leading Japanese trading company, that provides a public data network services in Japan.

Network Interface (NI) The physical point (e.g., connecting block, terminal strip, jack) at which terminal equipment or facilities (including station wiring) can connect to the network.

network interface machine A device used to interface a X.25 packet network with non-packet terminals. More commonly known as a Packet Assembler/Disassembler (PAD).

Network Interface Processor (NIP) A term used to identify the function of the Telenet Processor 3000 (TP3) product line.

network job entry facility (JES2) In IBM's SNA, a facility which provides for the transmission of selected jobs, operator commands, messages, SYSOUT data, and accounting information between communicating job entry nodes that are connected in a network either by binary synchronous communication (BSC) lines or by channel-to-channel (CTC) adapters.

network layer The third layer in the OSI model. Responsible for addressing and routing between subnetworks.

Network Logical Data Manager (NLDM) An IBM program product that collects and correlates LU–LU session-related data and provides the user with on-line access to the information. It runs as an NCCF communication network management application program.

network management 1. Functions which permit the operation of a network to be controlled and monitored. 2. The systematic approach to the planning, organizing, controlling and evolving of a communications network while optimizing the cost–performance relationship.

Network Management Productivity Facility (NMPF) An IBM product that adds several features to that vendor's NetView product to include a browse facility, a help desk facility, and a library of command lists (CLISTs).

Network Management System (NMS) A minicomputer system located in a Network Control Center (NCC) and connected to a network as a host.

Network Management Vector Transport (NMVT) In IBM's SNA, a solicited or unsolicited record containing alert data and issued by certain SNA resources to the host system.

network manager A person using network management.

network name 1. In IBM's SNA, the symbolic identifier by which end users refer to a network addressable unit (NAU), a link station, or a link. 2. In a multiple-domain network, the name of the APPL statement defining an ACF/VTAM application program is its network name and it must be unique across domains.

network node Synonym for node (IBM's SNA).

network operator In IBM's SNA: 1. A person or program responsible for controlling the operation of all or part of a network. 2. The person or program that controls all the domains in a multiple-domain network.

network operator console In IBM's SNA, a system console or terminal in the network from which an operator controls a communication network.

Network Packet Switching Interface (NPSI) An IBM X.25 product which permits that vendor's communications controllers to be interfaced to packet networks.

Network Performance Monitor (NPM) An IBM program host product that uses VTAM. It records performance data collected for various devices in a network.

network port A set of interfaces for receiving data from the outside world and sending data to the outside world.

Network Problem Determination Application (NPDA) A host IBM program that aids a network operator in interactively identifying network problems from a central point.

Network Processing Supervisor (NPS) A software program resident in Honeywell Datanet 355 or 6600 front-end processors that controls communications.

Network Processor (NP) A computer that controls data communications facilities. It performs functions such as routing, switching, concentrations, link/terminal control and, sometimes, information processor interfacing.

network relay Synonym for router.

Network Routing Facility (NRF) An IBM program product that resides in the NCP, which provides a path for messages between terminals, and routes messages over this path without going through the host processor.

Network Service Access Point (NSAP) An addressable point at which the network service is made available.

Network Services (NS) In IBM's SNA, the services within the network-addressable units (NAUs) that control network operations via sessions to and from the host system services control point (SSCP).

network services header In IBM's SNA, a 3-byte field in an FMD request/response unit (RU) flowing in an SSCP–LU, SSCP–PU, or SSCP–SSCP session. The network services header is used primarily to identify the network services category of the RU (for example, configuration services, session services) and the particular request code within a category.

Network Services Procedure Error (NSPE) In IBM's SNA, a request unit that is sent by an SSCP to an LU when a procedure requested by that LU has failed.

Network Services Protocol (NSP) A protocol operating in the DNA Transport layer.

Network Terminal Number (NTN) Number identifying the logical location of a DTE connected to a network. The NTN may contain a subaddress used by the DTE rather than by the network to identify equipment or circuits attached to it. The NTN can be up to 10 digits long.

Network Terminal Option (NTO) IBM program that enables an SNA network to accommodate non-SNA asynchronous and BSC devices via the NC-driven controller.

network terminating wire PBX-type service wire used to connect the intrabuilding network cable to the network interface.

network termination In ISDN, the device at the end-user's premises that receives signals from the provider's network and delivers it to the terminal equipment. The network termination can comprise two separate parts, NT1 and NT2; in this case, NT1 converts the network signal from the two-wire local loop to the four-wire end-user interface, and NT2 supplies the necessary demultiplexing, switching, and local signaling intelligence. The interface between NT1 and NT2 is the T-reference point; the interface between NT2 and the terminal equipment is the S-reference point. Outside the United States, the NT can be part of the provider's network; within the United States, the NT is defined as customer premises equipment and cannot be considered part of the provider's network. In this case, the interface between the NT and the network is the U-reference point.

Network Termination Unit (NTU) A network termination unit is a piece of equipment forming part of a network. Typically it has a keyboard, to select (dial) calls, and lamps or a display to indicate call progress signals.

network topology Describing the physical and logical relationship of nodes in a network.

network transparency The property of a communications system that makes it independent of the characteristics of the terminals and computers.

Network User Identification (NUI) In X.25 packet-switched networks, a combination of the network user's address and the corresponding password. Replaces the NTN in newer networks.

network virtual terminal A concept where DTEs with different data rates, protocols, codes, and formats are handled on the same network.

Network-Addressable Units (NAU) An SNA term referring to special program code segments that are used to represent software programs and hardware devices to a network. Not the actual programs or devices but a portion of a control program that represents the programs or devices.

networking In an SNA, multiple-domain network, communication among domains.

networking multiplexer A T1 multiplexer capable of channel bypass, drop-and-insert multiplexing, and intelligent channel-by-channel switching of T1 circuits. Also called a switching multiplexer.

neutral current loop Same as single-current version of current loop. In double-current version, the no-current condition is illegal and indicates a system failure.

neutral transmission Method of transmitting teletypewriter signals, whereby a mark is represented by current on the line and a space is represented by the absence of current. By extension to tone signaling, neutral transmission is a method of signaling employing two signaling states, one of the states representing both a space condition and also the absence of any signaling. Also called unipolar transmission.

Newsnet An on-line information service which provides full text of over 300 speciality newsletters. For additional information telephone 1-800-345-1301.

NEXT Near End CrossTalk.

NFS Network File System.

NI Network Interface.

NIB Node Initialization Block (IBM's SNA).

NIB list In IBM's SNA, a series of contiguous node initialization blocks.

nibble The first or last half of an 8-bit byte.

NIC Near Instantaneous Companding.

nickname A locally assigned name used to refer to a global name.

NIM Network Interface Machine.

Nippon Telegraph and Telephone Corporation (NTT) The dominant Japanese telephone company which became a public company during 1985.

NLDM Network Logical Data Manager (IBM's SNA).

NM Network Management.

NMC Network Management Center.

NMPF Network Management Productivity Facility.

NMS Network Management System.

NMVT Network Management Vector Transport (IBM's SNA).

no response In IBM's SNA, a value in the form-of-response-requested field of the request header (RH) indicating that no response is to be returned to the request, whether or not the request is received and processed successfully.

nodal processor A specialized computer system used to control the flow of traffic at a network node, often involving protocol conversion, buffering, multiplexing, and similar network control functions.

node 1. In general, a point of interconnection to a network. 2. In multipoint networks, a unit that is polled. 3. In LAN technology, a unit on a ring. Often used as a synonym for station. 4. In packet switched networks, one of the switches forming the network's backbone. 5. In IBM's SNA, a junction point in the network that contains a physical unit (PU). A node also contains other network addressable units, path control components and data link control components, and may contain boundary function. 6. In IBM's ACF/VTAM, a point in a network defined by a symbolic name. Synonymous with network node.

node entity The top-level entity in the management hierarchy of a system.

Node Initialization Block (NIB) In IBM's ACF/VTAM, a control block associated with a particular node or session that contains information used by the application program to identify the node or session and to indicate how communication requests on a session are to be handled by ACF/VTAM.

node name In IBM's ACF/VTAM, the symbolic name assigned to a specific major or minor node during network definition.

node type In IBM's SNA, the node type is the classification of a network device based upon the protocols it supports and the network-addressable units (NAUs) it can contain. Currently, Type 1 and Type 2 nodes are peripheral nodes; Type 4 and Type 5 are subarea nodes.

noise Random electrical signals, generated by circuit components or by natural disturbances, that corrupt the data by introducing errors.

noise suppressor Filtering or digital signal-processing circuit that eliminates or reduces noise.

noise weighting In telephone communications, a rating system developed to identify the characteristics of the type of receiving element in a telephone handset.

non-blocking Pertaining to a characteristic of some circuit switching systems, such as PABXs, in which sufficient internal switching paths are provided that no matter the call demand, no call will be blocked for lack of sufficient switching capacity.

non-deterministic network A network in which the access delay cannot be predicted with certainty, because it is based on a probability function. Also called stochastic network.

non-erasable Data storage that is unalterable; read-only.

non-impact printer A device that does not use mechanical strikes to print. Also called thermal, laser, and electrostatic.

non-interactive system A system where no interaction takes place between the user terminal and the computer during program execution. Usually offline.

non-linear distortion A type of distortion resulting from attenuation of a signal level. Also called clipping.

non-native network In IBM's SNA, any network attached to a gateway NCP that does not contain that NCP's resources.

non-persistent In LAN technology, pertaining to a CSMA LAN in which the stations involved in a collision do not try to retransmit immediately, even if the network is quiet.

Non-Return-to-Zero (NRZ) A unipolar digital signal which has only two states, high and low, where low is typically zero volts and high is a positive voltage. The signal stays at the level (high or low) for the total clock period.

Non-Return to Zero Inverted (NRZI) In SDLC, a binary encoding technique in which a change in state represents a binary 0 and no change in state represents a binary 1. Also known as invert-on-zero coding.

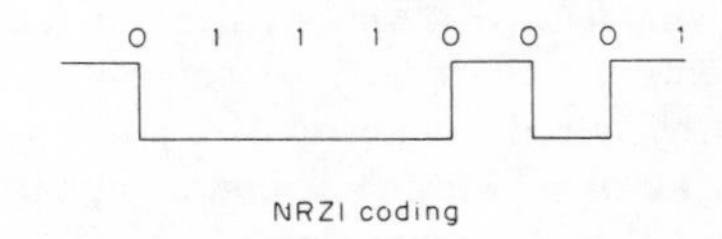

Non-Return to Zero Inverted (NRZI)

non-SNA terminal In IBM's SNA, a terminal that does not use SNA protocols.

non-switched data link In IBM's SNA, a connection between a link-attached device and a communication controller that does not have to be established by dialing.

non-switched line Leased line; a telecommunications line on which connections do not have to be established by dialing.

non-transparent mode A transmission mode of operation in which control characters and control character sequences are recognized through the examination of all transmitted data. Mainly used with byte-oriented protocols, such as IBM's Bisync.

non-volatile A term used to describe a data storage device (memory) that retains its contents when power is lost.

non-volatile storage Any storage medium or circuitry where the contents are not lost when power is turned off or interrupted.

NORDUNET NORDic University NETwork.

normal contacts Contacts of an access jack normally closed when a plug is not inserted into the jack and are opened by a plug.

normal flow In IBM's SNA, a data flow designated in the transmission header (TH) that is used primarily to carry end user data. The rate at which requests flow on the normal flow can be regulated by session-level pacing. *Note*: The normal and expedited flows move in both the primary-to-secondary and secondary-to-primary directions. Requests and responses on a given flow (normal or expedited) usually are processed sequentially within the path, but the expedited-flow traffic may be moved ahead of the normal-flow traffic within the path at queueing points in the half-sessions and for half-session support in the boundary functions.

normal mode An HDLC operational mode used in DNA over half-duplex links.

North American Presentation Level Protocol Syntax (NAPLPS) An ANSI and Canadian standard protocol for videotex graphics and screen formats.

notify In IBM's SNA, a network services request unit that is sent by an SSCP to an LU to inform the LU of the status of a procedure requested by the LU.

NP Network Processor.

NPDA Network Problem Determination Application.

NPDU A Network layer Protocol Data Unit.

NPM Network Performance Monitor (IBM's SNA).

NPS Networking Processing Supervisor.

NPSI X.25 NCP Packet Switching Interface (IBM's SNA).

NRF Network Routing Facility (IBM's SNA).

NRZ Non-Return to Zero.

NRZI Non-Return to Zero Inverted.

NS Network Services (IBM's SNA).

NSAP Network Service Access Point.

NSAP address The address of a Network Service Access Point.

NSAP selector In Digital Equipment Corporation Network Architecture (DECnet), the last byte of an NSAP address, which selects a particular NSAP and hence a Network layer user, within a system identified by the preceding fields of the address.

NSFNET National Science Foundation NETwork.

NSN NASA Science Network.
NSP Network Services Protocol.
NSPE Network Services Procedure Error (IBM's SNA).
NT Network Termination.
NT1 (Network Termination 1) In ISDN, the portion of the Network Termination that converts the physical signal from the two-wire local loop to the four-wire customer premises circuit. NT1 resides on the network side of the T-reference point. In the U.S., NT1 resides on the user side of the U-reference point.
NT2 (Network Termination 2) In ISDN, the portion of the Network Termination that provides local multiplexing, switching, and signaling intelligence. NT2 resides on the network side of the S-reference point and the user side of the T-reference point.
NTIA National Telecommunications and Information Agency.
NTN Network Terminal Number.
NTO Network Terminal Option (IBM's SNA).
NTPF Number of Terminals Per Failure.
NTSC signal National Television System specified signalling and display format. *De facto* standard governing the format of television transmission.
NUA Network Users Association.
Nucleus Initialization Program (NIP) The program that initializes the resident control program. It allows the operator to request last-minute changes to certain options specified during system generation.
NUI Network User Identification.
NUL An ASCII code ("nothing") which is used as a fill character in some communications formats.
null character A character (with all bits set to mark) used to allow time for a printer's mechanical actions, such as return of carriage and form feeding, so that the printer will be ready to print the next data character. Same as idle character.
null modem A device that connects two DTE devices directly by emulating the physical connections of a DCE device.

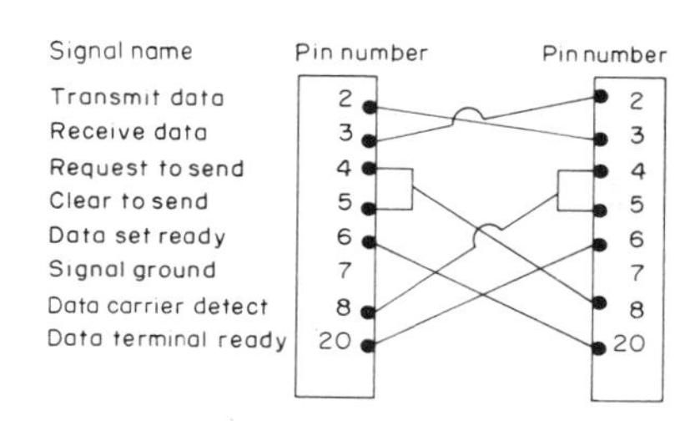

null modem

Numero Vert France's toll-free service which includes call forwarding during predefined time windows or during traffic overload as well as international access.
NVT Network Virtual Terminal.
Nyquist sampling rate In communication theory, the rate at which a band-limited signal must be sampled in order to accurately reproduce the signal. Equal to twice the highest frequency component of the signal.
Nyquist theorem A rule in information theory, Nyquist's Theorem states that in order to reproduce a signal of a given frequency accurately through digital sampling, the sampling frequency must be at least twice the frequency of the signal being sampled. Used to derive the bandwidth necessary for PCM signals.

O

OA Office Automation.

OA & M Operations, Administration, and Maintenance.

object code Executable machine code resulting from programs that were compiled or assembled.

object entry In Digital Equipment Corporation Network Architecture (DECnet), a Naming Service entry which contains the attributes of a network object.

OC-1 Optical carrier level 1. Lowest optical rate in the Sonet standard (51.84 Mb/s).

OC-3 Optical carrier level 3. Next highest optical rate in the Sonet standard (155.52 Mb/s).

OCC Other Common Carriers.

OCCF Operator Communication Control Facility (IBM's SNA).

OCR Optical Character Recognition.

octal A digital system with eight states, 0 through 7.

Octal	*Decimal*
0	0
1	1
2	2
3	3
4	4
5	5
6	6
7	7
10	8
11	9
12	10

octet A logical association of eight consecutive bits. In packet switched networks, a grouping of 8 bits similar, but not identical, to byte. In T1 transmission, the octet is the unit of transmission for an end user channel. Each octet in a D4 or ESF frame carries one channel.

octet interleaving A multiplexing technique in which the low-speed channels are combined onto the high-speed channel 8 bits at a time. The DS1 format uses octet interleaving.

OCU Office Channel Unit.

ODP Originator Detection Pattern.

OEM Original Equipment Manufacturer.

off-hook In a telephone environment, a condition indicating the active state of a subscriber's telephone circuit. A modem automatically answering a call on the dial network is said to go "off-hook."

off-line A condition in which a user, terminal, or other device is not actively transmitting data. Originally, the term referenced not in the line loop. In telegraph usage, paper tapes frequently are punched "off-line" and then transmitted using a paper tape transmitter.

off-loading Process of programming communications processing devices to alleviate processing requirements of other more expensive devices in the network, e.g., a FEP off-loads a host. A terminal may off-load a concentrator.

off-net (DDS) A location beyond the primary serving area of Digital Data System.

Office Automation (OA) A term used to describe the process of making wide use of the latest data processing and data communications technology—electronic mail, word processing, file and peripheral sharing, and electronic publishing—in the office environment, usually involving the installation of LAN.

Office Channel Unit (OCU) A telephone company device used at a Central Office to reciprocate the functions of a channel service unit or data service unit on a digital circuit.

office class The functional ranking of a telephone company network switching center based upon its transmission requirements and its hierarchical relationship to other switching centers.

Office of Telecommunications (OFTEL) A regulatory agency in the United Kingdom.

OFTEL Office of Telecommunications.

OIC Only-In-Chain (IBM's SNA).

OLU Origin Logical Unit (IBM's SNA).

Omninet A combination of hardware and software from Corvus Systems that uses a dedicated hard disk as a disk server on a local area network.

ones density The requirement that a digital signal must carry a certain number of logical "one" or "mark" pulses to maintain timing through repeaters. Digital circuits now in use must maintain a ones density of 1 "one" bit per 8 bits transmitted.

one-way function A function which it is reasonably easy to calculate but for which the inverse is so difficult to compute that, for all practical purposes, it cannot be done.

one-way trunk A trunk between a PBX and a Central Office, or between central offices, where traffic originates from only one end.

on-hook Deactivated condition of a subscriber's telephone circuit, in which the telephone circuit is not in use. A modem not in use is said to be "on hook."

on-line In data processing, pertaining to equipment or devices which are directly connected and controlled by a central processor. Originally referred to directly in the line loop. In telegraph usage, transmitting directly onto the line rather than, for example, perforating a tape for later transmission.

on-line computer A computer used for on-line processing in which the input data enter the computer directly from their point of origin and/or output data are transmitted directly to where they are used. The intermediate stages such as punching data into cards or paper tape, writing magnetic tape, or off-line printing, are largely avoided.

on-line processing A method of processing data in which data is input directly from its point of origin and output directly to its point of use.

Only-In-Chain (OIC) In IBM's SNA, a request unit whose request header (RH) begin chain indicator and RH end chain indicator are both on.

ONM Open Network Management.

on-net (DDS) A location within the primary serving area of a Digital Data System.

OnTyme A registered trademark of McDonnell Douglas' Tymnet public data network as well as an electronic mail system offered by Tymnet.

OOF Out Of Frame.

OPD Originator Detection Pattern.

open architecture An architecture that is compatible with hardware and software from any of many vendors.

open circuit A discontinuity in an electrical conductor or piece of equipment.

Open Network Management (ONM) An IBM network management architecture which allows users and vendors to incorporate non-IBM and non-SNA resource management under SNA.

Open System Interconnection (OSI) The Open System Interconnection (OSI) is a seven-layered ISO compliant architecture for network operations that enables two or more OSI devices to communicate with each other. The philosophy guiding OSI is that an open system of interconnection is poss-

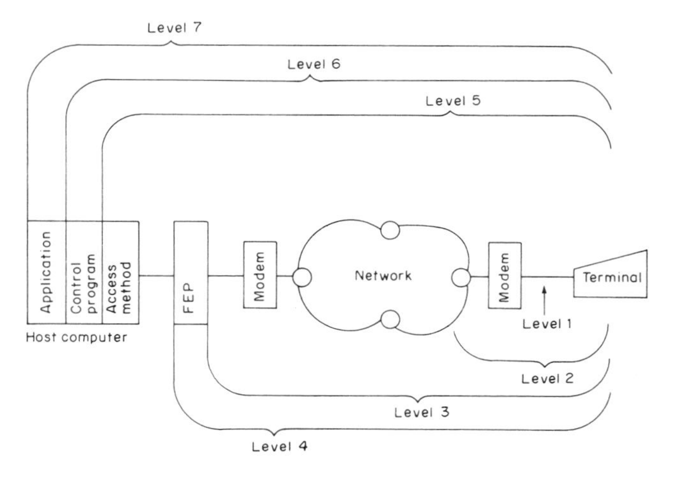

Open System Interconnection (OSI)

ible if an encompassing set of standards is created. With the implementation of standardized protocols, computers and related devices can exchange information despite a diversity in makes and models.

open wire A conductor separately supported above the surface of the ground—i.e., supported on insulators.

open wire line Parallel bare conductors that are strung on electrical insulators, mounted on the cross arms of telephone poles. Two wires constitute a line.

open-air transmission Radio frequency communications technique, including microwave, FM radio, and infrared.

operating environment The combination of computer software to include the operating system, telecommunications access method, data base software and user applications.

operating system A program that manages a computer's hardware and software components. It determines when to run programs, and controls peripheral equipment such as printers.

Operating Telephone Company (OTC) The telephone company holding the local franchise to service a defined geographic area.

operating time The amount of time required for dialing the call, waiting for the connection to be established, and coordinating the forthcoming transaction with the personnel or equipment at the receiving end.

Operating Support Systems (OSS) Software programs and associated hardware designed to help manage specific business functions. They keep records, update inventory, and forecast usage and equipment needs. Many are interlinked.

operation code A code which represents specific operations.

operational mode In HDLC, the particular operational state or protocol being used.

operator, computer The person who operates a computer.

Operator Communication Control Facility (OCCF) In NCCF (IBM's network control communications software), a set of transactions that allow communication with and the operation of remote MVS or DOS systems.

operator orientation point The generic name given to the point in IBM's model 3800 3 printing process at which the data becomes visible to the operator, and is therefore the point at which all operator commands are directed. Also called the transfer station.

operator services Any of a variety of telephone services which require the assistance of an operator. For example, collect calls, third-party billed calls, person-to-person calls, Calling Card calls, and directory assistance.

Operator Station Task (OST) In IBM's Network Communications Control Facility (NCCF), a subtask that establishes and maintains the on-line session with the network operator. There is one operator station task for each network operator who logs on to NCCF.

optical cavity The part of a laser where light is amplified by bouncing it between mirrors.

Optical Character Recognition (OCR) The machine identification of printed characters through use of light-sensitive devices.

optical connectors Connectors designed to terminate and connect either single or multiple optical fibers. Optical connectors are used to connect fiber cable equipment and interconnect cables.

optical cross-connect panel A cross-connect unit used for circuit administration and built from modular cabinets. It provides for the connection of individual optical fibers with optical fiber patch cords.

optical disk A very high capacity information storage medium that uses light generated by a laser to read information.

optical fiber One of the glass strands—each of which is an independent circuit—in a fiber optic cable.

optical fiber cable A transmission medium consisting of a core of glass or plastic surrounded by a protective cladding, strengthening material, and outer jacket. Signals are transmitted as light pulses, introduced into the fiber by a light transmitter (either a laser or light emitting diode). Low data loss, high-speed transmission, large bandwidth, small physical size, light weight, and freedom from electro-

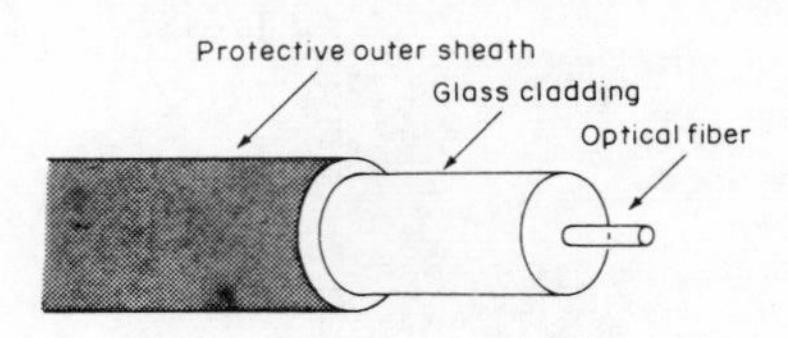

optical fiber cable

magnetic interference and grounding problems are some of the advantages offered by optical fiber cable. There are five common types: single, dual, quad, Lightpack, and ribbon.

optical interconnection panel An interconnection unit used for circuit administration and built from modular cabinets. It provides interconnection for individual optical fibers. Unlike the optical cross-connect panel, the interconnection panel does not use patch cords.

optical waveguide Dielectric waveguide with a core consisting of optically transparent material of low loss (usually silica glass) and with cladding consisting of optically transparent material of lower refractive index than that of the core. It is used for the transmission of signals with lightwaves and is frequently referred to as a fiber. In addition, there are planar dielectric waveguide structures in some optical components, like laser diodes, which also are referred to as optical waveguides.

OPX Off Premise Exchange.

orderly closedown In IBM's VTAM, the orderly deactivation of ACF/VTAM and its domain. An orderly closedown does not complete until all application programs have closed their ACBs. Until then, RPL-based operations continue; however, no new sessions can be established and no new ACBs can be opened.

orders of magnitude In order to understand some of the terms used to define measurements, it is necessary to understand the expressions used to delineate very large and very small numbers.

Terms	*Value*	*Prefix*	*Example*
pico	One trillionth (1/1 000 000 000 000)	p	picowatt
nano	One billionth (1/1 000 000 000)	n	nanosecond
micro	One millionth (1/1 000 000)		microsecond
milli	One thousandth (1/1 000)	m	milliwatt
kilo	One thousand (1 * 1000)	k	kilohertz
mega	One million times (1 * 1 000 000)	M	megahertz
giga	One billion times (1 * 1 000 000 000)	G	gigahertz

Examples: 1000 hertz = 1 kilohertz (1 kHz)
0.000 000 000 001 watt = 1 picowatt (1 pW)

Origin Logical Unit (OLU) In IBM's SNA, the logical unit from which data is sent.

Original Equipment Manufacturer (OEM) Manufacturer of equipment components built by another manufacturer.

Originator Detection Pattern (ODP) A string of characters sent by a V.42 error correcting protocol modem which permits the modems communicating with one another to determine if they have similar error-correcting protocols when they first connect.

Originator/Recipient Name (ORName) The specified form used to send messages to users who are part of the X.400 community.

ORName Originator/Recipient Name.

OS Operating System.

OS/2 An operating system jointly developed by Microsoft and IBM which allows programs to use memory in excess of 640 Kbytes.

OS/MFT IBM Operating System/Multi programming with Fixed number of Tasks.

OS/MVT IBM Operating System/Multi programming with Variable number of Tasks.

OS/VS1 IBM Operating System/Virtual Storage 1. Used on 370 machines.

OS/VS2 IBM Operating System/Virtual Storage 2. Used on 370 machines.

oscillator An electronic device used to generate repeating signals of a given amplitude or frequency.

OSI Open System Interconnection.

OSinet A test network sponsored by the National Bureau of Standards (NBS) to facilitate the testing of vendor products based on the OSI model.

OSN Operations System Network.

OSS Operating Support Systems.

OST Operator Station Task.

OSWS Operating System Work Station.

OTC Operating Telephone Company.

Other Common Carrier (OCC) Includes specialized common carriers (SCCs), domestic and international record carriers (IRCs), and domestic satellite carriers which are authorized to provide private line services in competition with the established telephone common carriers.

out-of-band signaling A technique in which call control signaling for one or more channels is carried in a separate channel dedicated only to signaling. Also called Common Channel Signaling. Common channel signaling allows the creation of clear channels. The current international standard for out-of-band signaling is Common Channel Signaling System No. 7.

Out-Of-Frame (OOF) An error condition on a framed, digital circuit, invoked when two of four framing bits are received in error.

outgoing access The capability of a user in one network to communicate with a user in another network.

output 1. Data that has been processed. 2. The state or sequence of states occurring on a specified output channel. 3. The device or collective set of devices used for taking data out of a device. 4. A channel for expressing the state of a device or logic element. 5. The process of transferring data from an internal storage to an external storage device.

output device Device which allows the user to receive data output from a computer system. For example, the screen on a CRT terminal.

Out-WATS See Out-Wide Area Telephone Service.

Out-Wide Area Telephone Service (Out-WATS) A WATS service that enables customers to call out on a line at a fixed charge, but does not allow incoming fixed charge calls.

overbooking ratio The ratio of the sum of the inputs to the link speed capacity. Usually associated with STDMs and is a result of the concentrating ability of an STDM.

overflow Switching equipment which operates when the traffic load exceeds the capacity of the regular equipment.

overhead Control characters, error control information, synchronization data, etc., sent along with user information in a data communication exchange.

overhead bit A non-data bit used in addressing control, error detection, error control, or synchronization.

overlay segment A portion of a program that can be loaded during execution to overlay another section of the program. An overlay segment is normally used when there is insufficient memory to accommodate all the code of a program.

overrun Loss of data because a receiving device is unable to accept data at the rate it is transmitted.

oversampling A TDM technique where each bit from each channel is sampled more than once.

overseas service conditioning AT&T conditioning for a two-point overseas service to meet CCITT recommended standards M.1020 or M.1025.

overspeed Condition in which the transmitting device runs slightly faster than the data presented for transmission. Overspeeds of 0.1% for modems and 0.5% for data PABXs are typical.

over-voltage protection Electrical protection provided by devices such as air gap discharge protectors and gas tube protectors and appropriate construction methods (fusing, bonding, grounding, shielding, etc.). Used where served by exposed outside plant.

P

PA Program Access (key).

PAB Process Anchor Block (IBM's SNA).

PABX Private Automatic Branch Exchange.

pacing A method for regulating traffic rates to ensure orderly handling of messages. In IBM's System Network Architecture, it refers to the number of path information units (PIUs) that can be sent before a response is received.

pacing group The number of data units (Path Information Units, PIUs) that can be sent before a response is received in IBM's SNA. Also IBM's term for window.

pacing group size In IBM's SNA, 1. The number of path information units (PIUs) in a virtual route pacing group. The pacing group size varies according to traffic congestion along the virtual route. Synonymous with window size. 2. The number of requests in a session-level pacing group.

pacing response In IBM's SNA, an indicator that signifies a receiving component's readiness to accept another pacing group; the indicator is carried in a response header (RH) for session-level pacing, and in a transmission header (TH) for virtual route pacing.

packet A group of bits—including information bits and overhead bits—transmitted as a complete package on a packet switched network. Usually smaller than a transmission block. Often called a message.

PACKET/74 An SNA Controller/Host PAD designed by PACKET/PC, Inc., of Farmington, CT, which permits communications between an IBM communications controller and a public or private X.25 network. PACKET/74 functions in conjunction with PACKET/3270 software operating on an IBM PC or compatible computer and provides such functions as DFT file transfer, CRC error checking, an Application Program Interface (API) at the PC and bit compressed communications.

packet assembly unit A user facility which permits non-packet-mode terminals to exchange data on the packet system. Also called packet control unit and network interface machine.

Packet Assembler/Disassembler (PAD) Equipment in a packet network used to convert the format of data from asynchronous or synchronous terminals for transmission over a packet network. A PAD permits terminals which do not have an interface suitable for direct connection to a packet switched network to access such a network as well as convert the terminal's usual data flow to and from packets. The PAD handles all aspects of call setup and addressing.

packet buffer Memory space reserved for storing a packet awaiting transmission or for storing a received packet.

packet data network Often used to mean packet switched network.

packet disassembly unit A user facility which enables packets to be delivered from the packet network to non-packet-mode terminals. Combined with a packet assembly unit to form a PAD (packet assembler/disassembler).

packet header In packet switched networks, the first three octets of an X.25 packet.

packet interleaving A form of multiplexing in which packets from various subchannels are interleaved on the line. The X.25 interface is an example.

packet level Level 3 of the CCITT X.25 Recommendation which defines the packet format and control procedures for the exchange of packets over a PSDN.

packet mode device Data Circuit-terminating Equipment (DCE) or Data Terminal Equipment (DTE) that can communicate using packets.

packet mode terminal DTE that can control and format, transmit, and receive packets.

packet radio Packet switching in which the transmission paths are radio links and a transmitted packet may be picked up by more than one station. May be used with mobile stations.

Packet Switch Exchange (PSE) A unit which performs packet switching in a network.

Packet Switch Stream (PSS) A value added carrier based upon packet switching operated by British Telecom.

packet switched data transmission service A service that transmits, assembles, disassembles, and controls data in the form of packets, e.g., Tymnet, Telenet.

packet switched network A data communications network that transmits packets. Packets from different sources are interleaved and sent to their destination over virtual circuits. The term includes PDNs and cable-based LANs.

packet switching A technique of switching data in a network whereby individual data blocks or "packets" of controlled size and format are accepted by the networks and routed to their destination. The sequence of packets is maintained and the destination is determined by the exchange of control information (also in packets) between the originating terminal and the network before data transfer starts. The equipment making up the network is shared by all users at all times, packets from different terminals being interleaved throughout the network.

Packet Switching Exchange (PSE) The top level in a network where each PSE is identified by a unique number and contains one or more packet switches. PSEs correspond to the major metropolitan centers within the U.S. and are interconnected with high-bandwidth, long-haul trunks. Area codes are assigned uniquely to the nearest PSE. The lower level in the hierarchy is the node level where each node is identified by a node number unique within its PSE. A X.25 Data Terminal Equipment (DTE) number is known only in the PSE serving that DTE.

Packet Switching Network (PSN) A network designed to assemble data into packets. Each packet is then routed to its destination by the most efficient means possible.

packet terminal A data terminal which can transmit and receive packets without a PAD.

packet type identifier In X.25 packet switched networks, the third octet in the packet header that identifies the packet's function and, if applicable, its sequence number.

Packetized Ensemble Protocol (PEP) A modem operation protocol in which the bandwidth of a dial-up line is split into up to 512 carriers, each of which is capable of supporting data transmission. PEP was developed by Telebit Corporation of Mountain View, CA, and is used in that vendor's series of Trailblazer modems.

packetized voice The formation of digitized voice into data packets.

PACNET PACific NETwork.

PAD Packet Assembler/Disassembler.

pad character A special character(s) usually sent at the beginning and/or end of a synchronous transmission to accomplish bit synchronization and timing.

page 1. A specified portion of main memory capacity used when allocating memory and for partitioning programs into units or control sections. 2. In IBM's SNA, the portion of a panel that is shown on a display surface at one time. 3. The unit of output from an IBM 3800 model 3 running with full function capability or an IBM 3820 printer.

page-mode data A type of data that can be formatted anywhere on a physical page. This data requires specialized processing such as is provided by the Print Services Facility for the IBM 3800-3 and 3820 printers.

page-mode environment checkpointing That process which preserves the information necessary to resume page mode printing.

page-mode printer A printer (such as the IBM 3800 model 3 and 3820) that can print page-mode data.

paging/paging system An audible, one-way communication through an on-site communications system for on-premises paging, or, through special equipment, a means of reaching someone remotely within a specified geographic area. With an internal paging system, you can be notified of important incoming calls even when you are not at your own desk or telephone. External paging systems provide a way to reach you when you are not in the office or at a previously specified telephone number.

Pair-Selected Ternary (PST) A pseudo-ternary code in which pairs of binary digits are coded together in such a way that the resultant signal has no long strings of zeros.

PAM Pulse Amplitude Modulation.

panel In IBM's SNA computer graphics, a predefined display image that defines the locations and characteristics of display fields on a display surface.

paper tape An input/output medium, on which data

can be recorded as a pattern of punched holes (five- or eight-channel).

PAR Peak to Average Ratio.

parallel interface An interface where an entire group of bits is transmitted at one time by sending each bit over a separate wire. The Centronics parallel Interface is a 36-pin, TTL level, byte-wide interface used for computer to parallel printer communications. The transmission of these data bits is controlled by the computer-supplied strobe pulse. Flow control is achieved by asserting or deasserting either the ACKNLG or BUSY leads or both. Pins 12, 13, 14, 15, 18, 31, 32, 34, 35, and 36 vary in function depending upon implementation by the manufacturer of the printer. Pins 16 and 17 are commonly used for logic ground and chassis ground respectively.

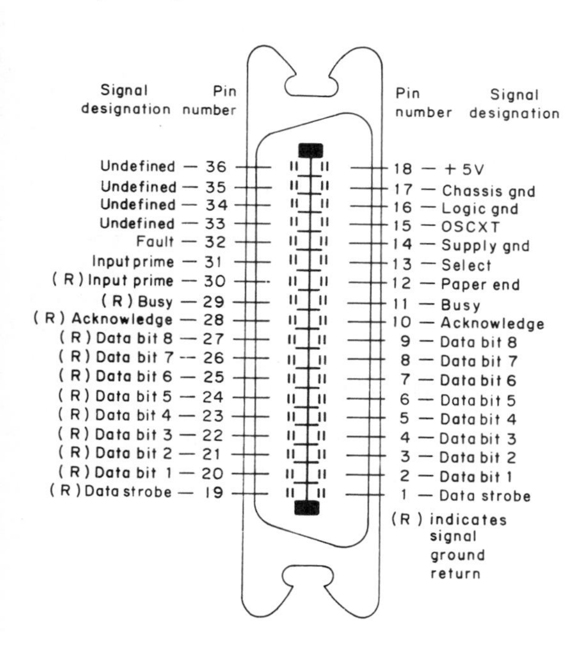

parallel interface

parallel interface extender A subsystem capable of remoting a parallel interface, such as a computer I/O channel. The parallel interface extender functions as a parallel to serial and serial to parallel converter, enabling such devices as line printers and card readers to be located remotely from a computer system.

parallel links In IBM's SNA, two or more links between adjacent subarea nodes.

parallel processing Concurrent execution of two or more programs, within the same processor.

parallel sessions In IBM's SNA, a parallel session is two or more concurrently active sessions between the same two logical units (LUs) using different network addresses. Each parallel session can have different transmission parameters.

parallel transmission A technique that sends each bit simultaneously over a separate line. Normally used to send data a byte (8 bits over eight lines) at a time to a high-speed printer or other locally attached peripheral.

Character 1
Character 2
Bit 1
Bit 2
Bit 3
Bit 4
Bit 5
Bit 6
Bit 7
Bit 8

parallel transmission

parameter A variable that takes on any value within a predefined range of values.

parity A method of checking for errors in data communications. An extra bit (either a "1" or "0"), called the parity bit, is added to the end of each ASCII character to make the final count of "1" bits in the character an even or odd number, according to a prearranged format. Some systems always use even parity, some always use odd parity, and some do not check for parity. Both terminal and system must be set for the same parity.

parity bit The bit which is set to 1 or 0 in a character to ensure that the total number of 1 bits in the data field is even or odd. Or may be fixed at 1 (mark parity), fixed at 0 (space parity), or ignored (no parity). Parity bits trap errors in the following manner: when the transmitting device frames a character, it tallies the number of 0s or 1s within the frame and attaches a parity bit. (The parity bit will vary according to whether the total is even or odd.) Then, the receiving end will count the 0s or 1s and compare the total to the "odd" or "even" recorded on the parity bit. If the receiving end finds a discrepancy, it can flag the data and request a retransmission.

Parity type	*Description*
odd	Eighth data bit is logical zero if total number of logical 1s in first seven data bits is odd.
even	Eighth data bit is logical zero if total number of logical 1s in first seven data bits is even.
mark	Eighth data bit always logical 1 (high/mark).
space	Eighth data bit always logical 0 (low/space).
none/off	Eighth data bit ignored.

parity check, horizontal A parity check applied to the group of certain bits from every character in a block. Also called longitudinal redundancy check.

parity check, longitudinal A parity check performed on a group of binary digits in a longitudinal direction for each track. Also called longitudinal redundancy check.

		Character ↓parity bit
Character	1	10110110
Character	2	01001010
	3	01101000
	4	10010010
	5	01111010
	6	10100001
	7	01011101
	.	01110011
	.	10001100
	.	01101011
Block parity character (LRC)		11101011

parity check, longitudinal

parity check, transverse A parity check performed on a group of binary digits in a transverse direction for each frame. Synonymous with transverse redundancy check.

parity check, vertical A parity check applied to the group which is all bits in one character. Also called vertical redundancy check.

parity checking A technique of error detection in which one bit is added to each data character so that the number of one bits per character is always even (or always odd).

parity error An error which occurs in data where an extra or missing bit is detected. Character parity cannot detect an even number of bits.

ASCII character R	1010010
Adding an even parity bit	10100101
1 bit in error	1Ø100101 (1 above)
2 bits in error	1Ø1Ø0101 (1 1 above)

parity error

parse In systems with time sharing, to analyze the operands entered with a command and build up a parameter list for the command processor from the information.

PART 68 FCC rules that allow registration of communications equipment provided they meet FCC requirements designed to ensure no harm to the telephone network.

Partitioned Data Base Management A Software-Defined Network (SDN) service feature offered by AT&T which permits a user to group locations into independent subnetworks.

Partitioned Emulation Programming Extension (PEP) IBM software package used with Network Control Program (NCP), allows a communications controller to operate in a split mode, controlling an SNA network while managing a number of non-SNA communications lines.

pass band filters Filters used to allow only the frequencies within the communications channel to pass while rejecting all frequencies outside the pass band.

Pass Change A network security service offered by Tymnet which enables a user to change his/her own password.

Passlife A network security service offered by Tymnet which enables an organization using this value added carrier to set a time limit on a user's password.

passthrough In general, a term used to describe the ability to gain access to one network element through another. It is also another name for channel bypass. A T1 multiplexing technique similar to drop-and-insert, used when some channels in a DS1 stream must be demultiplexed at an intermediate node. With passthrough, only those channels destined for the intermediate node are demultiplexed—ongoing traffic remains in the T1 signal, and new traffic bound from the intermediate node to the final destination takes the place of the dropped traffic. With drop-and-insert, all channels are demultiplexed, those destined for the intermediate node are dropped, traffic from the intermediate node is added, and a new T1 stream is created to carry all channels to the final destination.

password A word or character string that, when accurately presented, permits a user access to a system or computer program.

password protection A method of limiting login access to a network by requiring users to enter a password. Unless the password is entered correctly, access will be denied.

patch Machine language instructions added to a program to alter it or correct an error. Too many patches make it difficult to maintain programs.

patch cord A short length of wire or fiber cable with connectors on each end used to join communication circuits at a cross-connect.

patch panel A system of terminal blocks, patch cords, and backboards which facilitates administration of cross-connect fields for moves and rearrangements by non-technical end user personnel, thus enhancing and expanding desirability of PBX and Centrex systems.

patching jacks Series-access devices used to patch around faulty equipment by using spare units.

path In IBM's SNA: 1. The series of path control network components (path control and data link control) that are traversed by the information exchanged between two network-addressable units (NAUs). A path consists of a virtual route and its route extension, if any. 2. In defining a switched major node, a potential dial-out port that can be used to reach a physical unit.

path control layers In IBM's SNA, the network processing layer that handles routing of data units as they travel through the network and manages shared link resources.

path control network In IBM's SNA, the part of the SNA network that includes the data link control and path control layers.

path field In packet switching, a set of bits within a Call Request packet that indicates Telenet Processors (TPs) have routed or have attempted routing of the Call Request packet. The path field prevents the formation of loops in virtual circuit routes.

Path Information Unit (PIU) In IBM's SNA, a message unit consisting of a transmission header (TH) alone, or of a TH followed by a basic information unit (BIU) or a BIU segment.

path test In IBM's SNA, a test provided by NLDM Release 2 that enables a network operator to determine whether a path between two LUs that are currently in session is available.

Pathfinder A trademark of Ven-Tel Inc. of San Jose, CA, as well as a high speed modem designed for operation on the public-switched telephone network.

pathname A character string which provides the location of a file or directory on a disk.

PAX Private Automatic Exchange.

payload mapping The mapping of network services into synchronous payload envelopes.

PBX Private Branch Exchange.

PC 1. Personal Computer. 2. Phase Corrector. 3. Printed Circuit (board). 4. Path Control (IBM's SNA).

PC Network An IBM broadband local area network.

PC Pursuit A dial-up communications service marketed by Telenet Communications Corporation which allows users to communicate with personal computers, host computers, and information services during off-peak hours for a flat fee per month.

PCI Protocol Control Information.

PCM 1. Plug Compatible Machine. 2. Pulse Code Modulation.

PC-Term An asynchronous communications and terminal emulation program from Crystal Point of Kirkland, WA, that operates on IBM PC and compatible personal computers.

PCU Packet Control Unit. See packet assembly unit.

PDM Pulse Duration Modulation.

PDN 1. Packet Data Network. 2. Public Data Network.

PDS Premises Distribution System.

PDU Protocol Data Unit.

peak limiter A filter used to reduce the effect of noise on a signal by clipping off noise peaks above the desired peak level of a signal. Can be used on frequency-modulated (FM), frequency shift keyed (FSK), or pulse code modulated (PCM) signals.

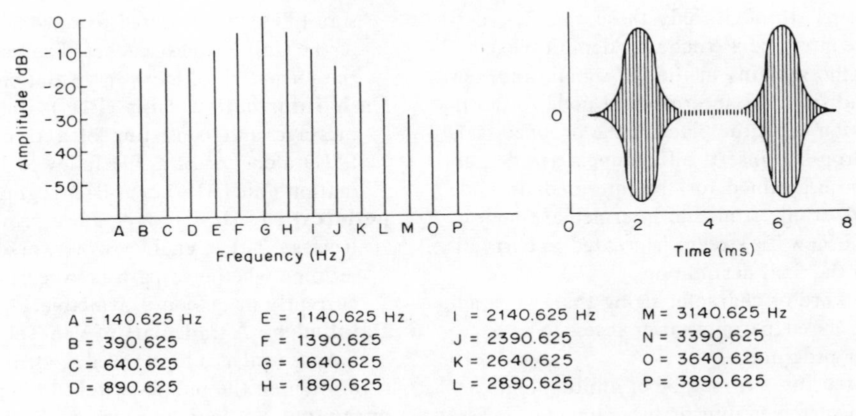

peak to average ratio test (P/AR)

peak to average ratio test (P/AR) The P/AR test is a signal fidelity measurement designed to be sensitive to envelope delay distortion and attenuation distortion and less sensitive to the normal steady interferences or impairments on a channel such as non-linear distortion, noise, and phase jitter. The measurement technique employs the signal transmission of a complex frequency spectrum. The receiver of the analog test set performs a calculation based on the charges to the original signal and generates a value of measurement called P/AR units. A P/AR value of 100 would indicate a channel with excellent fidelity; however, 75 is more of a practical reading to expect on a fairly good line.

pending active session In IBM's VTAM, the state of an LU–LU session recorded by the SSCP when it finds both LUs available and has sent a CINIT request to the primary logical unit (PLU) of the requested session.

penetration tap In an Ethernet local area network, a device used to connect a transceiver to the bus without requiring that the bus transmission be interrupted for the installation of fittings. This is accomplished by the use of a needle-like device which penetrates the insulation of the coaxial cable bus to reach the center of the coax conductor.

PEP 1. Partitioned Emulation Program. 2. Packetized Ensemble Protocol.

perforator A device used for the manual preparation of a perforated tape, in which telegraph signals are represented by holes punched in accordance with a predetermined code. Paper tape is prepared off-line with this device.

performance classifications CCITT G.821 has defined four error-rate performance categories as follows:

Available and Acceptable	Intervals of test time of at least one minute during which the error rate is less than one millionth.
Available but Degraded	Intervals in the test time of at least one minute during which the error-rate is between one thousandth and one millionth.
Available but Unacceptable	Intervals in the test time of at least one second but less than 10 consecutive seconds during which the error-rate is greater than one thousandth.
Unavailable	Intervals in the test time of at least 10 consecutive seconds during which the error-rate is less than one thousandth.

performance error In IBM's NPDA, a resource failure that can be resolved by error recovery programs. Synonym for temporary error.

periodic frames Segments of equal duration that are delineated by incorporation of fixed periodic patterns into the bit stream.

peripheral device A device which is connected to a computer to perform an input/output function such as data storage.

peripheral equipment Equipment that works in conjunction with a communications system or a computer, but not integral to them—printers and CRTs, for example.

peripheral interface A standard interface used between a computer and its peripherals so that new peripherals may be added or old ones changed without special hardware adaptation.

peripheral LU In IBM's SNA, a logical unit representing a peripheral node.

peripheral node In IBM's SNA, a node that uses local addresses for routing and therefore is not affected by changes in network addresses. A peripheral node requires boundary function assistance from an adjacent subarea node. A peripheral node is a type 1 or type 2 node connected to a subarea node.

peripheral PU In IBM's SNA, a physical unit representing a peripheral node.

permanent error In IBM's SNA, a resource error that cannot be resolved by error recovery programs.

Permanent Virtual Circuit (PVC) 1. A permanent virtual call existing between two Data Terminal Equipment (DTEs). It is a point-to-point, non-switched circuit over which only data, reset, interrupt, and flow-control packets can flow. 2. A logical channel that is maintained in a data transfer state between two user devices (typically a terminal and host) at all times when the network is operational. Supports only X.25 DTEs.

permissive (PE) arrangement A connection arrangement used to connect FCC registered equipment to the DDD network. This arrangement utilizes the type USOC RJ11C jack. The output signal level of the communications equipment is fixed at a maximum of 9 dBm. An assumption that at least 3 dB signal loss will occur on the local loop insures that the signal will not arrive at the central office at more than the maximum allowable level of −12 dBm.

permissive device A classification of registered modems with output limited to −9 dBm.

persistent In LAN technology, pertaining to a CSMA LAN in which the stations involved in a collision try to retransmit almost immediately. *p*-persistent where *p* (for probability) = 1 (hence, also called 1-persistent).

Personal Computer (PC) A microcomputer with an end-user-oriented application program (used by data processing professionals and non-professionals alike) for an assortment of functions.

Personal Identification Number (PIN) A special number assigned to a specific customer or user, to enable him/her to authenticate himself/herself to a system. A PIN is primarily used with automatic bank teller systems to permit a user to withdraw cash from his/her account.

Personal System/2 (PS/2) IBM's current family of microcomputers.

Perumtel The state owned network operator in Indonesia.

PF Program Function (key).

PFEP Programmable Front End Processor.

phantom telegraph circuit Telegraph circuit superimposed on two physical circuits reserved for telephony.

phase A measure of signal position with respect to a reference signal. The unit of measurement is the angular degree.

Phase Corrector (PC) A function of synchronous modems which adjusts the local data clocking signal to match the incoming receive signal.

phase equalizer, delay equalizer A delay equalizer is a corrective network which is designed to make the phase delay or envelope delay of a circuit or system substantially constant over a desired frequency range.

phase hit Undesired shifting in phase of an analog signal. Any case where the phase of a 1004-Hz test signal shifts more than 20 degrees.

phase inversion modulation A method of phase modulation, in which the two significant conditions differ in phase by 180 degrees.

phase jitter An analog line impairment caused by the shifting of phase of one part of the frequency tone relative to an earlier part of the tone. Line distortion is caused by the variation in phase or frequency of a transmitted signal from its reference timing position. This can cause data transmission errors particularly at high speeds. Phase jitter is caused by primary frequency supplies and power transients. These supplies are prevalent in telephone offices where ringing voltages are generated and from power supplies for frequency division multiplexed carrier systems. Phase jitter rarely occurs above 300 Hz and it is typically measured in two areas. These are called in-band and out-of-band jitter, or high and low frequency jitter. These encompass 20 to 300 Hz for the high band and 4 to 300 Hz for the low or out of band.

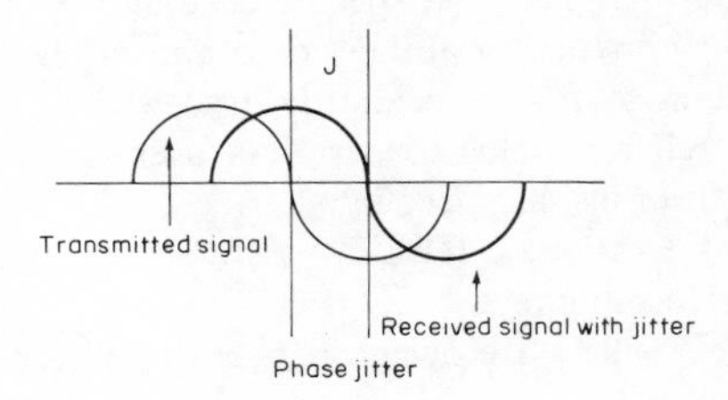

phase jitter

phase locked loop Device that compares the phase of two signals, i.e., a reference signal and a voltage controlled signal. A phase difference between the two signals produces an error voltage that locks the frequency of the voltage controlled signal to that of the reference signal.

phase modulation One of three basic ways (see also amplitude modulation (AM) and frequency modulation (FM)) to add information to a sine wave signal. The phase of the sine wave, or carrier, is modified in accordance with the information to be transmitted. With only discrete changes in phase, this technique is known as phase shift keying (PSK).

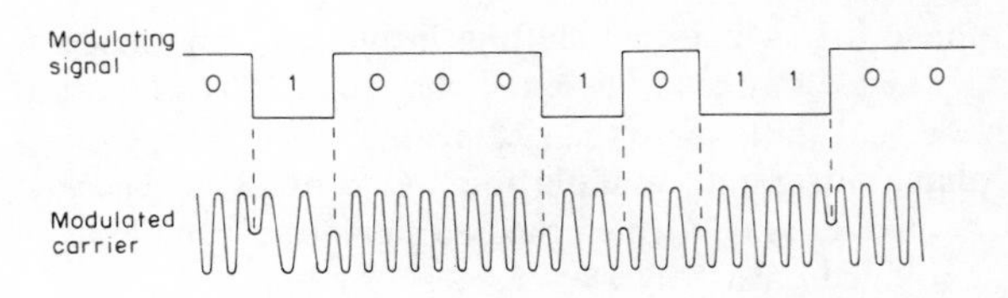

phase modulation

phase shift A change in the time or amplitude that a signal is delayed with respect to a reference signal.

Phase Shift Keying (PSK) A modulation technique in which the phase of the carrier is modulated by the state of the input signal.

phasor Temporary buffer storage that compensates for slight differences in data rate between TDM I/O ports and devices.

Philippine Long Distance Telephone Company The company based in Manila that has virtual control over all telecommunications services in the Philippines.

Phoenix 5500 A trademark of Phoenix Microsystems Corporation. The Phoenix 5500 is a widely used BERT tester, capable of testing high-speed digital circuits such as T1 lines.

phone mail A computer-controlled voice processing system which enables people to communicate when both parties are not available at the same time or if one party already has an on-going conversation.

photodetector In a lightwave system, a device that turns pulses of light into bursts of electricity.

photoelectric effect The emission of electrons by a material when it is exposed to light.

photon The fundamental unit of light and other forms of electromagnetic energy. Photons are to optical fibers what electrons are to copper wires; like electrons, they have a wave motion.

photonics The technology that uses light particles (photons) to carry information streams over hair-thin fibers of pure glass.

physical connection 1. The Physical layer communications path between two systems. 2. In IBM's VTAM, a point-to-point connection or multipoint connection.

physical layer The lowest (first) layer in the OSI model. Responsible for the physical signaling, including the connectors, timing, voltages, and other related matters.

physical record A single block of data transferred between an I/O device and main memory. May contain several logical records.

Physical Unit (PU) On an IBM SNA network, a type of network-addressable unit that represents hardware devices or nodes to the network. The program that resides in each node and provides the services required to manage and monitor that node's resources.

physical unit services In IBM's SNA, the components within a physical unit (PU) that provide configuration services and maintenance services for SSCP-PU sessions.

physical unit type In IBM's SNA, the classification of a physical unit (PU) according to the type of node in which it resides. The PU type is the same as its node type; that is, a type 1 PU resides in a type 1 node, and so forth.

PIC Plastic Insulated Conductors.

pico (p) A prefix for one trillionth of a unit.

picosecond One millionth of one millionth of a second, or one trillionth of a second.

pilot model A model of the system used for program testing purposes which is less complex than the complete model, e.g., the files used on a pilot model may contain a much smaller number of records than the operational files; there may be few lines and fewer terminals per line.

PIN 1. Personal Identification Number. 2. Positive, Intrinsic, Negative.

ping pong A method used to emulate full-duplex transmission on a half-duplex circuit where automatic line turnaround occurs when a receiving modem has data to transmit.

pipeline Slang for telephone cables or other telephone circuit bundles.

PIU Path Information Unit.

pixel An abbreviation for a picture element: the smallest unit into which an image can be divided, and to which can be assigned such characteristics as gray scale, color, and intensity.

PKS Public Key System.

pkt Packet(s).

PL/1 A high level programming language oriented toward both business and mathematical programs. PL1 stands for Programming Language 1.

Plain Old Telephone Service (POTS) A reference to the basic service provided by the public telephone network without any added facilities such as conditioning.

plaintext 1. In the context of cryptography, messages in their normal, readable form are called plaintext. 2. Synonym for clear data (IBM's SNA).

plant The physical means by which transmission services are provided. This can include copper wire, conduit, poles, microwave systems, fiber optic trunks, and switching systems.

plasma display A type of flat visual display in which selected electrodes in a gas-filled panel are energized, causing the gas to be ionized and light to be emitted.

plenum, return air The utilization of the false ceiling area on each floor to move air (for heating and air conditioning).

plenum cable Cable specifically designed for use in a "plenum" (the space above suspended ceiling used to circulate air back to the heating or cooling system in a building). Plenum cable has insulated conductors often jacketed with PVDF material to give them low flame spread and low smoke-producing properties. Its maximum allowable flame distance is five feet.

Plexar A trademark of Southwestern Bell Corporation as well as a service for an ISDN-based Centrex service.

plotter A type of computer peripheral printer that displays data in a two-dimensional graphics form.

PLS Private Line Service.

PLU Primary Logical Unit (IBM's SNA).

plug A device for connecting wires to a jack. It is typically used on one or both ends of equipment cords, and on wiring for interconnects and cross connects.

Plug-Compatible Machine (PCM) Term used to describe a device which can be directly substituted for an original manufacturer's device. The PCM device is usually an improvement over the original device—less expensive, more fully featured, or both.

PM Phase Modulation.

PMS Public Message Service.

PMX Packet Multiplexer.

PND Present Next Digit.

point of presence The point within a LATA where the local telephone company terminates subscribers' circuits for leased line or long distance dial up circuits.

Point Of Sale (POS) Transaction terminal used in retail.

Point Of Service (POS) The point at which a long distance call enters the long distance company's network from the local company's network.

Point Of Termination (POT) The point where the entrance cable or regulated riser cable meets the network interface. The POT is the location where the communications carrier's facility responsibility ends. POT is co-located at the Rate Demarcation Point (RDP).

pointer A word in a computer's store containing the address of another item of data. It is saying 'point' to that item. The manipulation of pointers may save many operations by avoiding the movement of larger items to which they point.

point-to-point Pertaining to a data channel which connects two, and only two, terminals.

point-to-point circuit A communication circuit, or system connecting two points through a telephone circuit, or line.

point-to-point line In IBM's SNA, a link that connects a single remote link station to a node. It may be either switched or non-switched.

point-to-point link A communications link connecting two stations.

point-to-point network A point-to-point network is one in which exactly two stations are connected. It may be a dial connection or a leased line.

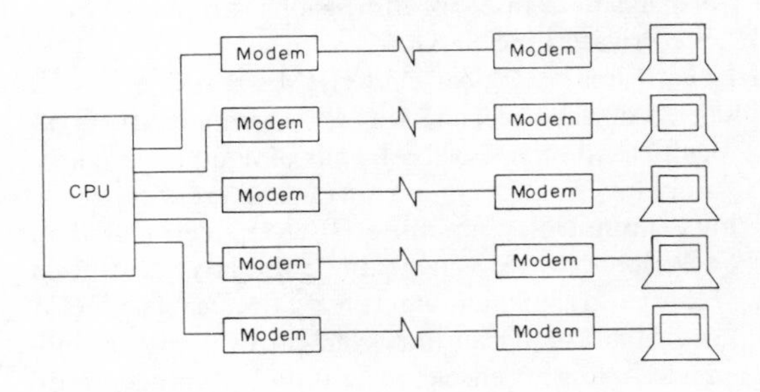

point-to-point network

polar non-return to zero signaling The polar non-return to zero signal uses positive current to represent a mark and negative current to represent space. As no transmission occurs between two consecutive bits of the same value, the signal must be sampled to determine the value of each received bit.

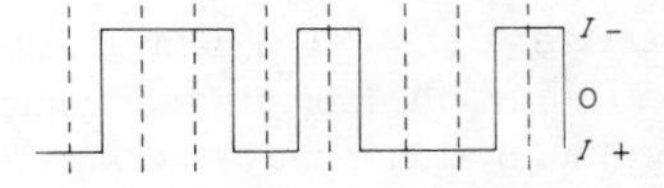

polar non-return to zero signaling

polar return to zero The polar return to zero signaling method uses positive current to represent a mark and negative current to represent a space. However, the signal returns to zero after each bit is transmitted. Since there is a pulse that has a discrete value for each bit, sampling of the signal is not required. Thus, the circuitry required to determine whether a mark or space has occurred is reduced.

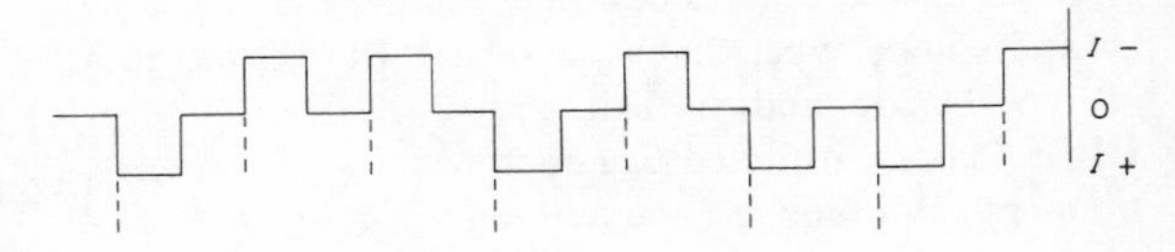

polar return to zero

polar transmission A method for transmitting teletypewriter signals, whereby marking signal is represented by direct current flowing in one direction and the spacing signal is represented by an equal current flowing in the opposite direction. By extension to tone signaling, polar transmission is a method of transmission employing three distinct states, two to represent a mark and a space and one to represent the absence of a signal. Also called bipolar.

polarity Any condition in which there are two opposing voltage levels or charges, such as positive and negative.

polarization A characteristic of electromagnetic radiation that occurs when the electric field vector of the energy wave is perpendicular to the main direction of the electromagnetic beam.

poll The process by which a computer asks a terminal associated with it if the terminal has any data for the computer.

poll character A unique character or sequence sent by the main computer to a device to check for availability for sending and receiving data.

polling A means of controlling terminals on a multipoint line. The computer, acting as the master station, sends a message to each terminal in turn saying, "Terminal A: have you anything to send?" If not, "Terminal B: have you anything to send?" and so on. Each such message is called a poll.

polling delay The elapsed time between successive polls to a given station which becomes the maximum delay after an operator is ready to send before transmission actually takes place.

polling list The order in which stations are polled and maintained in a list associated with each channel. A line polling list can also be used to provide priority in line service.

Poly-Star/240 Terminal emulation software from Polygon, Inc., which make a personal computer operate as a Digital Equipment Corporation (DEC) VT240 terminal.

polyvinyl chloride (PVC) A flame-retardant thermoplastic insulation material that is commonly used in the jackets of building cables.

POP Point of Presence.

port A computer interface capable of attaching to a modem for communicating with a remote terminal. The logical entrance and exit through which data traffic flows into and out of a network.

port concentrator A device that allows several terminals to share a single computer port. A concentrator link in which the port concentrator simplifies the software demultiplexing used in lieu of the demultiplexing normally performed by the computer-site concentrator.

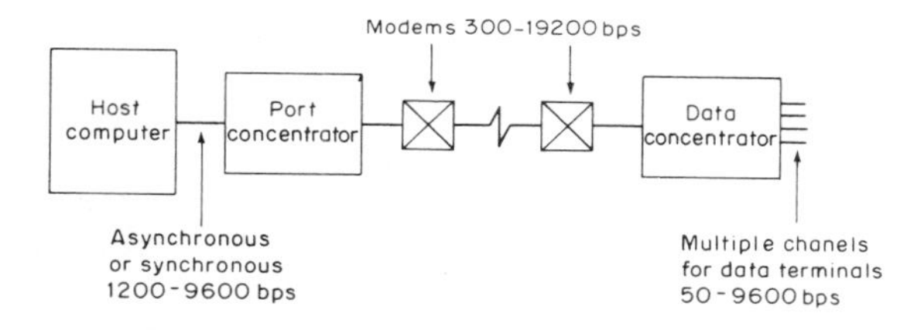

port concentrator

port contention The use of switching for incoming data calls to assign ports into the destination equipment on a first-come, first-served basis. This function is performed by a port selector or port concentrator.

port group Group of destination ports defined for a particular exchange.

port selector A switching device that extends the capability of a computer to handle more data traffic without more ports. It eliminates dedicated line-to-port interfaces, so fewer ports may handle more data lines. Also called a port concentrator and data PBX.

port sharing An arrangement by which a port in a concentrator, multiplexer, computer, or controller is used consecutively by two or more devices, such as terminals or printers.

port sharing device Digital device that treats several point-to-point lines as if they were a single multipoint line enabling the circuit capacity of a FEP to be exceeded. Located at FEP end of line. Also called port selector.

POS Point of Sale.

Positive, Intrinsic, Negative (PIN) A type of photodetector used to sense lightwave energy and convert it into electrical signals.

positive response In IBM's SNA, a response indicating that a request was received and processed.

Post, Telephone, and Telegraph Authority (PTT) The governmental agency that functions as the communications common carrier and administrator in many areas of the world.

postmortem Pertaining to the analysis of an operation after its completion.

POT Point of Termination.

POTS Plain Old Telephone Service.

power level The ratio of power at a given point to an arbitrary amount of power chosen as a reference. Usually expressed in decibels based on 1 milliwatt (dBm) or 1 watt (dBw).

power line connector A technique that uses a radio frequency carrier transmitted over the ac power line usually in a building.

power supply A hardware system that converts line current and voltage to a current and voltage that is suitable for circuit components. Most power supplies convert alternating current to filtered direct current.

PPDN Public Packet Data Network.

p-persistent In LAN technology, a term used to describe a CSMA LAN in which the stations involved in a collision try to retransmit almost immediately—with a probability of *p*.

PPM Pulse Position Modulation.

PRBS Pseudo-Random Binary-Pulse Sequence.

preamble A sequence of bits sent at the beginning of a transmission to condition the electronics at the receiving device.

preemption The act of interrupting a lower-priority call or message to use the same circuit to transmit an urgent message.

preform A solid glass rod that is heated and drawn out to form several kilometers of fiber lightguide.

premise distribution A scheme used to distribute services throughout a building complex. This may be a combination of twisted pairs and coaxial cables.

premises distribution system An AT&T intra-facility wiring scheme. The transmission network inside a building or group of buildings that connects various types of voice and data communications devices, switching equipment, and other information management systems together, as well as to outside communications network. It includes the cabling and distribution hardware components and facilities between the point where building wiring connects to the outside network lines back to the voice and data terminals in your office or other user work location. The system consists of all the transmission media and electronics, administration points, connectors, adapters, plugs, and support hardware between the building's side of the network interface and the terminal equipment required to make the system operational.

premises lightwave system An AT&T fiber optic intra-facility wiring scheme.

premises network Same as cable system.

presentation layer The sixth layer in the OSI model. Responsible for format and code conversion.

Prestel The term for the Videotex service provided by British Telecom in the United Kingdom.

PRI Primary Rate Interface.

primary application program In IBM's SNA, an application program acting as the primary end of an LU–LU session.

primary center A control center connecting toll centers. A class 3 office. It can also serve as a toll center for its local end offices.

primary end of a session In IBM's SNA, the end of a session that uses primary protocols. The primary end establishes the session. For an LU–LU session, the primary end of the session is the primary logical unit.

primary group The lowest level of a multiplexing hierarchy. The multiplexing of a large group of channels (for example telephone channels) is carried out by stages. The basic signals are first multiplexed into a primary group then a set of primary groups is multiplexed, and so forth. In the frequency division multiplexing of 4 kHz speech channels the primary group contains 12 channels and occupies 48 kHz. The primary group of PCM channels contains either 24 or 30 speech channels and uses approximately 1.5 or 2.0 Mbit/s of channel capacity, respectively.

primary half-session In IBM's SNA, the half-session that sends the session activation request.

primary link station In IBM's System Network Architecture (SNA), the link station on a link that is responsible for the control of that link. A link has only one primary link station. All traffic over the link is between the primary link station and a secondary link station.

Primary Logical Unit (PLU) In IBM's SNA, the logical unit (LU) that contains the primary half-session for a particular LU–LU session. *Note*: A particular logical unit may contain primary and secondary half-sessions for different active LU–LU sessions.

Primary Rate In ISDN, a high-bandwidth service offered to the end user. The Primary Rate signal comprises one H1 channel or any combination of B- and HO-channels possible within its aggregate bandwidth. In North America, the Primary Rate Service is offered at 1.544 Mbps; in Europe, the Primary Rate Service is offered at 2.048 Mbps. *Note*: The terms "23B+D" (North America) and "30B+D" (Europe) are often used interchangeably with "Primary Rate." Each of these terms describes only one possible Primary Rate channel configuration. In North America, there are 89 possible channel arrangements for the Primary Rate service.

Primary Rate Interface (PRI) 1. (*North American*) Twenty-three 64 Kbps B channels and one 64 Kbps D channel. 2. (*European*) Thirty 64 Kbps B channels and one 64 Kbps D channel (23B+D/30B+D).

primary route The path normally used to establish a virtual circuit to a destination. If this path is unavailable, other "secondary" routes are used.

primary station A station responsible for controlling a data link. Controls one or more secondary stations.

Prime Operating System (PRIMOS) The operating system for Prime computers providing time-shared access, segmented virtual address space, and a file system with user-implemented passwords and file-protection attributes.

Primex A private network switching service offered by British Telecom which is designed to permit users to benefit from international lease-circuit networks. Circuits interfaced can include lines routed to teleprinters, telex machines, word processors, video display units, personal computers and mainframe computers.

primitive name A name that denotes a single, unique object.

primitives The basic unit of machine instruction.

PRIMOS PRIMe Operating System.

print server In local area networks, a computer on the network that makes one or more of its attached printers available to other users. The computer normally uses a hard disk to spool the print jobs while they wait in a queue for the printer.

print services facility The program (code) that operates the IBM 3800 model 3 and 3820 printers. The print services facility operates as an a functional subsystem.

printed circuit board (PCB) A plastic board that supports electronic components interconnected by conductor deposits.

printer converter A coaxial converter that allows an asynchronous printer to emulate an IBM 3287 printer (as shown in the diagram following Type A Coax).

priority A condition denoting a level of relative urgency, importance, or value of a specific message over others and which serves as the basis for invoking special handling to ensure rapid delivery, or a higher level of protection.

priority indicators Such message priorities as urgent, rush, routine, and deferred are typically indicated by a code in the message header to define the sequence of transmission.

priority level A value that is assigned to a task or device to control processing.

privacy The techniques used for preventing access to specific system information from system users.

privacy feature Ports of packet networks to include private dial ports, and single-connection ports associated with Dedicated Access Facilities (DAFs) may be equipped with the privacy feature. This feature uses a privacy list established by the customer at restricted access host port/rotaries to block virtual connections to or from unauthorized Data Terminal Equipment (DTEs). A privacy list must be established by the customer for each access port/rotary for which the privacy feature is requested.

Private Automatic Branch Exchange (PABX) A private, automatic telephone exchange connected to the public telephone network that handles calls unattended.

Private Automatic Exchange (PAX) A telephone exchange which provides private telephone service within an organization, but not to/from the public switched network.

Private Branch Exchange (PBX) A manual, user-owned telephone exchange. Sometimes used in a general sense to include both PBXs and PABXs.

private dial ports Private dial ports are available on a leased basis for the exclusive use of a particular customer or his authorized users. Such ports provide a means of access via dial telephone networks and Foreign Exchange (FX) access channels. Two types of private dial ports are available: those equipped for inward dialing to the network and those equipped for outward dialing.

Private Exchange (PX) An exchange serving a subscriber's premises without connection to the public switched network.

private line A circuit which is 100% dedicated to the user. Unlike the circuits which are continually built up in the message telephone network, the facilities on the private line are permanently assigned to the customer and no one else uses the circuit. Private lines may be both two-wire and four-wire. They are also called leased lines and dedicated lines. Private lines are physically connected at the Central Office independent of public switching and signaling equipment. These are leased for a flat monthly rate based on line length and quality, regardless of use. These are the Bell Private Line offerings:

Series	*Examples of service*
1000	Low speed (narrowband) data, for example, private line telegraph, teletypewriter, and remote metering (telemetering).
2000	Voice
2001	This was originally developed for private line voice communications.
3000	Medium speed (voiceband) data.
3002	This is used for applications requiring dedicated lines for voice band data transmission.
4000	Telephoto/facsimile.
6000	Audio (music transmission).
7000	Television.
8000	High speed (broadband) data.

Private Line Service (PLS) Initially private line service was point-to-point telecommunications service over a channel dedicated to a particular customer's private user. FCC regulation now allows all parts of private line services, except access lines, to be used in common by many customers. Private line services are utilized by customers with high-volume or specialized requirements. However, most private lines are connected directly or indirectly (PBX) to the public switched telephone network.

Private Management Domain (PRMD) A management domain managed by a company or noncommercial organization (X.400 specific).

private network A network established and operated by a private organization for the benefit of members of the organization.

Private Transatlantic Telecommunications System The first private transoceanic fiber optic cable system between the United States and Europe.

private wire Same as leased circuit.

PRMD PRivate Management Domain.

PRN Pseudo-Random Noise.

process An activity in a software system organized as a set of self-contained but interacting activities. A process is most simply regarded as a "pseudo-processor" which may possess certain states such as "active" or "dormant" and which may execute a piece of program code.

process mode The mode in which sysout data exists and is to be processed by a JES output device. There are two IBM-defined process modes—line mode and page mode.

processing, batch A method of computer operation in which a number of similar input items are accumulated and grouped for processing.

processing, in-line The processing of transactions as they occur, with no preliminary editing or sorting of them before they enter the system.

Procomm A popular shareware communications program from Datastorm Technologies, Inc., of Columbia, MO, designed for use on the IBM PC and compatible computers.

Prodigy A videotext service provided by a joint venture of Sears, Roebuck and Co. and IBM. The initial name for this service was Trintex.

Professional Office System (PROFS) An IBM office automation software package.

profile In packet switched networks, refers to a set of parameter values, such as for a terminal, which can be defined and stored. The parameters can then be recalled and used as a group by identifying and selecting the appropriate profile.

PROFS PRofessional OFfice System.

program A sequence of step-by-step instructions that tell a computer what to do.

program operator In IBM's ACF/VTAM, an application program that is authorized to issue ACF/VTAM operator commands and receive ACF/VTAM operator awareness messages.

programmable arrangement A connection arrangement used to connect FCC registered equipment to the DDD network. This arrangement can employ either of two telephone company supplied data jacks—USOC RJ45S (programmable) or USOC RJ41S (universal). With this arrangement, the telephone company measures signal loss over the local loop between the subscriber's site and the Central Office. A "programming" resistor is selected and installed in the data jack to enable the communication equipment to transmit at a level that delivers the maximum −12 dBm signal at the Central Office.

programmable jack A jack that contains a resistor whose value can be changed to control the output level of a modem to meet a required signal loss level.

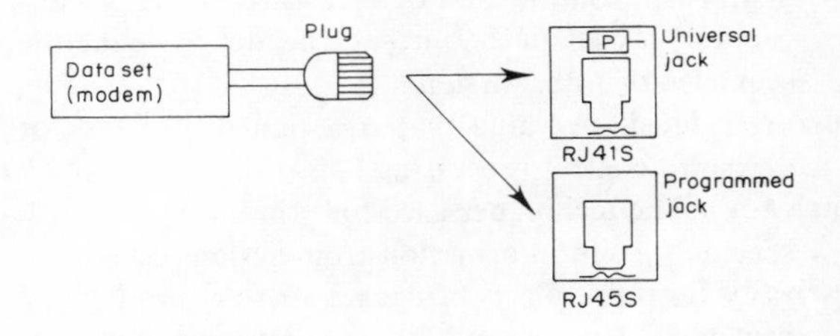

programmable jack

Programmable Read-Only Memory (PROM) Permanently stored data in a non-volatile semiconductor device. The semiconductor is electrically programmed by the equipment manufacturer and can only be changed with special equipment which erases the previous program.

programmable terminal A terminal that has processor capability. Also called intelligent terminal.

programmer, application A person who writes or maintains application programs.

programmer, systems A person who writes or maintains system programs such as operating systems or data base managers.

Project Victoria A service based upon a technology developed by Pacific Bell which enables a single, twisted copper pair in a telephone company's local loop to simultaneously transmit seven channels—two voice, one 9.6 Kbps data channel and four 1.2 Kbps data channels.

PROM Programmable Read-Only Memory.

prompt A symbol that alerts the user that a device is on-line and connected to a channel for transmission.

proof test In a fiber optic system, the process of applying constant stress to an entire length of fiber to determine and/or verify its minimum strength.

propagation delay The transit time for a signal to travel through a link, network, system, or piece of equipment.

propagation velocity The speed at which an electrical or optical signal travels through a transmission medium.

protected memory An independently established area in a storage module that protects stored programs from being inadvertently overwritten or destroyed by other programs sharing a core storage area. Other programs cannot write or transfer within the protected area.

protector 1. A device which is applied to a telephone line to protect the connected equipment from overvoltage and/or over-current. Excessive voltages and currents are shorted to ground. 2. A base of insulating material equipped with protector units, or carbon protector blocks, and sometimes fuses. Provides protection against over-voltage and over-current.

protector bypass Internal shorting or grounding of a protector due to the lack of physical separation between input and output fields, results in an overcurrent or over-voltage condition being passed into the building wiring without activating the protector unit.

protector unit A small device which screws or plugs into a protector to provide over-voltage or overcurrent protection.

PROTO A family of Telenet Processor 400 (TP4) software routines that interface the Packet Assembler/Disassembler (PAD) to the Switch and Link Access Procedure (LAP) software. PROTO provides protocol quality control, privacy screening, facilities processing, and packet reformatting. Other PROTO functions include rotary processing, autoconnects, and table support.

protocol A set of rules governing information flow in a communications system. Sometimes called data link control.

protocol control information Information sent between communicating protocol modules to coordinate their operation, as distinct from user data.

protocol conversion The conversion of data transmissions from one protocol to another, thus enabling compatibility between two dissimilar systems.

protocol converter A device that translated from one communications protocol into another, such as IBM SNA/SDLC to ASCII.

Protocol Data Unit (PDU) An ISO term which refers to the exchange of packet information between two Network layer entities.

protocol ID A five-byte field in the header of a SNAP frame on a LAN, used to identify the Data Link client at the receiving system that is to receive this frame.

protocol identifier A character string name for a protocol.

protocol intelligence The ability to decode, display, and transmit information about the contents of a bit stream as well as its format, especially about the data communications protocols in use on a circuit.

protocol sequence An ordered list of protocol identifiers.

protocol type A two byte field in the header of an Ethernet frame, used to identify the Data Link client at the receiving system that is to receive this frame.

PRTM Printing Response Time Monitor.

PRW Pseudo-Random Word.

PS/2 Personal System/2.

PSC Public Service Commission.

PSDN Public Switched Data Network.

PSDS Public Switched Digital Service.

PSE Packet Switch Exchange.

Pseudo-Random Noise (PRN) Another name for a pseudo-random pattern, used in BERT testing.

pseudo-random pattern A repeating bit pattern of

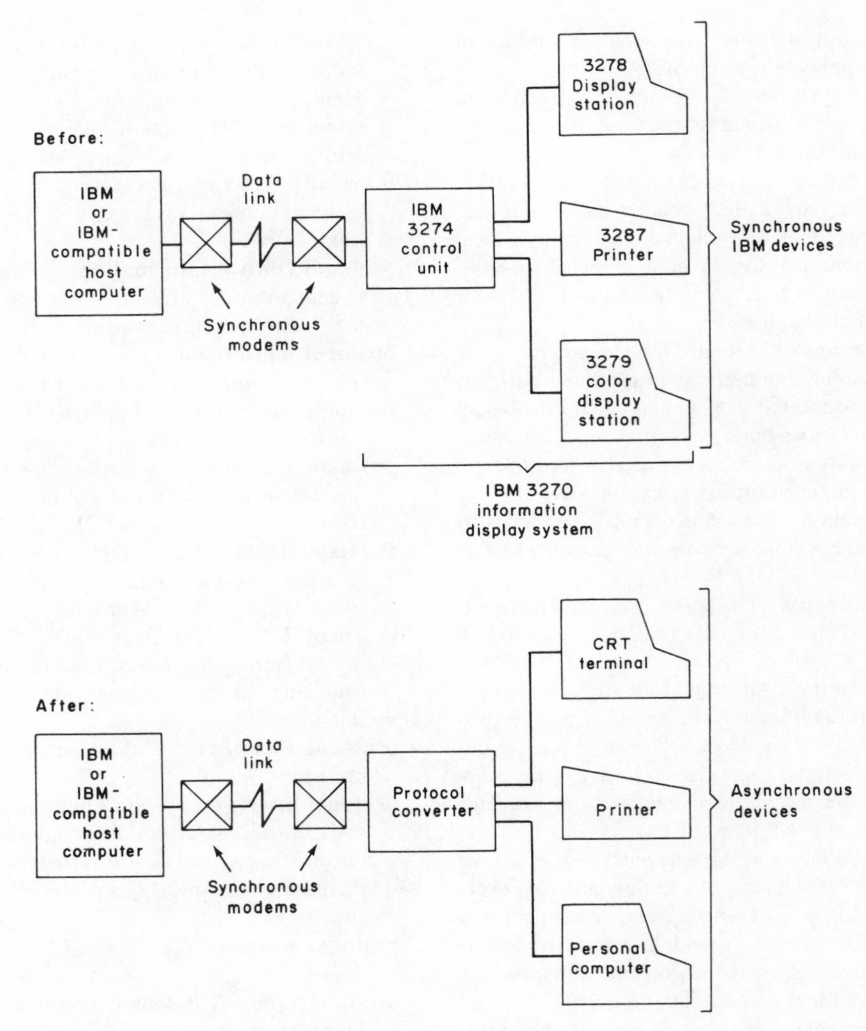

Typical protocol converter application

protocol converter

a specific length whose bit order appears to be random within the length of the pattern. Used in BERT testing to minimize the effects of regular or repeating sequences in the pattern on transmitting or receiving equipment.

Pseudo-Random Word (PRW) Specific bit pattern that is generated by a special algorithm that simulates random bit patterns associated with normal data transmission.

PSK Phase Shift Keying.

PSN 1. Packet Switching Network. 2. Public Switched Network.

psophometric weighting A type of telephone noise weighting established by the CCITT for use in a psophometer noise measuring set. Similar to F1A line weighting, psophometric weighting is used mostly in Europe.

PSS Packet Switch Stream.

PSTN Public Switched Telephone Network.

PSU Packet Switch Unit.

PTAT-1 Private TransAtlantic Telecommunications System.

PTR Printer.

PTT Post, Telephone, and Telegraph authority.

PU Physical Unit.

PU 2.1 In IBM's SNA, a low-level network entry point with limited routing facilities. Also referred to as a Single Node Control Point (SNCP).

PU 5 In IBM's SNA, ACF/VTAM in the host computer and ACF/NCP in the 37xx communications controller. Also referred to as a System Service Control Point (SSCP).

PU–PU flow In IBM's SNA, the exchange between physical units (PUs) of network control requests and responses.

PU type Physical Unit type.

PUBLIC Provided by a common carrier for use by many customers.

Public Data Network (PDN) A network established and operated by a PTT, common carrier, or private operating company for the specific purpose of providing data communications services to the public. May be a packet switched network or a digital network such as DDS. The Data Network Identification Code (DNIC) of major PDNs are listed in the following table:

Country	*PDN name*	*DNIC*
Africa:		
Egypt	ARENTO	6023
Gabon	GABONPAC	6282
Ivory Coast	SYTRANPAC	6122
Réunion 7DOMPAC	6470	
South Africa	SAPONET	6550
Sudan	via Bahrain	
Zimbabwe	ZIMNET	6482
Europe:		
Austria(1)	DATEX-P	2322
Belgium	DCS	2062
Denmark	DATAPAC	2382
England(1)	PSS	2342
Finland	DATAPAC	2442
France(1)	TRANSPAC	2080
West Germany(1)	DATEX-P	2624
Greece	HELPAC	2022
Hungary	via Austria	
Iceland	ICEPAC	2740
Ireland(1)	EIRPAC	2724
Italy(1)	ITAPAC	2222
Luxembourg	LUXPAC	2703
Netherlands	DATANET-1	2041
Norway	DATAPAC	2422
Portugal(1)	TELEPAC	2680
Spain(1)	IBERPAC	2145
Sweden	DATAPAC	2402
Switzerland	TELEPAC	2284
Middle East:		
Bahrain(1)	BAHNET	4263
Iraq	via Bahrain	
Israel	ISRANET	4251
Jordan	via Bahrain	
Kuwait	via Bahrain	
Qatar	via Bahrain	
Saudi Arabia	via Bahrain	
Turkey	TURPAC	2862
United Arab Emirates	EMDAM	4243
Asia:		
China	no name	4600
Thailand	IDAR	5200
Pacifica:		
Australia(1)	AUSTPAC	5052
French Polynesia	TOMPAC	5470
Guam	LSDS	5350
Hong Kong(1)	DATAPAC	4545
Indonesia	SKDP	5101
Japan(1)	DDX-P	4401
South Korea	DNS	4501
Malaysia	MAYPAC	5021
New Caledonia	TOMPAC	5460
New Zealand	PACNET	5301
Philippines	EASTNET	5156
Singapore	TELEPAC	5250
Taiwan(1)	PACNET	4872
North America:		
Antigua	AGANET	3443
Bahamas	BATELCO	4263
Barbados	IDAS	3420
Bermuda	BERMUDANET	3503
Canada(1)	DATAPAC	3020
Cayman Islands	IDAS	3463
Costa Rica	RACSAPAC	7120
Curaçao	UDTS	3620
Dominican Republic	no name	3701
French Antilles	DOMPAC	7420
Guatemala	GUATEL	7040
Honduras	HONDUTEL	7080
Jamaica	JAMANTEL	3380
Mexico	TELEPAC	3340
Panama	INTELPAQ	7141
Puerto Rico	PDIA	3301

Trinidad	TEXTEL	3745
United States(1)	various	various
Virgin Islands	UDTS	3320
South America:		
Argentina(1)	ARPAC	7222
Brazil(1)	RENPAC	7241
Chile(1)	ENTEL	7302
Colombia	DAPAQ	7320
French Guiana	DOMPAC	7420
Peru	no name	7160

(1) Indicates that the country has more than one PDN and operates as a gateway PDN to other countries.

public dial-in ports Public dial-in ports are available on a network on a continuous basis and may be used by the customer or authorized user upon demand. Local access ports provide a means of access via the local public telephone network. In-WATS access ports provide a toll-free means of access from any point within the continental United States via the telephone network.

public exchange Central Office.

Public Key System (PKS) Data Encryption An asymmetrical, two-key encryption algorithm that transforms data from plaintext to ciphertext with one key that is made public and converts ciphertext back to plaintext with a different key that remains secret. Also called Public Key Cryptosystem.

public network A network established and operated by communication common carriers or telecommunication administrations for the specific purpose of providing circuit switched, packet switched, and leased circuit services.

Public Service Commission (PSC) An agency charged with regulating communications services at the state level. Also known as the Public Utility Commission (PUC).

Public Switched Digital Service (PSDS) A service of Regional Bell Operating Companies (RBOC) and AT&T which permits full-duplex dial-up, 56 Kbps transmission on an end-to-end basis.

Public Switched Network (PSN) Any switching communications system—such as the Telex, TWX, or public telephone networks—that provides circuit switching to many customers.

public telephone network A telephone network which is shared among many users, any one of which can establish communications with any other user by use of a dial or pushbutton telephone. Includes DDD service. In the United Kingdom and some other countries, the network is known as the PSTN (public switched telephone network).

Public Utility Commission (PUC) See Public Service Commission.

PUC Public Utility Commission.

pulse A brief change of current or voltage produced in a circuit in order either to operate or relay, or to be detected by a logic circuit.

Pulse Amplitude Modulation (PAM) A voice sampling technique, used in digital telephony, in which the amplitude of the voice waveform is sampled 8000 times per second to produce discrete values for Pulse Code Modulation encoding.

Pulse Code Modulation (PCM) Modulation in which an analog signal is sampled and the sample quantized and coded. Standard North American sampling is 8000 times per second with 8 bits representing each sample pulse, giving a transmission rate of 64 Kbps. The resultant bit stream is sent down the line as interleaved data. The original analog signal, now in digital form, is less susceptible to noise. At the demodulator, the interleaved signals are separated and regenerated into an analog signal.

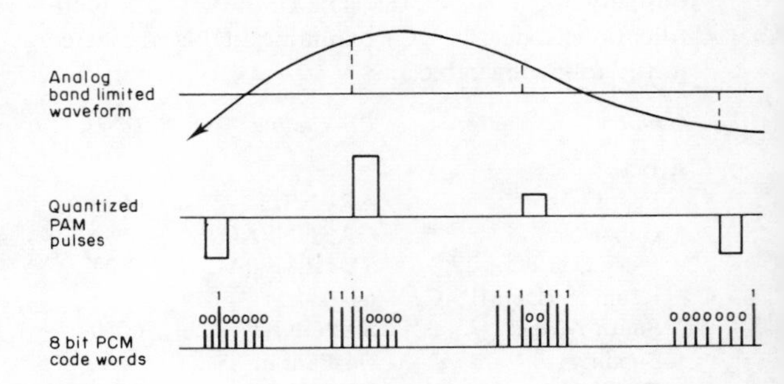

Pulse Code Modulation (PCM)

pulse dialing Older form of phone dialing, utilizing breaks in dc current to indicate the number being dialed.

pulse modulation Transmission of information by modulation of a pulsed, or intermittent, carrier. Pulse width, count, position, phase, and/or amplitude may be the varied characteristic.

pulse stuffing Same as bit stuffing.

pulse trap A device that monitors any RS-232 lead

for changes in logic levels (high to low or low to high).

Pulse-Duration Modulation (PDM) A form of pulse modulation in which the duration of pulses are varied. Also called pulse-width modulation and pulse-length modulation.

Pulsenet The packet switching network of Cincinnati Bell Telephone.

Pulse-Position Modulation (PPM) A form of pulse modulation in which the positions in time are varied, without modifying their duration.

pure code A program written so that none of the program (code) is altered during execution. All data and working storage is outside the program area. A process using the pure code can be stopped at any point and the program re-entered with a different process. Also called re-entrant code.

pushbutton dialing The use of keys or pushbuttons instead of a rotary dial to generate a sequence of digits to establish a circuit connection. The signal is usually multiple tones. Also called tone dialing, Touch-call, Touch-Tone.

PVC Permanent Virtual Circuit.

PVC PolyVinyl Chloride.

PWI Power Indicator.

PWM Pulse Width Modulation.

pyramid configuration A communications network in which the data link(s) of one or more multiplexers are connected to I/O ports of another multiplexer.

Q

Q.921 "ISDN User-Network Interface Data Link Layer Specification" is the CCITT recommendation in the Q-Series that describes LAP-D. Identical to Recommendation I.441.

Q.931 "ISDN User-Network Interface Data Link Layer 3 Specification" is the CCITT recommendation in the Q-Series that describes LAP-D. Identical to Recommendation I.451.

QAM Quadrature Amplitude Modulation.

Q-bit Qualifier bit.

QLLC Qualified Logical Link Control.

QLLC PAD Qualified Logical Link Control Packet Assembler/Disassembler.

QPSK Quadrature Phase-Shift Keying.

QRSS Quasi-Random Signal Source.

QTAM Queued Telecommunications Access Method.

quad A cable consisting of two twisted pairs of conductors.

Quadrature Amplitude Modulation (QAM) A modulation technique that combines phase modulation and AM techniques to increase the number of bits per baud. It can transmit at rates of 4800 and 9600 bps, and even higher. It is also capable of transmitting from one to seven bits per baud while keeping within the 3000 kHz limits of the phone line.

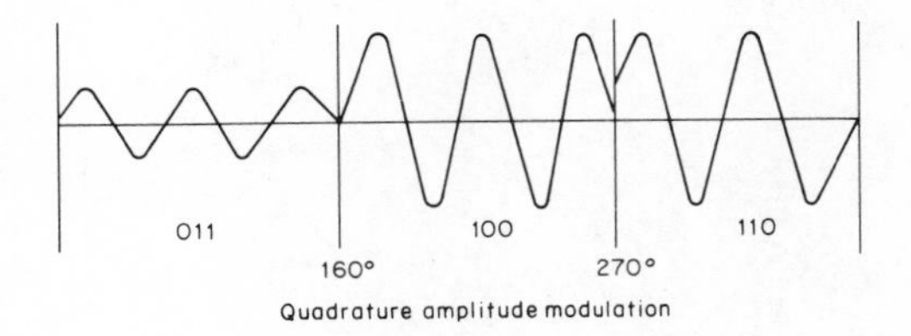

Quadrature Amplitude Modulation (QAM)

quadrature distortion Analog signal distortion frequently found in phase modulated modems.

Qualified Logical Link Control (QLLC) Logical link control procedures designed to provide Synchronous Data Link Control (SDLC) station communication through a Packet Switching Network (PSN). The QLLC protocol employs the Qualifier (Q) bit in X.25 data packets to identify unnumbered and supervisory commands and responses.

Qualified Logical Link Control Packet Assembler/Disassembler (QLLC PAD) A PAD that supports inputs from Synchronous Data Link Control (SDLC) devices and allows them to communicate with X.25 devices using QLLC, e.g., IBM's NPS2.

qualifier bit (Q bit) In X.25 packetswitched networks, bit 8 in the first octet of the packet header. It is used to indicate if the packet contains control information.

quantization error The difference between the signal level encoded as a digital value and its actual analog value.

quantizer A device used in digital communication systems to assign one of a discrete set of values to the amplitude of each successive sample of an analog signal.

Qasi-Random Signal Source (QRSS) A pseudorandom data pattern with properties similar to gaussian noise used for out-of-service, bit-error-rate testing of the ones density requirements of a digital service. Common QRSS data patterns are 2+E20−1 and 2+E15−1 defined by Bell and CCITT standards respectively.

quasi-random word One iteration of the Quasi-Random Signal Source (QRSS) pattern.

Qube Cable Network A videotex service offered by Warner Communications and American Express from 1977 to 1984.

query language A programming language that makes it relatively simple to engage in a conversational mode with a computer.

queue A line or list formed by items waiting for service, such as tasks waiting to be performed, stations waiting for connection, or messages waiting for transmission.

queued BIND In IBM's VTAM, a BIND request, sent from the primary logical unit (PLU) to the secondary logical unit (SLU), that has not yet been responded to by the SLU. This creates a pending active session at the SLU. When the SLU is a VTAM application program, it responds to a BIND by issuing an OPNSEC or SESSIONC macro instruction.

queued CINIT In IBM's VTAM, a CINIT request, sent from an SSCP to an LU, that has not yet been responded to by the LU. This creates a pending active session at the LU. A VTAM application program responds to a CINIT by issuing an OPNDST ACCEPT or a CLSDST macro instruction.

queued session In IBM's VTAM, pertaining to a requested LU–LU session that cannot be started because one of the LUs is not available. If the session-initiation request specified queueing, the SSCP(s) will record the request and later continue with the session-establishment procedure when both LUs become available.

Queued Telecommunications Access Method (QTAM) IBM teleprocessing access method that provides the capabilities of BTAM plus the capability of queued messages on direct access storage devices.

queueing A feature that allows transactions to be "held" at the origination point, node, or delivery point, while waiting for a facility to become available.

quick closedown In IBM's VTAM, a closedown in which any RPL-based communication macro instruction is terminated (posted complete with an error code) and no new sessions can be established and no new ACBs can be opened.

Quicktran A data compression program marketed by Eidolan Technologies of New York City, NY, which compresses files in the background as they are sent and received.

quiesce protocol In IBM's VTAM, a method of communicating in one direction at a time. Either the primary logical unit (PLU) or the secondary logical unit (SLU) assumes the exclusive right to send normal-flow requests, and the other node refrains from sending such requests. When the sender wants to receive, it releases the other node from its quiesced state.

R

R interface (ISDN) The two-wire physical interface which is used for a single customer termination between the TE2 and TA.

rack Same as cabinet.

rack-mount Designed to be installed in a cabinet.

radial wiring Wiring in which all cable runs from a common point to the point requiring service by the most direct means possible.

radio frequency (RF) That portion of the electromagnetic spectrum between 10 kHz and 300 MHz where propagation occurs without a guide in free space.

radio frequency (RF) noise Noise caused by an electronic spark developed across relay contacts or electronic motor brush contacts. Usually suppressed by a resistor in series with a capacitor.

radio telephone Telephone which operates over radio frequencies.

radio wave Electromagnetic waves of frequencies between 30 kHz and 3000 000 MHz, propagated without guide in free space.

radioactive isotopes Radioactive gases that are used in some gas tube surge protection devices.

Radiodetermination Satellite Service (RDSS) A satellite-based system that allows two-way communications between moving vehicles and a base station. In addition, the system permits the base station to pinpoint the exact location of a vehicle on a real-time basis.

Raduga satellites A class of Soviet geostationary communications satellites that were first launched in December 1975.

RAM Random Access Memory.

random access The process of obtaining information from or placing information in storage, where the time required for such access is independent of the location of the information most recently obtained from or placed in storage.

Random Access Memory (RAM) A storage device into which data can be entered (written) and read. Usually but not always a volatile semiconductor memory.

random noise Distortion due to the aggregate of a large number of elementary disturbances that occur at random.

random retry time In an Ethernet local area network, the time a station waits after a collision prior to attempting to transmit again.

RANGKOM RANGkaian KOmputer Malaysia.

RAS Reliability, Availability, and Serviceability.

raster A scanning pattern used for image representation in which successive horizontal lines are followed in presenting, or detecting, individual picture elements (pixels) of the image. Commonly used in television, facsimile, and electronic (laser) printing.

rate center A defined geographic point used by telephone companies in distance measurements for inter-LATA mileage rates.

Rate Demarcation Point (RDP) Also referred to as Demarc, this point has been defined by the FCC as: "The point of interconnection between telephone company communications facilities and equipment, protective apparatus or wiring at a subscriber's premises. The network interface or demarcation point shall be located on the subscriber's side of the telephone company's protector, or the equivalent thereof in cases where a protector is not employed." (First Report and Order, CC Docket 81-216, May 18, 1984.) The RDP is the point where tariffed charges end and deregulated charges begin.

ratio, signal to noise A relative measurement of the power of a signal to the power of the noise on a communications channel. Expressed in decibels (dB). Used in measuring channel quality and specifying channel or equipment characteristics.

RBOC Regional Bell Operating Company.

RBS Robbed-Bit-Signaling.

RBT Remote Batch Terminal.

RCA Global Communications An international record carrier that provides international telex, facsimile, leased private-line channels and data transmission services that were purchased by MCI Communications Corporation.

RCAC Remote Computer Access Communications Service.

RCD Receiver Carrier Detect.

RCI Remote Computer Interface.

RCV ReCeiVer.

RD Receive Data.

RDC Remote Data Concentrator.

RDP Rate Demarcation Point.

RDR ReDirectoR.

RDSS Radiodetermination Satellite Service.

RDT Resource Definition Table (IBM's VTAM).

reactance Frequency sensitive communications line impairment causing loss of power and phase shifting.

read-only A teleprinter receiver without a transmitter.

read-only bulletin board A bulletin board to which messages may be read but no messages can be sent unless the user is the designated owner of the board.

Read Only Memory (ROM) A non-volatile memory device manufactured with predefined contents that can only be read.

read-only replica In Digital Equipment Corporation Network Architecture (DECnet), a replica which responds only to lookup requests.

Read/Write (R/W) A memory that allows data to be read from it and written to it. Read/Write memory is generally a Random Access Memory (RAM).

real name In IBM's SNA, the name by which a logical unit (LU), logon mode table, or class of service (COS) table is known within the SNA network in which it resides.

real network address In IBM's SNA, the address by which a logical unit (LU) is known within the SNA network in which it resides.

real-time Responding to requests for service on demand, in contrast to time sharing, in which all requests for service are responded to on a round-robin basis in a predetermined time sequence.

real-time system An on-line computer that generates output nearly simultaneously with the corresponding inputs. Often, a computer system whose outputs follow its inputs with only a very short delay.

rearrangeable switch A circuit switch which accommodates extra connections by rearrangement of some of the connections it already carries, so that they follow different paths through the switch. By such rearrangement a switch may become nonblocking.

reasonableness checks Tests made on information reaching a real-time system or being transmitted from it to ensure that the data is within a given range. It is a means of protecting a system from data transmission errors. Also called limit check.

reassembly The process of reconstructing a complete user data message from the received segments.

reassignment In Digital Equipment Corporation Network Architecture (DECnet), when operating over the CONS, the process of transferring a Transport Connection to use a new Network Connection when the original has failed for some reason.

rebooting The process of reinitializing an operating system.

Receive Clock (RxC) An interface timing signal which synchronizes the transfer of Receive Data (RxD), provided by a DCE device.

receive pacing In IBM's SNA, the pacing of message units that the component is receiving.

Received Data (RD) An RS-232 data signal received by DTE from DCE on pin 3.

Received Line Signal Detector (RLSD) Modem interface signal, defined in RS-232, that indicates to the attached data terminal equipment that it is receiving a signal from the distant modem.

Receive-Only (RO) Of device, usually a printer, that can receive transmissions but cannot transmit.

receiver Any device that is capable of receiving a transmitted message.

RECMS Record Maintenance Statistics (IBM's SNA).

Recommendation X.21 (Geneva 1980) A Consultative Committee on International Telegraph and Telephone (CCITT) recommendation for a general-purpose interface between data terminal equipment and data circuit equipment for synchronous operations on a public data network.

Recommendation X.25 (Geneva 1980) A Consultative Committee on International Telegraph and Telephone (CCITT) recommendation for the interface between data terminal equipment and packet switched data networks.

reconfiguration The process of changing the quantity, types, or arrangement of hardware and/or software.

record A collection of logically related fields.

Record Formatted Maintenance Statistics (RECFMS) In an IBM SNA network, a RECFMS is sent to a focal point as a solicited reply unit in response to a Request Maintenance Statistic (REQMS) request.

Record Maintenance Statistics (RECMS) In IBM's NPDA, an SNA error event record built from an NCP or line error and sent unsolicited to the host.

Record mode In IBM's ACF/VTAM, the mode of data transfer in which the application program can communicate with logical units (LUs).

Record Separator (RS) A control character.

recovery The necessary actions required to bring a system to a predefined level of operation after a failure.

Red Alarm An alarm condition on a T1 circuit. A Red Alarm is in effect whenever an Out-Of-Frame condition persists for 2.5 seconds. A Red Alarm condition at one end of a circuit causes the hardware at the other end to transmit the Yellow Alarm signal.

Red Book The 1984 compilation of CCITT Standards for international telecommunications.

redirect In packet switching, a function that routes a call to an alternate network address if the communications line to the original address is blocked. Redirect occurs at the endpoint switches.

redirect address In packet switching, the 12- or 14-digit X.25 address to which a call is routed if the primary address is unavailable.

redirect NPDU In Digital Equipment Corporation Network Architecture (DECnet), an NPDU issued by a router when it forwards a data NPDU onto the same subnetwork from which it was received. It includes the subnetwork address to which the NPDU was forwarded. This indicates to the sender of the original data NPDU that it can send subsequent NPDUs destined for the same NSAP address directly to the indicated subnetwork address.

redirector A module of DOS 3.1 and later versions of the operating system that allows a user to access the resources of a remote file or print server as if those resources were attached directly to the user's computer.

Reduced Instruction Set Computing (RISC) A computer-designed architecture where the number of processing instructions is reduced to enable most instructions to execute faster.

redundancy In data transmission, the portion of characters and bits that can be eliminated without losing information. Also refers to duplicate facilities.

redundancy checking A technique of error detection involving the transmission of additional data related to the basic data in such a way that the receiving terminal can determine to a certain degree of probability whether an error has occurred in transmission.

redundant code A code using more signal elements than necessary to represent the intrinsic information. For example, five-unit code using all the characters of International Telegraph Alphabet No. 2 is not redundant. Five-unit code using only the figures in International Telegraph Alphabet No. 2 is redundant. Seven-unit code using only signals made of four "space" and three "mark" elements is redundant.

reentrant code/program Same as pure code.

reentrant routine A software routine that does not alter itself during execution. A reentrant routine can be entered and reused at any time by any number of callers.

R & D Research and Development.

reference clock The master timing source of a synchronous system.

reference noise (dBrn) The level of noise that is equal to 1 picowatt (-90 dBm) of power at 1000 Hz.

reference pilot A reference pilot is a different wave from those which transmit the telecommunication signals (telegraphy, telephony). It is used in carrier systems to facilitate the maintenance and adjustment of the carrier transmission system such as automatic level regulation and synchronization of oscillators.

reference point In the CCITT model for ISDN, a reference point is an abstract location between two proposed devices. In other words, a reference point is the potential location of a physical and logical interface. The CCITT identifies reference points by single, capital letters from the end of the alphabet. Thus, the R-reference point resides between a nonstandard terminal and a terminal adapter; the S-reference point stands between a terminal device (standard terminal or terminal adapter) and Network Termination 2; the T-reference point stands between Network Termination 2 and Network Termination 1; and the U-reference point (in the U.S.) stands between Network Termination 1 and the provider's network. *Note*: In common practice, the physical and logical interface proposed for a given reference point is named for that reference point, thus, the interface at the S-reference point is known as the S-interface.

reflection The abrupt change in direction of a light beam at an interface between two dissimilar media so that the light beam returns into the medium from which it originated.

refraction The bending of a beam of light at an interface between two dissimilar media or in a medium whose refractive index is a continuous function of position (graded-index medium).

refractive index The ratio of the speed of light in a vacuum to its speed in a given material such as glass. The larger the ratio, the more the light entering the material is bent.

refresh rate In CRT displays, the rate per unit of time that a displayed image is renewed in order to appear stable. Typically 50 times per second, or 50 Hz, in Europe or 60 times per second, or 60 Hz, in the United States.

regeneration A technique used to restore a digital signal on a cable pair. The digital signal is reconstructed and transmitted to the next regenerator. The regenerator is the device which makes digital transmission superior to analog transmission. Because the signal is reconstructed, the effects of noise and attenuation distortion are minimized.

regenerative repeater Telegraph repeaters which are speed- and code-sensitive, and are used to retime and retransmit a received signal at its original strength.

regenerator Equipment that restores the shape, timing, and amplitude to digital signals that may have been distorted during transmission. In a lightwave system, an electronic tonic for pooped pulses. A photodetector collects the pulses and converts them to electricity. An electronic circuit reconverts the electrical signals to light pulses, and a laser speeds them on their way. A costly part of lightwave systems.

Régie des Télégraphes ET DES Téléphones The Belgium PTT.

Regie van Telegrafie en Telephone The Belgian PTT.

Regional Bell Operating Company (RBOC) One of seven holding companies (Ameritech, Nynex, Bell Atlantic, BellSouth, Pacific Telesis, Southwestern Bell, and US West) set up as a result of the AT&T divestiture. Publicly traded, each is responsible for owning various Bell operating companies.

regional center A control center (class 1 office) connecting sectional centers of the telephone system together. Every pair of regional centers in the U.S. has a direct circuit group running from one center to the other.

register A storage device having a specified storage capacity, such as a bit, a byte, or a computer word. Usually intended for a special purpose.

regulatory agency An agency which controls common and specialized carrier tariffs, such as the Federal Communications Commission and the State Public Utility Commissions.

REJ Reject.

relative file organization Organization of a file whose records are to be accessed sequentially or directly by their record position relative to the beginning of a file.

relative humidity The amount of water present in the air as measured at 72 degrees Fahrenheit.

relative record number A number representing the position of a record relative to the beginning of a file. The initial record in a file is relative record number 1.

relative transmission level The ratio, expressed in decibels, of the test-tone signal power at one point in a circuit to another circuit point selected as a reference.

relay An electronic device that utilizes variation in the condition or strength of a circuit to affect the operation of the same or another circuit.

Relay Gold A full featured communications program from VM Personal Computing of Danbury, CT, for use on the IBM PC and compatible computers. It is marketed by Relay Communications, Inc., of Danbury, CT. The program is noted for its file transfer capabilities, terminal emulation, text editing and script language.

release In IBM's ACF/VTAM, resource control, to relinquish control of resources (communication controllers or physical units).

reliability A measure of the impact of line failures and the ability to recover. Enhanced by ability to reconfigure or by redundancy.

relocatable address A reference to a storage location that has a fixed displacement from the program origin, but whose displacement from absolute memory location zero depends upon the loading address of the program.

remote Physically distant from a local computer, terminal, multiplexers etc. Synonym for link-attached.

remote access The ability of a transmission station to gain access to a computer from which it is physically removed.

remote analog loopback An analog loopback test that forms the loop at the line side (analog output) of the remote modem.

remote batch processing A batch process in which the source data is prepared and collected at a remote station. The station then uses remote access to enable the computer to retrieve and process the data.

Remote Batch Terminal (RBT) A collection of input/output devices such as a card reader, line printer, and console that is controlled by a single processing unit which uses a single communications line.

remote channel loopback A channel loopback test that forms the loop at the input (channel side) of the remote multiplexer.

remote composite loopback A composite loopback test that forms the loop at the output (composite side) of the remote multiplexer.

Remote Computer Interface (RCI) The communications procedure used to operate the link between a Honeywell front end processor, and a Level 6 computer. RCI supports one logical stream on each physical line.

remote digital loopback A digital loopback test that forms the loop at the DTE side (digital input) of the remote modem. No modem operator is required at the looping modem. With RDL a signal is sent down the communications line which instructs the remote modem to place itself in digital loopback mode.

Remote File Service (RFS) A distributed file system network protocol developed by AT&T and adopted by other vendors as part of UNIX V. The protocol permits one computer to use the files and peripherals of another as if they were local.

Remote Job Entry (RJE) Entering batch jobs from remote terminals usually including a card reader and printer. Also called remote batch entry.

Remote Network Processor (RNP) A network that is not directly connected to an information processor.

remote processing Moving part of the main (host) computer's processing responsibility to a computer at a remote location. This remote (mini) computer may be connected to and supervised by the host computer.

remote station Any device attached to a controlling unit by a data link.

Remote Terminal (RT) 1. The terminating equipment for a digital line furthest from the Central Office. 2. A terminal attached to a system through a data link. 3. In telephony, a terminal attached through a trunk or tieline.

Remote Terminal Access Method (RTAM) A facility that controls operations between the job entry subsystem (JES2 or JES3) and remote terminals.

Remote Terminal Supervisor A software program originally developed by General Electric and modified by Honeywell which controls communications on the Honeywell Datanet 355 and 6600 series front end processors.

REN Ringer Equivalence Number.

repair The restoration or replacement of parts or components as necessitated by wear, tear, damage, or failure.

repairables Parts or items that are economically and technically repairable.

repeater 1. In a lightwave system, an optoelectric device or module that receives an optical signal, converts it to electrical form, amplifies it (or, in the case of a digital signal, reshapes, retimes, or otherwise reconstructs it), and retransmits it in optical form. 2. In digital transmission, a device used to extend transmission ranges/distance by restoring signals to their original size or shape. Repeaters function at the Physical layer of the OSI model. 3. In a local area network (LAN), a device that repeats all messages from one LAN segment to another or connects two or more separate LAN segments and logically joins them into a single LAN, increasing the maximum length of the LAN connection.

repeater, regenerative Normally, a repeater utilized in telegraph applications. Its function is to retime and retransmit the received signal impulses restored to their original strength.

repeater, telegraph A device which receives telegraph signals and automatically retransmits corresponding signals.

reperforator (receiving perforator) A telegraph instrument in which the received signals cause the code of the corresponding characters or functions to be punched in a tape.

Reperforator/Transmitter (RT) A teletypewriter unit consisting of a perforator and a tape transmitter, each independent of the other. It is used as a relaying device and is especially suitable for transforming the incoming speed to a different outgoing speed, and for temporary queueing.

repertory dialing The ability of a PABX to dial a complete external telephone number upon receipt of a predetermined abbreviation code for that number from an internal station.

replica In Digital Equipment Corporation Network Architecture (DECnet), a copy of a directory stored in a particular clearinghouse.

Request For Information (RFI) A general notification of an intent of an organization to purchase computer or communications equipment. An RFI is normally sent to potential suppliers to determine their interest and solicit information concerning their products.

Request for Price Quotation (RPQ) A document which is used for the solicitation of pricing for a hardware device, software product, service, or system.

Request for Proposal (RFP) A document sent to interested vendors which defines the requirements of an organization so the vendor can configure and price their product(s).

Request Header (RH) In IBM's SNA, control information preceding a request unit (RU).

Request Parameter List (RPL) In IBM's VTAM, a control block that contains the parameters necessary for processing a request for data transfer, for establishing or terminating a session, or for some other operation.

Request/Response Header (RH) In IBM's SNA, control information, preceding a request/response unit (RU), that specifies the type of RU (request unit or response unit) and contains control information associated with that RU.

Request/Response Unit (RU) In IBM's SNA, a generic term for a request unit or a response unit.

Request Unit (RU) In IBM's SNA, a message unit that contains control information such as a request code or FM headers, end-user data, or both.

Request-To-Send (RTS) An RS-232 modem interface signal (sent from the DTE to the modem on pin 4) which indicates that the DTE has data to transmit.

required cryptographic session In IBM's SNA, a cryptographic session in which all outbound data is enciphered and all inbound data is deciphered. Synonymous with mandatory cryptographic session.

reseller 1. A company that resells products made by another company. 2. A company that buys or leases transmission lines and then resells them to its customers.

reset A term referring to system initialization.

residual error rate, undetected error rate The ratio of the number of bits, unit elements, characters, or blocks incorrectly received but undetected or uncorrected by the error-control equipment, to the total number of bits, unit elements, characters, or blocks sent.

resistance The opposition to the flow of electrons in a conductor, measured in ohms. Resistance causes a voltage applied at the transmitting end of a circuit to drop.

resistor A circuit component designed to resist electrical current flow. A resistor can be used to provide a desired voltage drop for use in a circuit.

resource 1. Any data processing device in a network. 2. In IBM's SNA, any facility of the computing system or operating system required by a job or task, and including main storage, input/output devices, the processing unit, data sets, and control or processing programs. 3. In IBM's NPDA, any hardware or software component that provides function to the network or to locally attached environments.

resource class In LAN technology, a collection of computers or computer ports that offer similar facilities, such as the same application program. Each can be identified by a symbolic name.

Resource Definition Table (RDT) In IBM's VTAM, a table that describes the characteristics of each node available to VTAM and associates each node with a network address. This is the main VTAM network configuration table.

resource hierarchy In IBM's VTAM, the relationship among network resources in which some resources are subordinate to others as a result of their position in the network structure and architecture. For example, the LUs of a peripheral PU are subordinate to that PU, which, in turn, is subordinate to the link attaching it to its subarea node.

resource level In IBM's NPDA, the hardware (and software contained in the hardware) configuration of a data processing system. For example, a first-level resource would be the communication controller, and the second-level resource would be the line connected to it.

resource takeover In IBM's VTAM, action initiated by a network operator to transfer control of resources from one domain to another.

resource types In IBM's NPDA, a concept to describe the organization of data displays. Resource types are defined as central processing unit, channel, control unit, and I/O device for one category; and communication controller/adapter, link, cluster controller, and terminal for another category. Resource types are combined with data types and display types to describe NPDA display organization.

responded output In IBM's VTAM, a type of output request that is completed when a response is returned.

response An answer to an inquiry. In IBM's SNA, the control information sent from a secondary station to the primary station under SDLC.

Response Header (RH) In IBM's SNA, a header, optionally followed by a response unit (RU), that indicates whether the response is positive or negative and that may contain a pacing response.

response time The elapsed time between the generation of the last character of a message at a terminal and the receipt of first character of the reply (often an echo). It includes all propagation delays.

Response Time Monitor (RTM) In IBM's NLDM, a feature available with the 3274 control unit to measure response times.

Response Unit (RU) In IBM's SNA, a message unit that acknowledges a request unit. It may contain prefix information received in a request unit. If positive, the response unit may contain additional information (such as session parameters in response to a Bind Session), or if negative, contains sense data defining the exception condition.

restorer A trademark of Atlantic Research Corporation of Springfield, VA, as well as a system which uses the direct dial network to backup multidrop circuits.

restricted resource group In a switching network, a group of ports that can only be called by other ports within the same group.

restricted use cable Cable designed for use in non-concealed spaces where the exposed cable length does not exceed 10 feet. This type of cable can also be used in an enclosed raceway or non-combustible tubing, or when it is less than 0.25 inches in diameter and installed in one- or multiple-family dwellings.

Restructured Extended Executive (REXX) An interpreter that supports structured coding on an IBM mainframe computer system.

retransmission The procedure of transmitting a message for a second or subsequent time. Performed when it is assumed that the previous copy of the message was not successfully delivered.

retransmissive start An optical fiber component that permits the light signal on an input fiber to be retransmitted on multiple output fibers.

retry The process of retransmitting a block of data a prescribed number of times.

return address The address of an instruction in a program to which control is returned after a call to a subroutine occurs.

return loss The measurement of reflected power at various frequencies.

Return to Zero (RZ) Encoding information so that after each encoded bit, voltage returns to zero level.

REUNIR RÉseaux des UNIversités et de la Recherche.

Revenue Volume Pricing Plan (RVPP) An AT&T pricing plan for large 800 service customers. Under RVPP the billing for both domestic and international inbound 800 service can be added to obtain a discount whenever the monthly volume exceeds a revenue threshold.

reverse channel A feature provided on some

modems which provides simultaneous communication from the receiver to the transmitter on a two-wire channel. It may be used for circuit assurance, circuit breaking, error control, and network diagnostics. Also called a backward channel.

Reverse Interrupt (RVI) In the bisynchronous protocol, a reverse interrupt is a control character sequence sent by a receiving station to request the premature termination of a transmission in progress.

revisable form document An electronic document with its formatting information intact, making it readable and modifiable.

REX Route EXtension (IBM's SNA).

REXX REstructured eXtended eXecutive.

RF Radio Frequency.

RFI 1. Radio Frequency Interference. 2. Request For Information.

RFP Request For Proposal.

RFS Remote File Service.

RH 1. Request Header. 2. Request/Response Header (IBM's SNA).

RI Ring Indicator.

ring In LAN technology, a closed loop network topology.

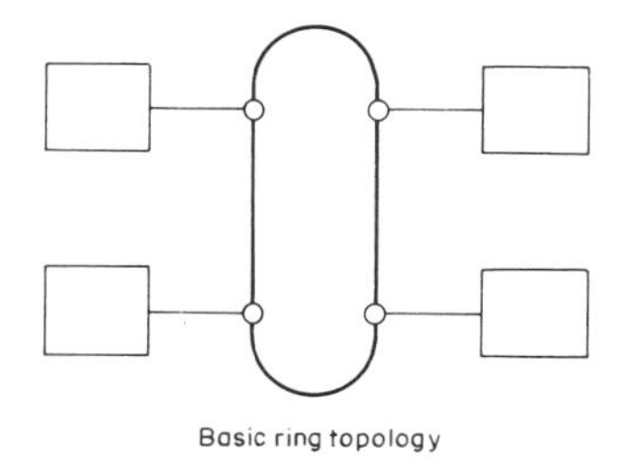

ring

Ring Indicator (RI) An RS-232 modem interface signal (sent from the modem to the DTE on pin 22) which indicates that an incoming call is present.

ring network A network in which there is not a central computer, but a series of computers which communicate with one another.

ringdown A method of signaling subscribers and operators using either a 20-cycle ac signal, a 135-cycle ac signal, or a 100-cycle signal interrupted 20 times per second.

Ringer Equivalence Number (REN) A number which indicates the quantity of ringers (or products) which may be connected to a single telephone line and still ring, for example, the number 1.0B. The total of all RENs connected to a single line may not exceed the value of 5 or some or all of the ringers may not work. The letter indicates the frequency response of the ringer. The REN is on the "FCC Registration" label located on the bottom of analog phone products.

R-interface An ISDN reference point which is the interface for older, non-ISDN equipment accessing an ISDN network.

RISC Reduced Instruction Set Computing.

riser cable A cable designed to be used in vertical shafts (risers) and not for inside plenums unless it is enclosed in non-combustible tubing. Riser cable has fire-resistant characteristics that are designed to prevent it from spreading fire between floors.

riser closet Same as backbone closet.

RJ11C The modular jack connector used on dial lines (two-wire) and is the connector most commonly used in the office for telephone connections. The RJ11 is an optional connector for four-wire private lines. For dial line operation, modems are designed to transmit at a level of −9 dBm.

RJE Remote Job Entry.

RJ41S The universal "data jack" can connect modems classified either as programmable devices or fixed loss loop devices. This type of jack can function like the RJ45S programmable jack, or allow the modem to transmit at the fixed level of -9 dBm. With a fixed loss loop, the phone company installs the dial line with a "fixed loss" from their central office to your premises. This insures a much closer to optimum level of loss in your dial-up lines. If your modem cannot use the RJ45S programmable jack, but you wish to install a dial data line that is maintained at close tolerances, then the RJ41S and corresponding fixed loss line is a feature to consider. With this type of line and jack, the telephone

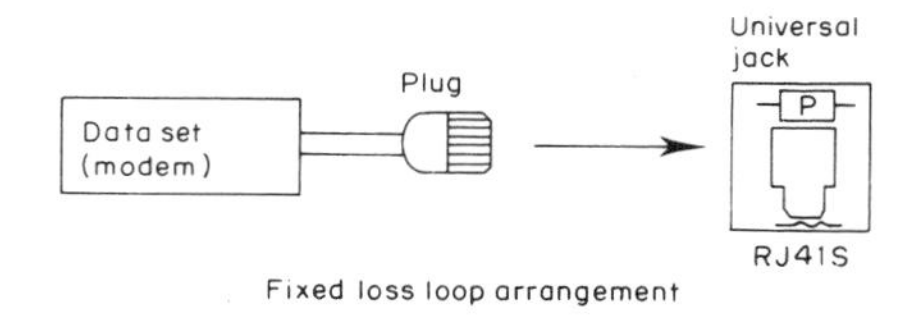

RJ41S

company is more able to assist you with data communications troubleshooting.

RJ45S The data modular jack that can be programmed by inserting a resistor of the proper value. Used with modems classified as "programmable devices." This jack is used with modems that are engineered to vary their output signal, based on the programmable function of the jack. The jack "programs" the transmit level of the modem, rather than the modem using the default transmit level of -9 dBm. This feature allows the modem to transmit to the central office at a higher level. If you are experiencing problems with dial-up connections, the problem can exist in the link between you and your telephone company central office, as this is the only portion of the dial-up network that is used in every connection. This jack is only used with dial-up lines.

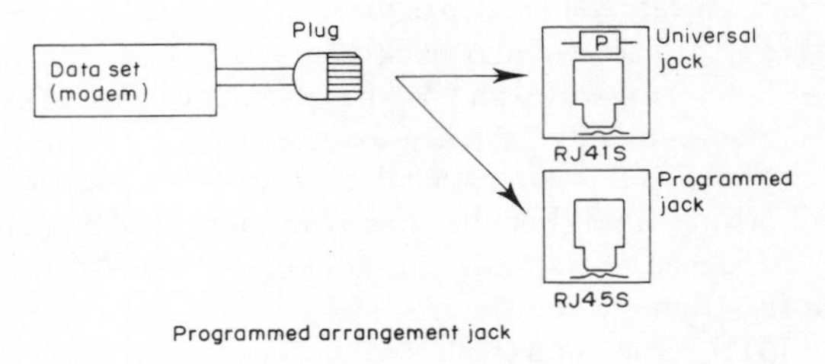

Programmed arrangement jack

RJ45S

RJ36X The modular jack used with a single line telephone when employed for alternate voice data use. In this case, the RJ16X modular jack is used for the modem.

RLSD Received Line Signal Detector.

RMS Root Mean Square.

RMT generation Generation of remote work stations for remote job entry.

RNP Remote Network Processor.

RNR Receive Not Ready.

RO 1. Read-Only. 2. Receive-Only.

robbed bits The units of AB or ABCD signaling, in which the least significant bit of each channel octet in the 6th and 12th frames of a 12-frame Superframe, or the 6th, 12th, 18th, and 24th frames of a 24-frame Superframe, are removed from the useful bandwidth for on-hook/off-hook signaling.

Rockwell Semiconductor Products A subsidiary of Rockwell International located in Newport Beach, CA, which manufactures chip sets used in a majority of modems and facsimile machines.

ROM Read-Only Memory.

ROM bootstrap loader A firmware routine in a computer or intelligent terminal that reads the first record from a designated storage device into memory.

root directory The base of the directory structure of a disk.

rotary An arrangement of a group of lines, such as telephone or data PABX lines, that are identified by a single symbolic name or number. Upon request, connection is made to the first available (free) line. Also called destination group.

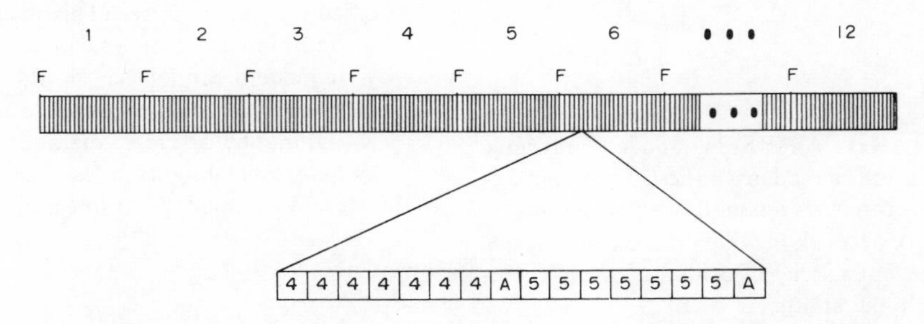

Bit robbing: A 12-frame Superframe. Two octets from the 6th frame have been enlarged to illustrate robbed-bit signaling. The least significant bit of each octet in frames 6 and 12 is "robbed" to provide signaling. Robbed bits in frame 6 are "A" bits; those in frame 12 are "B" bits

robbed bits

rotary dial Calling device that generates pulses for establishing connection in a telephone system.

rotary feature Ports on packet networks associated with Dedicated Access Facilities (DAFs) may be equipped with the rotary feature. The customer must specify which ports are to operate under rotary control. All ports on a rotary list designated by the customer are assigned a single network address, and a virtual connection initiated by a distant Data Terminal Equipment (DTE) to the group is established to the first available port in the list.

rotary hunt An arrangement which allows calls placed to seek out an idle circuit in a prearranged multicircuit group and find the next open line to establish a through circuit.

ROTR Receive Only Typing Reperforation.

round robin retraining A method of training in which the receiving modem asks for a training pattern by sending a training pattern.

round trip delay The amount of time it takes for an electrical signal to travel from one end of a transmission medium to the other and back.

route The process of directing a message to the appropriate line and terminal, based on information contained in the message header.

Route Extension (REX) In IBM's SNA, the path control network components, including a peripheral link, that make up the portion of a path between a subarea node and a network-addressable unit (NAU) in an adjacent peripheral node.

Route Table Generator (RTG) An IBM-supplied field developed program that assists the user in generating path tables for SNA networks.

router In local area networking, an inter-networking device that dynamically routes frames based upon the quality of the service required and the amount of traffic in the network.

routing The assignment of the communications path by which a message or telephone call will reach its destination.

routing, alternate Assignment of a secondary communications path to a destination when the primary path is unavailable.

routing code Third group of digits dialed to place an international call, or the area code for a domestic long distance call. Also called area code.

routing domain In Digital Equipment Network Architecture (DECnet), a collection of end systems, intermediate systems, and subnetworks which operate according to the same routing procedures and which is wholly contained within a single administrative domain.

routing indicator The bits or characters in a message header that identify the destination of the message to the routing mechanisms.

routing table A table which contains predefined information concerning the connections or port routing for a network or device.

RPL Request Parameter List (IBM's SNA).

RPL exit routine In IBM's VTAM, an application program exit routine whose address has been placed in the EXIT field of a request parameter list (RPL). VTAM invokes the routine to indicate that an asynchronous request has been completed.

RPL-based macro instruction In IBM's VTAM, a macro instruction whose parameters are specified by the user in a request parameter list.

RPM Revolutions Per Minute.

RPOA Recognized Private Operating Agency.

RPQ Request to Price Quotation.

RR Receive Ready.

RS 1. Recommended Standard. 2. Record Separator.

RS-232, RS-232C An EIA recommended standard (RS). The most common standard for connecting data processing devices. RS-232 defines the electrical characteristics of the signals in the cables that connect DTE with DCE. It specifies a 25-pin connector (the DB-25 connector is almost universally used in RS-232 applications), and it is functionally identical to the CCITT V.24/V.28. Of the 25 pins in the interface, 20 are specified for routine system operating. Of the remaining five pins, two (pins 9 and 10) are reserved for modem testing and three (pins 11, 18, and 25) are unassigned. For pin assignments see the diagram and table.

RS-422 An EIA recommended standard for cable lengths that extended the RS-232 50-foot limit. Although introduced as a companion standard with RS-449, RS-422 is most frequently implemented on unused pins of DB-25 (RS-232) connectors. Electrically compatible with CCITT recommendation V.11.

RS-423 An EIA recommended standard for cable lengths that extended the RS-232 50-foot limit. Although introduced as a companion standard with RS-422, RS-423 is not widely used. Electrically

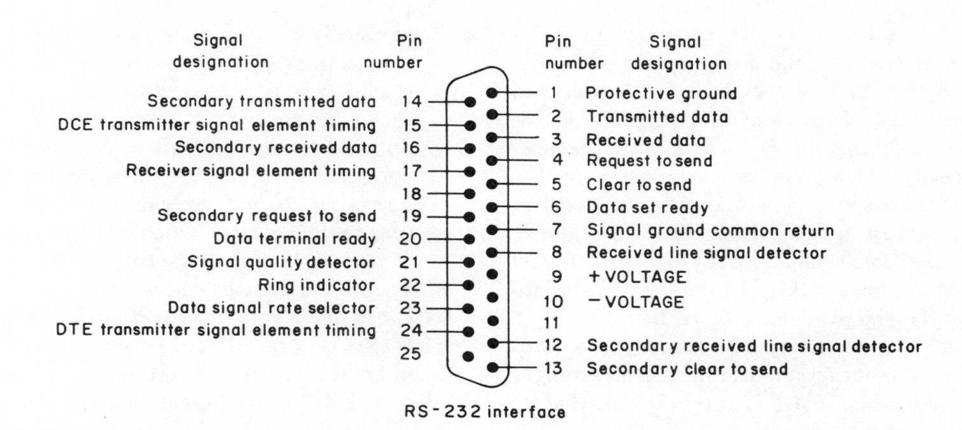

RS-232 interface

Pin	Interchange circuit	CCITT equivalent	Description	Gnd	Data		Control		Timing	
					From DCE	To DCE	From DCE	To DCE	From DCE	To DCE
1	AA	101	Protective ground	X						
7	AB	102	Signal ground/common return	X						
2	BA	103	Transmitted data			X				
3	BB	104	Received data		X					
4	CA	105	Request to send					X		
5	CB	106	Clear to send				X			
6	CC	107	Data set ready				X			
20	CD	108.2	Data terminal ready					X		
22	CE	125	Ring indicator				X			
8	CF	109	Received line signal detector				X			
21	CG	110	Signal quality detector				X			
23	CH	111	Data signal rate selector (DTE)					X		
23	CI	112	Data signal rate selector (DCE)				X			
24	DA	113	Transmitter signal element timing (DTE)							X
15	DB	114	Transmitter signal element timing (DCE)						X	
17	DD	115	Receiver signal element timing (DCE)						X	
14	SBA	118	Secondary transmitted data			X				
16	SBB	119	Secondary received data		X					
19	SCA	120	Secondary request to send					X		
13	SCB	121	Secondary clear to send				X			
12	SCF	122	Secondary received signal detector				X			

RS-232 circuit summary with CCITT equivalents

RS-232 interface

compatible with CCITT recommendation V.10.

RS-449 interface This EIA interface specifies the functional and mechanical characteristics of the interconnection between DTE and DCE and compliance to EIA electrical interface standards RS-422 and RS-423. RS-449 specifies a 37 position connector for all interchange circuits with the exception of secondary channel circuits which are accommodated in a separate nine position connector. This 9-pin connector is only necessary when the secondary channel capability is implemented in the interface. On the 37-pin connector, pins 3 and 21 are undefined. Because RS-449 can support an electrically balanced interface (RS-422), provisions are made on the connector for a return line for the various leads that are balanced. In the illustrations, the RS-449 table shows all return leads under the "B" column and the 37-pin connector shows its respective pin outs.

RSA An encryption algorithm developed by Rivest, Shamir, and Adleman that uses a public key.

RS-366-A The RS-366-A interface is employed to connect terminal devices to automatic calling units. This interface standard uses the same type 25-pin connector as RS-232. However, the pin assignments are different. The RS-366-A interface is illustrated. Note that each actual digit to be dialed is transmitted as parallel binary information over circuits 14 through 17.

RT 1. Remote Terminal. 2. Reperforator/transmitter.

RTG Route Table Generator (IBM's SNA).

RTM Response Time Monitor (IBM's SNA).

RTS Request To Send.

RTZ Return To Zero.

RU Request Unit or Response Unit.

RU chain In IBM's SNA, a set of related request/response units (RUs) that are consecutively transmitted on a particular normal or expedited data flow. The request RU chain is the unit of recovery—if one of the RUs in the chain cannot be

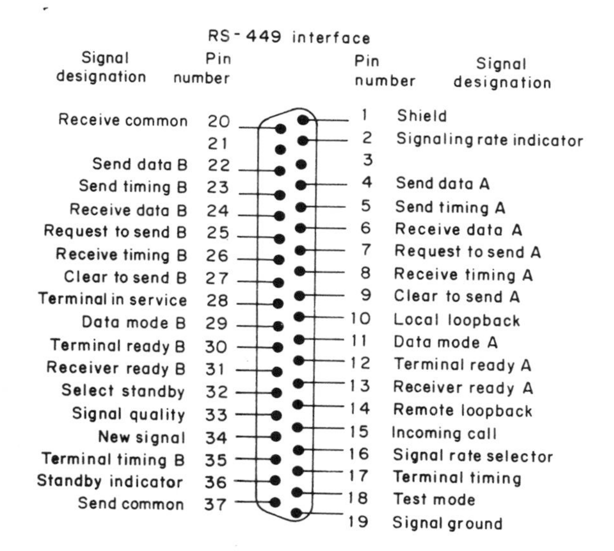

9 pin	37 pin A	37 pin B	RS-449 Circuit	Function	To DTE	To DCE
1	1		Shield	Frame ground		
5	19		SG	Signal ground		→
9	37		SC	Send common		→
6	20		RC	Receive common	←	
	4	22	SD	Send data		→
	6	24	RD	Receive data	←	
	7	25	RS	Request to send		→
	9	27	CS	Clear to send	←	
	11	29	DM	Data mode	←	
	12	30	TR	Terminal ready		→
	15		IC	Incoming call	←	
	13	31	RR	Receiver ready	←	
	33		SQ	Signal quality	←	
	16		SR	Signaling rate		→
	2		SI	Signaling rate indicator	←	
	17	35	TT	Terminal timing		→
	5	23	ST	Send timing	←	
	8	26	RT	Receive timing	←	
3			SSD	Secondary send data		
4			SRD	Secondary receive data		
7			SRS	Secondary request to send		
8			SCS	Secondary clear to send		
2			SRR	Secondary receiver ready		
	10		LL	Local loopback		→
	14		RL	Remote loopback		→
	18		TM	Test mode	←	
	32		SS	Select standby		→
	36		SB	Standby indicator	←	
	16		SF	Select frequency		→
	28		IS	Terminal in service		→
	34		NS	New signal		→

RS-449 table

RS-449 interface

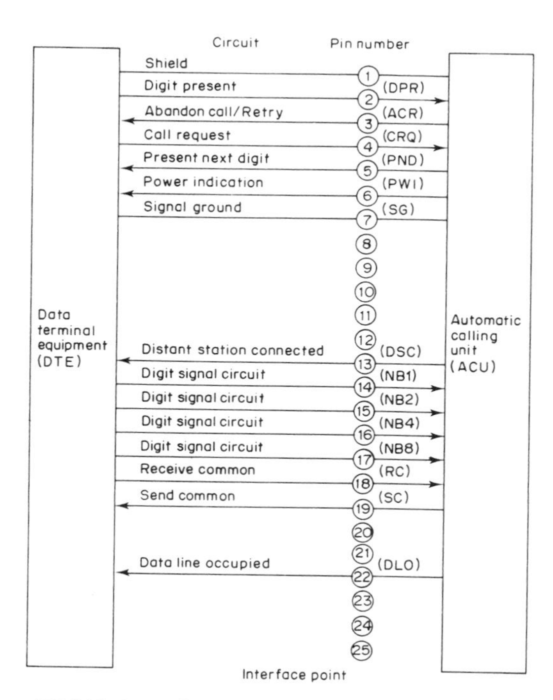

RS-366-A interface

processed, the entire chain is discarded. Note: Each RU belongs to only one chain, which has a beginning and an end indicated via control bits in request/response headers within the RU chain. Each RU can be designated as first-in-chain (FIC), last-in-chain (LIC), middle-in-chain (MIC), or only-in-chain (OIC). Response units and expedited-flow request units are always sent as only-in-chain.

run-length encoding A data compression technique in which lengths of identical symbols are encoded in terms of the actual symbol value and the number of units of symbols. The "run" of identical characters is replaced by a special character indicating the presence of this compression technique, a chracter which represents the "run" character, and a count character which indicates the length of the run.

RVI ReVerse Interrupt.

RVPP Revenue Volume Pricing Plan.

RXD Received Data.

RZ Return to Zero.

S

SAA System Application Architecture.

SABM Set Asynchronous Balanced Mode.

Sabre The airline reservation system operated by American Airlines.

SAC System Status Alarm and Control System.

SAFENET Survivable Adaptable Fiber Embedded NETwork.

SAM Status Activity Monitor.

same-domain lu–lu session In IBM's SNA, an LU–LU session between logical units (LUs) in the same domain.

sampling The process of obtaining a group of measurements representative of a universe in order to make inferences about that universe.

Samsung Data Systems A joint venture between IBM Korea and Samsung Electronic Devices.

SARM Set Asynchronous Response Mode.

satellite Transmission from earth station to antennas in earth orbit and broadcast back to earth stations. Signal is delayed approximately 275 milliseconds between earth and satellite.

Satellite Business Systems (SBS) A U.S. satellite carrier originally formed by IBM and now owned by MCI Communications.

satellite closet A walk-in or shallow wall closet that supplements a backbone closet by providing additional facilities for connecting backbone subsystem cables to horizontal wiring subsystem cables from information outlets. Also referred to as "satellite location."

satellite communications The utilization of geostationary orbiting satellites to relay the transmission received from one earth station to one or more other earth stations.

satellite microwave radio Microwave or beam radio system using geosynchronously orbiting communications satellites.

satellite relay An active or passive repeater in geosynchronous orbit around the earth which amplifies the signal it receives before transmitting it back to earth.

Satstream A private digital communications service using satellites and small dish earth terminals offered by British Telecom International.

saturation testing The testing of systems with large volumes of messages intended to expose errors that occur very infrequently and can be triggered by such rare coincidences as simultaneous arrival of two messages. Also called volume testing.

Saturn (R) The registered trademark of Siemens for a series of office communications equipment to include PABXs.

save and repeat A PBX or telephone set feature which allows the caller to store the last number dialed into the phone's memory so that a pushbutton can be pressed later when retrying the same number.

SBS Satellite Business Systems.

SC Session Control (IBM's SNA).

Scanner Interface Trace (SIT) In IBM's SNA, a record of the activity within the communication scanner processor (CSP) for a specified data link between a 3725 Communication Controller and a resource.

scattering The cause of lightwave signal loss in optical fiber transmission. It is caused by the glass itself, which bends some of the light so sharply that it leaves the fiber. The second leading cause of scattering is impurities in the glass.

SCC 1. Satellite Communications Controller. 2. Specialized Common Carrier.

scheduled output In IBM's VTAM, a type of output request that is completed, as far as the application program is concerned, when the program's output data area is free.

SCIP Exit Session Control In-bound Processing Exit (IBM's VTAM).

SCP Service Control Point.

SCPC Signal Channel Per Carrier.

scrambler A device which transforms data into an apparently random bit sequence. At the receiving locations, the data is unscrambled.

scrolling Moving the contents of a display screen up or down.

SCS An SNA character string consisting of EBCDIC control characters optionally mixed with user data that is carried within an SNA request/response unit.

SCTO Soft Carrier Turn Off.

SD Send Data.

SD-80 A Siemens telephone system with features for both small businesses and hotel/motel applications.

SD 192 A Siemens microprocessor-controlled telephone switching system for small-to-medium-sized businesses.

SD 232 A Siemens microprocessor-controlled telephone system with special features for hotel/motel industry.

SDLC Synchronous Data Link Control.

SDN System Development Network.

SDNS Software Defined Network Services.

SDU Service Data Unit.

secondary application program In IBM's SNA, an application program acting as the secondary end of an LU–LU session.

secondary channel A data channel derived from the same physical path (telephone lines) as the main data channel but completely independent from it. Running at a low data rate, it carries auxiliary information such as device control, diagnostics, or a simultaneous data stream.

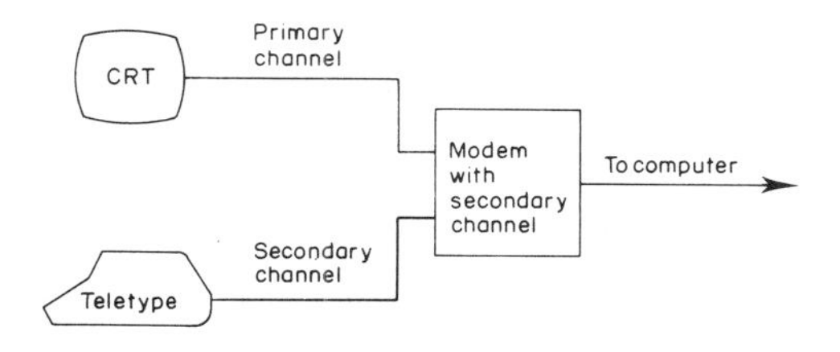

secondary channel

secondary end of a session In IBM's SNA, that end of a session that uses secondary protocols. For an LU–LU session, the secondary end of the session is the secondary logical unit (SLU).

secondary half-session In IBM's SNA, the half-session that receives the session-activation request.

secondary link station In IBM's SNA, any link station on a link, using a primary–secondary protocol, that is not the primary link station. A secondary link station can exchange data only with the primary link station. No data traffic flows from one secondary link station to another.

Secondary Logical Unit (SLU) In IBM's SNA, the logical unit (LU) that contains the secondary half-session for a particular LU–LU session. *Note*: An LU may contain secondary and primary half-sessions for different active LU–LU sessions.

Secondary Logical Unit (SLU) Key In IBM's SNA, a key-encrypting key used to protect a session cryptography key during its transmission to the secondary half-session.

secondary route In packet switching, a path taken to establish a virtual circuit if the "primary" path is not available.

secondary station A remote or tributary station that is under the control of a primary station.

second-party maintenance Maintenance performed by an original equipment manufacturer (OEM) owned and operated service department or division, authorized dealer, distributor, or system integrator.

sectional center A control center connecting primary centers. A class 2 office.

sector The smallest unit of data written onto or retrieved from a track on a disk.

Securities Industry Automation Corporation (SIAC) A company formed to provide data processing services to the New York and American Stock Exchanges.

security The techniques used for preventing physical access to information. May involve use of encryption.

seed Specific hex characters placed at the beginning of a pseudo-random word. The seed is used by test sets in determining the beginning of the pseudo-random word.

seek A mechanical movement involved in locating a record in a random-access file. This can, for example, be the movement of the arm and head mechanism that is necessary before a read instruction can be given to read data in a certain location in the file.

segment In a local area network using a bus topology, a segment is an electrically continuous piece of the bus. Segments can be connected to one another via repeaters.

segmentation The process of breaking a large user data message into multiple, smaller messages for transmission.

segmenting of BIUs In IBM's SNA, an optional function of path control that divides a basic information unit (BIU) received from transmission control into two or more path information units (PIUs). The first PIU contains the request header (RH) of the BIU and usually part of the RU; the remaining PIU or PIUs contain the remaining parts of the RU. *Note*: When segmenting is not done, a PIU contains a complete BIU.

selecting port A programmable port type which can communicate with any switched port in the network.

selection The process of identifying the destination terminal for a message.

selective call ringing A service of Signaling System 7 (SS7) which results in calls from important numbers being given a distinctive ring.

selective call rejection A service of Signaling System 7 (SS7) which results in calls from a preselected list of telephone numbers being diverted to a recording rather than sent to the subscriber.

selective calling The ability of the transmitting station to specify which of several stations on the same line is to receive a message.

selective cryptographic session In IBM's VTAM, a cryptographic session in which an application program is allowed to specify the request units to be enciphered.

selector channel An input/output (I/O) channel designed to operate with only one I/O device at a time. Once the I/O device is selected, complete records are transferred in one byte intervals.

selector lightpen An instrument that can be attached to the display station as a special feature. When pointed at a portion of the display station's image on the screen and then activated, the selector lightpen identifies that portion of the displayed screen for subsequent processing.

self-checking numbers Numbers which contain redundant information so that an error in them, caused, for example, by noise on a transmission line, may be detected.

self-test A diagnostic test mode in which the modem is disconnected from the telephone facility and its transmitter's output is connected to its receiver's input to permit the looping of test messages (originated by the modem test circuitry) through the modem to check the performance of the modem.

semaphore A mechanism for the synchronization of a set of cooperating processes. It is used to prevent two or more processes from entering mutually critical sections at the same time.

semiconductor A material (like silicon, germanium, and gallium arsenide) with properties between those of conductors and insulators. Used to manufacture solid state devices—diodes, transistors, integrated circuits, injection lasers, and light-emitting diodes.

semiconductor laser A device made of two or more semiconductors (materials with electrical properties between those of conductors and insulators) that produces an intense, very pure beam of light when stimulated by electricity. Their small size, high speed, and minimal power consumption make these lasers a good choice for lightwave systems.

send data Data from DTE to DCE.

send pacing In IBM's SNA, pacing of message units that a component is sending.

sequence number A field which indicates the sequential ordering of successive blocks or frames.

sequencing The process of dividing a user's message into smaller blocks, frames, or packets for transmission, with each subdivision of data given a sequence number to insure the correct reassembly of the complete message at its destination.

sequential access The method of reading or writing a record in a file by requesting the next record in sequence.

sequential file organization A file on disk or magnetic tape whose records are organized for access in consecutive order.

SERCnet Science Engineering Research Council Network.

serial communication The standard method of ASCII character transmission where bits are sent, one at a time, in sequence. Each 7-bit ASCII character is preceded by a start bit and ended with a parity bit and stop bit.

serial interface Mechanical and electrical components that enable data to be transmitted sequentially bit-by-bit over a transmission medium.

serial networks A group of SNA networks connected in series by gateways.

serial port The communications port to which devices such as a modem or a serial printer can be attached.

serial transmission A technique in which each bit of information is sent sequentially on a single channel, rather than simultaneously as in parallel transmission. Serial transmission is the normal mode for data communications. Parallel transmission is often used between computers and local peripheral devices.

server Any node on a LAN that allows other nodes to access its resources. Dedicated servers are totally dedicated to providing resources to other nodes on the LAN. Non-dedicated servers are used both as servers and as work stations.

Server Message Block (SMB) A distributed file system developed by Microsoft Corporation and adopted by IBM and other vendors. SMB allows one computer on a network to use the files and peripherals of another as if they were local.

SERVICE A set of functions provided by a layer to its client.

Service 800 An international toll-free facsimile service marketed by International 800 Telecom Corporation.

service access point In the OSI model, the destination or source address at a layer boundary.

Service Control Point (SCP) A centralized node in a communication carrier's network that contains service control logic.

service data unit A distinct unit of data passed by a client of a service, for transmission to the remote client. A message.

service element That part of an entity which performs the primary functions of the entity.

service entrance The point at which network communications lines ("telephone company" lines) enter a building.

Service Management System (SMS) A centralized operations system in a communication carrier's network used for creating and establishing services.

service point In an IBM SNA network, the location which converts native vendor protocol to an SNA format and then transmits the resulting data to the focal point.

service provider In the ISO Model, all the lower-level layers that provide a service to a service user.

service request A request for service signal as defined by CCITT Recommendation X.28.

Service Switching and Control Point (SSCP) A centralized node in a communication carrier's network which has integrated service switching point (SSP) and service control point (SCP) functionality.

Relationship of carrier services providers

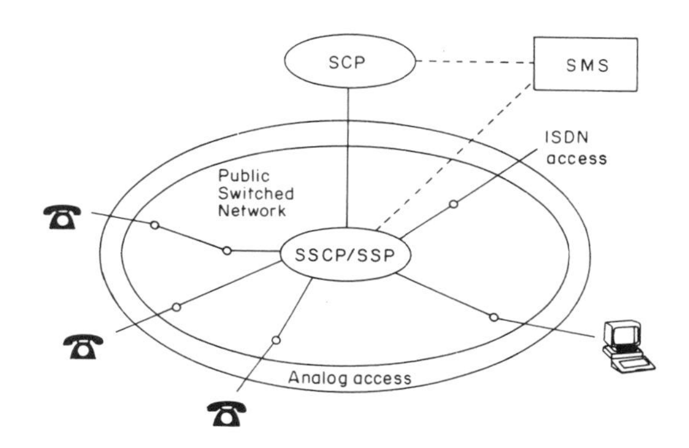

Service Switching and Control Point (SSCP)

Service Switching Point (SSP) A distributed switching node in a communication carrier's network that processes calls by interacting with the service control points (SCPs) in the network.

service terminal The equipment needed to terminate a channel and connect to station equipment.

service user In the ISO Model, the recipient of services provided by lower-level functions.

serving area Geographic area handled by a telephone facility. Generally a LATA.

Serving Test Center (STC) A test location established to control and maintain circuit layout records (CLR).

session 1. In SNA communications protocol, a logical network connection between two addressable units for the exchange of data. For example, a 3278 display station could be a logical unit in a session with a software application. 2. Another name for "connect time."

session activation request In IBM's SNA, a request that activates a session between two network addressable units (NAUs) and specifies session parameters that control various protocols during session activity, for example, BIND and ACTPU. Synonymous with generic BIND.

session address space In IBM's VTAM, an ACB address space or an associated address space in which an OPNDST or OPNSEC macro instruction is issued to establish a session.

session awareness data In IBM's SNA, data relating to sessions that is collected by NLDM and that includes the session type, the names of session partners, and information about the session activation status. It is collected for LU–LU, SSCP–LU, SSCP–PU, and SSCP–SSCP sessions and for non-SNA terminals not supported by NTO.

session control 1. A module providing functions for system-dependent process-to-process communication, name to address mapping, and protocol selection. 2. In IBM's SNA, one of the components of transmission control. Session control is used to purge data flowing in a session after an unrecoverable error occurs, to resynchronize the data flow after such an error, and to perform cryptographic verification. 3. In IBM's SNA, an RU category used for requests and responses exchanged between the session control components of a session and for session activation/deactivation requests and responses.

Session Control In-bound Processing Exit (SCIP) In IBM's VTAM, a user exit that receives control when certain request units (RUs) are received by VTAM.

session cryptography key In IBM's SNA, a data encrypting key used to encipher and decipher function management data (FMD) requests transmitted in an LU–LU session that uses cryptography.

session data In IBM's SNA, data relating to sessions that is collected by NLDM and that consists of session awareness data and session trace data.

session deactivation request In IBM's SNA, a request that deactivates a session between two network-addressable units (NAUs), for example, UNBIND and DACTPU. Synonymous with generic UNBIND.

Session Information Retrieval (SIR) In IBM's SNA, the function allowing the operator to enable/disable session information retrieval for a particular gateway or all gateway sessions. When a gateway session ends, trace information will be passed back to all SSCPs that have enabled session information retrieval (SIR) for that or all sessions. Trace information can also be requested for a particular gateway session to be passed back to the requesting host.

Session layer The fifth layer in the OSI Model responsible for establishing, managing, and terminating connections for individual application programs.

session limit 1. In IBM's SNA, the maximum number of concurrently active LU–LU sessions a particular logical unit can support. *Note*: VTAM application programs acting as logical units have no session limit. Device-type logical units have a session limit of one. 2. In the network control program (NCP), the maximum number of concurrent line-scheduling sessions on a non-SDLC, multipoint line.

session management exit routine In IBM's SNA, an installation-supplied VTAM exit routine that performs authorization, accounting, and gateway path selection functions.

session parameters In IBM's SNA, the parameters that specify or constrain the protocols (such as bracket protocol and pacing) for a session between two network-addressable units.

session partner In IBM's SNA, one of the two network-addressable units (NAUs) having an active session.

session seed Synonym for initial chaining value in IBM's SNA.

session sequence number In IBM's SNA, a sequentially incremented identifier that is assigned by data flow control to each request unit on a particular normal flow of a session, typically an LU–LU session, and is checked by transmission control. The identifier is carried in the transmission header (TH) of the path information unit (PIU) and is returned in the TH of any associated response.

session services In IBM's SNA, one of the types of network services in the system services control point (SSCP) and in the logical unit (LU). These services provide facilities for an LU or a network operator to request that the SSCP initiate or terminate sessions between logical units.

session trace In IBM's NLDM, the function that collects session trace data for sessions involving specified resource types or involving a specific resource.

session trace data In IBM's SNA, data relating to

sessions that is collected by NLDM whenever a session trace is started and that consists of session activation parameters, access method PIU data, and NCP data.

session-establishment macro instructions In IBM's VTAM, the set of RPL-based macro instructions used to initiate, establish, or terminate LU–LU sessions.

session-establishment request In IBM's VTAM, a request to an LU to establish a session. For the primary logical unit (PIU) of the requested session, the session-establishment request is the CINIT sent from the SSCP to the PLU. For the secondary logical unit (SLU) of the requested session, the session-establishment request is the BIND sent from the PLU to the SLU.

session-initiation request In IBM's SNA, an Initiate or logon request from a logical unit (LU) to a system services control point (SSCP) that an LU–LU session be activated.

session-level pacing In IBM's SNA, a flow control technique that permits a receiving connection point manager to control the data transfer rate (the rate at which it receives request units) on the normal flow. It is used to prevent overloading a receiver with unprocessed requests when the sender can generate requests faster than the receiver can process them.

session-termination request In IBM's VTAM, a request that an LU–LU session be terminated.

Set Asynchronous Balanced Mode (SABM) A command sent over a communications line to establish two-way, link-level communication. A SABM is generated by Link Access Procedure Balanced Mode (LAPBM) software.

Set Asynchronous Response Mode (SARM) A command sent over a communications line to establish link-level communication in one direction. SARM, a command generated by the link access procedure (LAP) software, must be sent in both directions of the call.

setup The preparation of a computing system to perform a job or job step. Setup is usually performed by an operator and often involves performing routine functions, such as mounting tape reels and loading card decks.

severely errored second An error condition on a T1 circuit. A severely errored second is any second of time in which 320 or more out-of-frame conditions or CRC error occur, or which has a bit error rate worse than 1×0.00010. Ten consecutive severely errored seconds constitute a failed signal state.

SG Signal Ground.

SGND Signal GrouND.

SH Switch Hook.

shadow resource In IBM's VTAM, an alternate representation of a network resource that is retained as a definition for possible future use.

shadowing The process of maintaining a duplicate file whereby modifications made to the primary file are also made to the duplicate file as changes occur.

Shannon equation A formula for determining the maximum rate of transmission for binary digits based on signal-to-noise ratio and bandwidth.

Shannon limit The theoretical maximum data rate capable of being transmitted over a band-limited channel with a given signal to noise ratio, for a given error rate.

share limit In IBM's SNA, the maximum number of control points that can concurrently control a network resource.

shared In IBM's SNA, pertaining to the availability of a resource to more than one user at the same time.

shared access In LAN technology, an access method that allows many stations to use the same (shared) transmission medium. Contended access and explicit access are two kinds of shared access methods.

shared-control gateway In IBM's SNA, a gateway consisting of one gateway NCP that is controlled by more than one gateway SSCP.

shared-tenant services A PABX feature which permits multiple-tenant organizations to share the use of a communications switch that is normally cost effective for large organizations.

sheath The outside, opaque, concentric layer in a fiber optic cable.

SHF Super High Frequency.

shield The metallic layer that surrounds insulated conductors in "shielded" cable. The shield may be the metallic sheath of the cable or the metallic layer inside a non-metallic sheath.

shielding Protective covering that eliminates electromagnetic and radio frequency interference.

shift In Baudot transmission, the prefix code that defines the interpretation of the following character.

Shift-In (SI) A control character used in conjunction with SO (Shift-Out) and ESC (Escape) to extend the graphic character set of the code.

Shift-Out (SO) A control character used in conjunction with SI (Shift-In) and ESC (Escape) to extend the graphic character set of the code.

ship to shore telephone Marine telephones.

short circuit An electrical system in which current flows directly from one conductor to another without passing through the device that is supposed to receive the current.

short-haul Pertaining to a transmission distance typically of less than 50 miles.

short-haul modem A signal converter which conditions a digital signal to ensure reliable transmission over dc continuous private line metallic circuits without interfering with adjacent pairs in the same telephone cable.

show cause In IBM's SNA, the reason code in the RECMS indicating to VTAM or NPDA the threshold that was exceeded and whether or not the threshold has been dynamically altered.

shunt 1. An alternate path in parallel with a part of a circuit. 2. A bypass or parallel path which diverts current from a circuit.

SI Shift-In.

SIAC Securities Industry Automation Corporation.

sideband The frequency band on either the upper or lower side of the carrier frequency within which fall the frequencies produced by the process of modulation.

SIGINT SIGnal INTelligence.

signal Aggregate of waves propagated along a transmission channel and intended to act on a receiving unit.

signal, analog A nominally continuous electrical signal that varies in some direct correlation to an impressed signal.

signal, digital A nominally discontinuous electrical signal that varies in some direct correlation to an impressed signal.

signal conditioning The amplification or modification of electrical signals to make them more appropriate for conveying information.

signal constellation An arrangement for graphically depicting the possible phase and amplitude combinations of a complex modem transmission scheme. The following table lists the relative signal element amplitude of V.29 modems, based upon the value of the first bit in the quadbit and the absolute phase which is determined from bits two through four. Thus, a serial data stream composed of the bits 1 1 0 0 would have a phase change of 270° and its signal amplitude would be 5. The resulting signal constellation pattern of V.29 modems is illustrated in the figure.

Absolute phase	*1st bit*	*Relative signal element amplitude*
0,90,180,270	0	3
	1	5
45,135,225,315	0	$\sqrt{2}$
	1	$3\sqrt{2}$

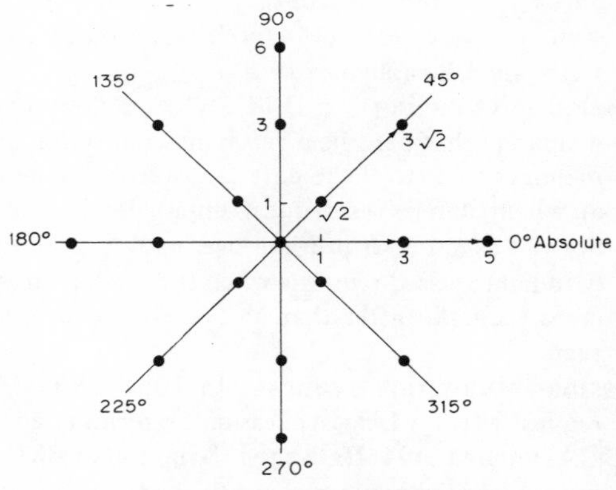

V.28 signal constellation pattern

V.29 signal amplitude construction

signal converter Device that received input signal information and outputs it in another form. Also called modem or data set.

signal distance Same as "Hamming distance."

signal element Each part of a digital signal, distinguished from others by its duration, position, or sense. Can be a start, information, or stop element.

signal ground The common ground reference potential for all circuits except protective ground.

signal intelligence (SIGINT) The monitoring and analysis of communications traffic.

signal level The strength of a signal, generally expressed in units of either voltage or power.

signal mode An optical waveguide that is designed to propagate light of only a single wavelength. An optical fiber that allows the transmission of only one light beam, and is optimized for a particular lightwave frequency.

signal reflection A condition that results from an electrical signal passing through a transmission medium encountering a point where the impedance changes, causing a portion of the signal to return (reflect) back to its origin.

Signal Transfer Points (STP) Tandem packet switches in a carrier network which route common channel signaling #7 (CCS#7) messages among service control points (SCPs) and service switching points (SSPs).

signaling In telephony, the process of controlling calls, specifically the communication between stations of such information as whether a station is available (on-hook) or unavailable (off-hook). Digital telephony offers a broader range of services than analog telephony, and thus requires a more complex form of signaling.

signaling synchronization sequence A logical sequence of six bit values carried in the framing bits of the 2nd, 4th, 6th, 8th, 10th, and 12th frames of a D4 Superframe, or the 2nd, 6th, 10th, 14th, 18th, and 22nd frames of an ESF Superframe. In combination with the frame synchronization sequence, this sequence indicates the locations in the bit stream of the 6th and 12th frames, which contain A and B signaling bits. The signaling synchronization sequence is the pattern 0 0 1 1 1 0.

Signaling System No. 7 CCITT Common Channel Signaling System No. 7. The current international standard for out-of-band call control signaling on digital networks.

Signaling Terminal Equipment (STE) The CCITT Term that defines an X.75 gateway switch.

signal to noise ratio (S/N) The ratio between the signal and the actual noise bed as illustrated. dBrn is a measurement of noise: −90 dBm = 0 dBrn.

sign-on character The first character sent on an ABR circuit; used to determine the data rate.

silicon A dark gray, hard, crystalline solid. Next to oxygen, the second most abundant element in the earth's surface. It is the basic material for most integrated circuits and semiconductor devices.

SIM 3278 A software program from Simware, Inc., of Ottawa, Canada, which operates on an IBM mainframe and makes asynchronous terminal devices appear to the host as 3278 full-screen terminals.

simple gateway In IBM's SNA, a gateway consisting of one gateway NCP and one gateway SSCP.

simplex Loosely, a communications system or equipment capable of transmission in one direction only. Permits transmission in only one direction from "A" to "B" but not from "B" to "A."

simplex circuit 1. A circuit permitting the transmission of signals in either direction, but not in both directions simultaneously (CCITT). 2. A circuit permitting transmission of signals in one direction only (common usage).

simplex mode Operation of a communication channel in one direction only, with no capability for reversing.

simplex transmission Transmission in only one direction.

simulated logon In IBM's VTAM, a session-initiation request generated when an ACF/VTAM application program issues a SIMLOGON macro instruction. The request specifies an LU with which the application program wants a session in which the requesting application program will act as the PLU.

simulation A technique through which a model of the working system is built in the form of a computer program. Special computer languages are available for producing this model, and the program is run on a separate computer. A complete system can be described by a succession of different models. These models can then be easily adjusted, and the system that is being designed or monitored can be used to test the effect of any proposed changes.

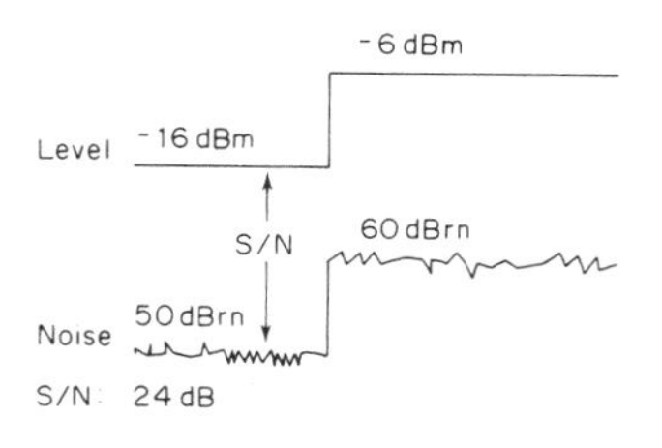

signal to noise ratio (S/N)

simulator A program in which a mathematical model represents an external system or process.

sine wave A periodic waveform with amplitude proportional to the sine of a linear function of time, space, or both.

SINET Schlumberger Information NETwork.

single fiber cable A plastic-coated fiber surrounded by an extruded layer of polyvinyl chloride, encased in a synthetic strengthening material and enclosed in an outer polyvinyl chloride sheath.

Single Message-unit Rate Timing (SMRT) A U.S. telephone company tariff under which local service is measured and calls timed in increments of 5 minutes or less—with a single message unit charge applied to each increment.

single threading A program which completes the processing of one message before starting another message.

single-current transmission A form of telegraph transmission effected by means of unidirectional currents.

single-domain network In IBM's SNA, a network with one system services control point (SSCP).

single-mode fiber A fiber lightguide with a slender core that confines light to a single path. It offers low dispersion and enormous information-carrying capacity.

single-phase Pertaining to the simplest form of alternating current, where one outward and one return conductor are required for transmission.

single-sideband transmission Transmission where one sideband of the carrier signal is transmitted while the other is suppressed.

single-thread application program In IBM's ACF/VTAM, an application program that processes requests for multiple sessions one at a time. Such a program usually requests synchronous operations from ACF/VTAM, waiting until each operation is completed before proceeding.

single-wire circuit A single conductor that uses the ground return path to complete a transmission circuit.

sink In data communications, a receiver.

sink tree For a given destination (sink) the paths taken by packets sent to it from all points of the network form a tree when fixed routing tables are used. This is the sink tree.

S-interface In ISDN, the two-wire physical interface used for single customer termination at the S-reference point, between NT2 and Tetminal equipment. The S-interface is the standard, end user interface for ISDN's basic rate service. According to current ISDN standards, the S- and T-interfaces are physically and logically identical.

SIO Scientific and Industrial Organization.

SIP Societa Italiana per L'Esercizio Telefonico pA.

SIR Session Information Retrieval (IBM's SNA).

SIT Scanner Interface Trace (IBM's SNA).

SITA Société Internationale de Télécommunications Aeronautiques.

skewing The time delay or offset between any two signals.

Skylin X.25 A satellite-based shared X.25 network service offered by Scientific Atlanta, Inc., of Atlanta, GA. Skylin X.25 supports interactive communications between multiple VSAT locations and a shared network hub via a Ku-band satellite.

Skynet Digital transmission service offering from AT&T Communications, featuring on-site earth station facilities for wideband satellite transmission, with Accunet Reserved 1.5 circuits. Also called high-capacity satellite digital service (HCSDS) and Skynet 1.5.

Skyphone A direct-dial telephone service for airline passengers offered by British Telecom International.

slave A called unit under the control of commands and signals from a master (calling) unit.

slave station In point-to-point circuits, the unit controlled by the master station.

SLCSAT Submarine Laser Communications SATellites.

SLIC Subscriber Line Interface Card.

slicing level A voltage or current level of a digital signal where a one or zero bit can be determined or not. Also called threshold.

slip A defect in timing that caused a single bit or a sequence of bits to be omitted or read twice. Slips are primarily caused by improper synchronization resulting from wander and improper distribution of the network reference frequency.

slot A unit of time in a TDM frame where a subchannel bit or character is carried.

slot time A local network parameter that describes the contention behavior of a media access control. Nominally equal to twice the propagation delay.

slotted ALOHA A packet broadcast system like ALOHA in which packets are timed to arrive at the center in regular time slots, synchronized for all stations.

SLU Secondary Logical Unit.

SM Statistical Multiplexer.

Small-Scale Integration (SSI) A term used to describe a multifunction semiconductor device with a sparse density (10 circuits or less) of electronic circuitry contained on a single silicon chip.

smart terminal A smart terminal has semiconductor-based memory. Its features include the ability to be polled, to store data in blocks, and to perform error checking. Functions such as editing can be performed on a smart terminal, but specific capabilities are built into the terminal and cannot be changed. A smart terminal is not user-programmable.

Smartcom A series of communications programs marketed by Hayes Microcomputer Products, Inc., of Atlanta, GA, for use on personal computers.

Smarterm 240 Terminal emulation software from Persoft, Inc., which makes a personal computer operate as a Digital Equipment Corporation (DEC) VT240 terminal.

Smartnet A trademark of Telematics International of Calabasas, CA, as well as a family of wide area networking products that permit personal computers and work stations to access SNA, BSC and X.25 networks.

SMB Server Message Block.

SMBASE Statistical Multiplexer BASE card.

SMDR Station Message Detail Recording.

SMEXP Statistical Multiplexer EXPansion feature.

SMF System Management Facilities (IBM's SNA).

SMN System Message Number.

SMP System Modification Program (IBM's SNA).

SMP/E System Modification Program Extended (IBM's SNA).

SMRT Single Message-unit Rate Timing.

SMS Service Management System.

S/N Signal-to-Noise ratio.

SNA Systems Network Architecture.

SNA Character String (SCS) In IBM's Systems Network Architecture, a data stream composed of EBCDIC controls, optionally intermixed with end-user data, that is carried within a request/response unit.

SNA distribution services In IBM's Systems Network Architecture, the facilities permitting multiple SNA networks to be interconnected.

SNA network In IBM's Systems Network Arlchitecture, the part of a user-application network that conforms to the formats and protocols of Systems Network Architecture. It enables reliable transfer of data among end users and provides protocols for controlling the resources of various network configurations. The SNA network consists of network addressable units, boundary function components, and the path control network.

SNA Network Interconnection (SNI) An IBM product which permits multiple-domain Systems Network Architecture networks.

SNA terminal A terminal that supports Systems Network Architecture protocols.

SNADS SNA Delivery System.

SNAP Standard Network Access Protocol.

snap shot The capture of the state of a protocol analyzer during operation. This information content may include memory contents, register status, the flag status, etc.

SNBU Switched Network BackUp (IBM's SNA).

SNDCF SubNetwork Dependent Convergence Function.

sneak currents Low-level currents that are insufficient to trigger electrical surge protectors and therefore able to pass them undetected. These currents may result from contact between communications lines and ac power circuits or from power induction, and may cause equipment damage from overheating.

SNI 1. SNA Network Integration. 2. SNA Network Interconnection.

SNR Signal-to-Noise Ratio.

SNRM Set Normal Response Mode.

SNUMB (submittal number) A job sequence number used to identify a particular job.

SO Shift-Out.

soak A term referring to a means of uncovering problems in software and hardware by running them under operating conditions while they are closely supervised by their developers.

Societa Italiana per l'Esercizio Telefonico pA (SIP) The Italian state telephone agency.

Société Internationale de Télécommunications Aeronautiques (SITA) Sponsor of an international data communication network used by many airlines.

soft copy A visual display on a CRT screen that provides no permanent record of the information that is displayed.

soft link An alternate name for an object or directory in a namespace. Soft links allow users to view names as forming an acyclic directed graph rather than a pure tree.

soft turn-off A soft carrier frequency transmitted by a modem operating on the switched telephone network to prevent transients at the end of a message being misinterpreted as spurious space signals at the remote modem.

Softerm PC A communications program marketed by Softronics, Inc., for use on the IBM PC and compatible personal computers. The program is known for its inclusion of over 50 exact terminal emulations and seven file transfer protocols.

software A computer program or set of computer programs held in some kind of storage medium and loaded into read/write memory (RAM) for execution.

software defined network A network constructed by a long distance carrier, in which a user is given a means to define a "virtual network" that operates in some respects as a private network, yet is assembled from conventional switching and trunking components of the carrier's public switched network.

Software Defined Network Services (SDNS) An AT&T provided service that gives customers control of their own special call-routing programs stored in AT&T's network. This will provide fast and cost-effective network flexibility.

software demultiplexing In multiplexer applications, the connection of a computer directly to a multiplexer trunk without an intervening multiplexer to permit the remotely multiplexed transmission to be demultiplexed by software running within the computer.

software engineering A broadly defined discipline that integrates the many aspects of programming, from writing code to meeting budgets, in order to produce affordable software that works.

software maintenance The continual improvements and changes required to keep programs up to date and working properly.

SOH Start Of Header.

solicited message In IBM's ACF/VTAM, a response from VTAM to a command entered by a program operator.

SOM Start Of Message.

SONET Synchronous Optical NETwork.

source In data communications, the originator of a message.

source code A file or listing of a high-level program in the original language in which it was coded.

source of data, source A term used to describe any device capable of supplying data in a controlled manner.

source rotary A subscription option for defining symbolic mail and a list of X.121 addresses associated with a name. The Packet Assembler/Disassembler (PAD) attempts to connect to the first address on the list and continues with other addresses until connection is made or the list is exhausted.

Source Service Access Point (SSAP) The one byte field in a LLC frame on a LAN that identifies the sending data link client protocol.

Southernnet An Atlanta-based long-distance communications carrier.

SP SPace Character.

space 1. An ASCII or EBCDIC character that results in a one-character-wide blank when printed. 2. In single-current telegraph communications, the open circuit or no-current-flowing condition. 3. In data communications, represents a binary 0.

space-division switching A switching technology where a separate physical path through the switch is maintained for each connection.

space-hold The normal no-traffic line condition whereby a steady space is transmitted.

space-to-mark transition The transition, or switching, from a spacing impulse to a marking impulse.

span 1. A T1 circuit connecting two endpoints with no intermediate switching, multiplexing, or demultiplexing; a T1 link. 2. Space Physics Analysis Network.

span line A repeated T1 line section between two central offices.

spanned record A variable length record that is segmented and that spans one or more control intervals. Spanned records can occur on disk-resident sequential files.

spanning tree A logical topology used for bridge forwarding in an extended LAN. It includes all bridges in the extended LAN but has no loops.

speaker phone Telephone equipment which allows anyone in the room to hear the telephone conversation and to participate in the conversation through a microphone and speaker. Speaker phone options may be part of the telephone set itself, or provided as ancillary equipment. Speaker phones eliminate the need to repeat entire conversations to the rest of the staff.

SPEARNET South Pacific Education And Research NETwork.

Special Prefix Information Delivery Service (SPIDS) A service of Southwestern Bell Telephone Company which provides access to unique prefix telephone numbers on a subscription basis.

Specialized Common Carrier (SCC) A company providing special or value-added communication services.

specific-mode In IBM's ACF/VTAM: 1. The form of a RECEIVE request that obtains input from one specific session. 2. The form of an accept request that completes the establishment of a session by accepting a specific queued CINIT request.

specifications Literature which clearly and accurately describe essential requirements for equipment, systems or services including the procedures which determine that the requirements have been met.

spectral density In T1, the specification concerning the space/mark ratio. AT&T specifies that no more than 15 consecutive zeros are allowed.

spectrum 1. A continuous range of frequencies, usually wide in extent, within which waves have some specific common characteristic. 2. A graphical representation of the distribution of the amplitude (and sometimes phase) of the components of a wave as a function of frequency. A spectrum may be continuous or, on the contrary, contain only points corresponding to certain discrete values.

spectrum roll-off Applied to the frequency response of a transmission line, or a filter, it is the attenuation characteristic at the edge of the band.

speech plus Technique used to combine voice and data on the same line by assigning the top part of the normal voice bandwidth to data.

speed Same as data rate.

speed calling A telephone feature that permits caller to reach frequently called numbers by using abbreviated codes.

speed conversion The changing of transmission speeds to enable two devices with unequal transmission speeds to communicate.

speed dial A PBX or telephone set feature which allows users to preprogram phone numbers so that a pushbutton can be accessed to dial them. Individual speed dialing uses designated buttons on the specific phone. System speed dialing logs the number into a central memory buffer in the PBX.

speed dialing Process of using short sequences of digits to represent complete telephone numbers.

speed number A one-, three-, or four-digit number that replaces a seven- or ten-digit telephone number. These numbers are programmed into the switch in the carrier's office or in a PBX.

SPIDS Special Prefix Information Delivery Service.

spike A sudden surge of current in an electronic circuit. Spike is a transient characterized by a short rise-time.

spin data set A data set that is deallocated (available for printing) when it is closed. Spin off data set support is provided for output data sets just prior to the termination of the job that created the data set.

spine network A local area network in which the user's computers and network servers connect to access networks, with the access networks in turn connected via gateways to the spine.

Spiral Redundancy Checking (SRC) A validity-checking technique for transmission blocks where the information sent for receiver checking is accumulated in a spiral bit position fashion (IBM 2780).

spiral-4 cable A cable used in many transmission line circuits which has two pairs of wires that are insulated with polyethylene and constructed in standard quarter-mile lengths.

splice 1. In lightwave systems, the joining of two fiber lightguides, end-to-end. Mechanical splices use coupling devices to connect lightguides; fusion splices use heat to melt and join them; and bonded splices "glue" lightguides together with an epoxy. 2. In wire systems, the joining of two ends of copper wire.

splitter An analog device for dividing one input signal into two output signals or combining two input signals into one output signal. Used to achieve tree topologies in CATV or broadband local area networks.

spoofing The process of introducing spurious transmission into a communications system.

SPOOL Simultaneous Peripheral Operation On Line.

spooled data set A data set written on an auxiliary storage device.

spooling A technique whereby output is stored on disk or tape for subsequent printing.

spot beam In satellite transmission, a narrow and focused down-link transmission.

spread spectrum communications The process of modulating a signal over a significantly larger bandwidth than is necessary for the given data rate for the purpose of lowering the bit error rate (BER) in the presence of strong interference signals.

Sprint A U.S. long distance communications carrier owned 80 percent by United Telecommunications and 20 percent by GTE Corporation. Sprint was one of the first long distance communications carriers to install a nationwide fiber optic network.

spying In a communications environment, the process of line tapping.

SQD Signal Quality Detector.

Squeeziplexer A trademark of Astrocom Corporation as well as a coaxial cable multiplexer designed to replace up to 32 individual cables that normally connect to ports on an IBM control unit.

SRC Spiral Redundancy Checking.

SRDM SubRate Data Multiplexer.

SRV SeRVer.

SS No. 7 CCITT Common Channel Signaling System No. 7.

SSAP Source Service Access Point.

SSCP 1. Service Switching and Control Point. 2. System Services Control Point (IBM's SNA).

SSCP ID In IBM's SNA, a number uniquely identifying a system services control point (SSCP). The SSCP ID is used in session activation requests sent to physical units (PUs) and other SSCPs.

SSCP rerouting In IBM's SNA network interconnection, the technique used by the gateway SSCP to send session-initiation RUs, by way of a series of SSCP–SSCP sessions, from one SSCP to another, until the owning SSCP is reached.

SSCP–LU session In IBM's SNA, a session between a system services control point (SSCP) and a logical unit (LU). The session enables the LU to request the SSCP to help initiate LU–LU sessions.

SSCP–PU session In IBM's SNA, a session between a system services control point (SSCP) and a physical unit (PU). SSCP–PU sessions allow SSCPs to send requests to and receive status information from individual nodes in order to control the network configuration.

SSCP–SSCP session In IBM's SNA, a session between the system services control point (SSCP) in one domain and the SSCP in another domain. An SSCP–SSCP session is used to initiate and terminate cross-domain LU–LU sessions.

SSI Small-Scale Integration.

SSP 1. Service Switching Point. 2. Advanced Communications Function for the System Support Programs.

stack A collection of items which can be thought of as arranged in sequence. One end of the sequence is the "top" of the stack. New items are added at the top, and items are also read and removed from the top of the stack. A stack can be stored as a list. Sometimes pointers in parts of the stack other than the top are retained.

star 1. In LAN technology, a network topology where the central control point is connected individually to all stations. 2. A multi-node network topology that consists of a central multiplexer node with multiple nodes feeding into and through the central node. The outlying nodes are usually interconnected with one another such that more than one path exists to any other node in the network.

star network A network topology in which each device is directly connected to a central node.

star topology A point-to-point wiring system in

star topology

which all network devices are connected to a central server.

starlan A local area network specification of 1 Mbps baseband data transmission over two twisted pair (IEE 802.3).

start bit In asynchronous communication, the start bit is attached to the beginning of a character so that bit and character synchronization can occur at the receiver. Always a "0" or space condition.

start element The first element of a character in certain serial transmissions, used to permit synchronization. In Baudot teletypewriter operation, it is one space bit.

Start Of Header (SOH) A communication character used to identify the beginning of the header field in a message block.

Start Of Message (SOM) A control character or group of characters transmitted by a station to indicate to other stations on the line that what follows are the addresses of stations to receive the message.

Start Of Text (STX) A control character used to indicate the beginning of a message. It immediately follows the header in transmission blocks.

start option In IBM's ACF/VTAM, a user-specified or IBM-supplied option that determines certain conditions that are to exist during the time an ACF/VTAM system is operating. Start options can be predefined or specified when ACF/VTAM is started.

start/stop A system in which each code combination is preceded by a start signal which serves to prepare the receiving mechanism for the reception of a character and is followed by a stop signal which serves to bring the receiving mechanism to rest. The start and stop signals (bits) are known as "machine information" or synchronizing bits.

start–stop envelope A form of digital signal which is anisochronous and comprises a short group of signal elements, fixed in number and length. For example, 11 elements at 300 elements per second is a common format. The envelope can begin whenever the line is clear of the previous envelope and the timing starts with the first signal element, the start element. The following elements carry the information and the envelope concludes with a stop element or elements.

start–stop transmission Asynchronous transmission such that a group of signals representing a character is preceded by a start bit and followed by a stop bit.

state In packet switching, information about the condition of a virtual call used to select appropriate processing routines.

static routing The use of manually entered routing information to determine routes.

station A logically addressable end point in a communications network.

station clock An internal oscillator and circuit that provide a clock source for an electronic device such as a multiplexer or transmission facilities management system.

station controller Processor that controls communications to and from a terminal device. Provides control and error handling functions.

Station Message Detail Recording (SMDR) A PABX feature that provides magnetic tape records of calling information data which the end customers use to help analyze billing or calling patterns.

station protector A device mounted on the customer's premise which prevents excessive voltage on the outside plant (lightning, power contacts, power induction, ground potential rise, etc.) from damaging equipment or harming personnel inside the building.

statistic In IBM'S NPDA, a resource-generated data base record that contains recoverable error counts, traffic, and other significant data about a resource.

statistical multiplexer A statistical multiplexer divides a data channel into a number of independent data channels greater than that which would be indicated by the sum of the data rates of those channels. It does this on the basis that not all the channels will want to transmit simultaneously, thus freeing up capacity to accommodate additional channels. The microprocessor in the statistical multiplexer scans the channels for input, formats the data into "blocks" and transmits it over the "composite" output channel. Buffers hold temporary bursts of data that exceed composite channel capacity.

Statistical multiplexers reduce the high-speed channel idle time by "concentrating" more channels together using microprocessors, buffers and synchronous protocols.

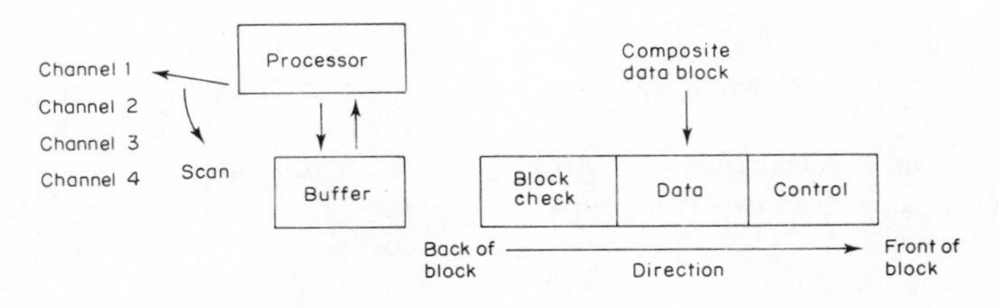

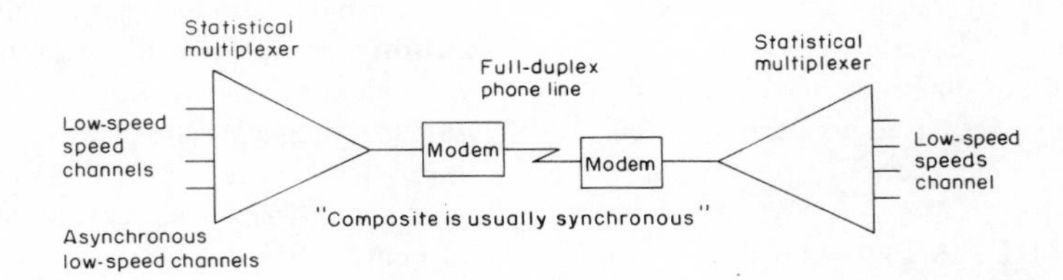

statistical multiplexer operation

Status Activity Monitor (SAM) A testing device that provides full break-out box functions in addition to LED monitoring for indicating low or high lead activity.

Status Alarm and Control system (SAC system) Collects Telenet Processor (TP) cabinet and cage environmental information (e.g., power, temperature, fan, and door).

status information Information about the logical state of a piece of equipment. Status information is one kind of control signal. Examples are: (1) A peripheral device reporting its status to the computer. (2) A network terminating unit reporting status to a network switch.

status maps Tables which give the status of various programs, devices, input–output operations, or the status of the communication lines.

STC Serving Test Area.

STD Subscriber Trunk Dialing.

STE Signaling Terminal Equipment.

steamer A trademark of Datagram of East Greenwich, RI, as well as a device which performs data compression.

step-by-step An automatic switching system in which a call is extended progressively, step-by-step.

step-index A type of waveguide (optical fiber) preferred for single-mode operation and long-distance transmission.

step-index fiber A fiber lightguide in which the refractive index changes in one big jump between core and cladding, and not gradually as in graded-index fibers. When the core is small, high transmission speeds over relatively long distances are possible.

stop bit In asynchronous transmission, the last bit used to indicate the end of a character. Normally a mark condition which serves to return the line to its idle or rest state.

stop element The last element of a character in asynchronous serial transmissions, used to ensure recognition of the next start element. In Baudot teletypewriter operations, it is 1.42 mark bits.

storage Memory. A device into which data can be entered, held, and be retrieved at a later time. Any device that can store data.

store and forward 1. The operation of a data network in which the interruption of data flow from the originating terminal to the designated receiver is accomplished by storing the information en route and forwarding it at a later time. 2. A technique in which a message is received from the originator and held in storage until a circuit to the addressee becomes available. 3. The handling of messages or packets in a network by accepting them completely into storage before sending them forward to the next switch.

stored program control Electronic switching equipment that can be programmed to perform a variety of functions.

STP Signal Transfer Point.

STR Synchronous Transmit Receive.

straight-through pinning RS-232 and RS-422 configuration that matches DTE to DCE, pin for pin (e.g., pin 1 with pin 1, pin 2 with pin 2).

straight tip connector An optical fiber connector used to join single fibers together at interconnects or connect them to optical cross-connects.

strain rate In a fiber optic system, the rate at which strain is increased during a fiber tensile test.

strapping Options built into a modem and other devices that can be activated by the user for a variety of functions. Strapping options include switching from half- to full-duplex operation, going from two- to four-wire transmission, or converting a point-to-point modem into a multipoint modem and vice versa.

stream encryption Encryption of a data stream (such as a stream of characters) with no hold-up of data in the system. Thus a single character can pass through the encryption, transmission and decryption processes without waiting for other characters. The encryption may depend on previous data in the stream.

streaming A condition of a modem when it is sending a carrier signal on a multipoint communication line when it has not been polled.

string code A technique for combining several sequential occurrences of the same character or bits.

structured programming The practice of organizing a program into modules that can be designed, prepared, and maintained independently of each other. Makes programs easier to write, check, read, and modify.

stub (tip) cable The short (usually 25 feet or less) cable which terminates on or is terminated on a cable terminal, protector, block, etc., and extends the connection into the cable.

stub-in/stub-out Where both the input and the output sides of a cable terminal, protector, block, etc., are prewired, eliminating the need for infield splicing or punch downs.

stunt box 1. A device to control the non-printing function of a teletypewriter terminal, such as carriage return and line feed. 2. A device to recognize line control characters and to cause specific action to occur when those characters are recognized.

STX Start of TeXt.

STX System Network Architecture-to-X.25.

SU Signalling Unit.

SUB SUBstitute Character.

subaddress An additional piece of addressing information for use by a DTE, beyond that needed to identify a particular subscriber line or group of lines.

subarea In IBM's SNA, a portion of a network containing a subarea node, any attached peripheral nodes, and their associated resources. Within a subarea node, all network-addressable units, links, and link stations that are addressable within the subarea share a common subarea address and have distinct element addresses.

subarea address In IBM's SNA, a value in the subarea field of the network address that identifies a particular subarea.

subarea link In IBM's SNA, a link that connects two subarea nodes.

subarea LU In IBM's SNA, a logical unit in a subarea node.

subarea node In IBM's SNA, a node that uses network addresses for routing and whose routing tables are therefore affected by changes in the configuration of the network. Subarea nodes can provide boundary function support for peripheral nodes. Type 4 and type 5 nodes are subarea nodes.

subarea PU In IBM's SNA, a physical unit in a subarea node.

subarea/element address split In IBM's SNA, the division of a 16-bit network address into a subarea address and an element address.

subchannel The result of subdividing a communications channel into narrower bandwidth channels.

Submarine Laser Communications Satellites (SLCSAT) A proposed array of satellites that will use blue laser technology to relay high data rate messages to submerged U.S. Navy submarines.

sub-multiframe In European T1, a group of eight consecutive frames. Half of a 16-frame multiframe.

subnet In local area networking, the portion of a network that is partitioned by a router or another device from the remainder of the network.

subnetwork The communication subsystem of a computer network. A collection of equipment and physical media which forms an autonomous whole and which can be used to interconnect systems for the purposes of communication. A public data communication service could function as the subnet. For example, an X.25 packet switched network or an HDLC datalink.

Subnetwork Access Protocol (SNAP) A form of

LLC frame used on LANs where multiplexing is done using a five byte Protocol ID field. This allows higher layer protocols that are not national or international standards to be addressed.

subnetwork dependent convergence function The set of functions required to enhance the service provided by a particular subnetwork to that assumed by the Connectionless-mode Network Protocol (ISO 8473).

subrate On a digital circuit, a bit rate less than DS0 (64 Kbps). The subrates most commonly used are 9600 bps and 56 Kbps.

subrate (DDS) A data speed that is either 9600 bps, 4800 bps, or 2400 bps.

Subrate Data Multiplexer (SRDM) A unit that combines a number of data streams at or below some basic rate (2400, 4800, 9600 Kbps) into a single 64 Kbps time division multiplexed signal.

subrate multiplexing The multiplexing of subrate data (bit rates less than 64 Kbps) into DS0 channels or DDS circuits. With current technologies, subrate multiplexing imposes limits on the number and kinds of subrate channels that can be multiplexed.

subroutine A procedure that alters data in an area which is common to both the subroutine and its caller.

subscriber Anyone who pays for and/or uses the services of a communications system. A subscriber can also be called a party, station, customer, or user.

subscriber line module Circuitry that provides the interface between a PABX and the telephone line.

Subscriber Trunk Dialing (STD) The European version of Direct Distance Dialing.

subscriber's line The telephone line connecting the exchange to the subscriber's station.

subscriber's loop Same as local loop.

subset A subscriber set of equipment, such as a telephone. A modulation and demodulation device.

subsplit A method of allocating available bandwidth in a single cable broadband system.

Substitute Character (SUB) A control character used in the place of a character that has been found to be invalid or in error.

subsystem 1. A conceptual grouping of programs by function (X.400 specific). 2. In IBM's SNA, a secondary or subordinate system, usually capable of operating independently of, or asynchronously with, a controlling system.

subvoice-grade channel A channel of bandwidth narrower than that of voice-grade channels. Such channels are usually subchannels of a voice-grade line.

SUNET Swedish University NETwork.

Super High Frequency (SHF) The portion of the electromagnetic spectrum in the microwave region with frequencies ranging from approximately 2 to 20 GHz.

Superframe A unit of consecutive frames on a multiplexed digital circuit. In D4 framing format, a Superframe contains 12 frames; in ESF, a Superframe contains 24 frames. In CCITT terminology, a Superframe is called a Multiframe. A European Multiframe contains 16 frames.

supergroup The assembly of five 12-channel groups (60 circuits), occupying adjacent bands in the spectrum, for the purpose of simultaneous modulation or demodulation.

supermastergroup 600 circuits processed as a unit in a carrier system.

supervisory port An interface on communications equipment to an operator console that can monitor and control communications functions.

supervisory programs Those computer programs designed to coordinate service and augment the machine components of the system, and coordinate and service application programs. They handle work scheduling, input–output operations, error actions, and other functions.

supervisory signals 1. Signals used to indicate the various operating states of circuit combinations. 2. Signals, such as "on-hook" or "off-hook," which indicate whether a circuit or line is in use.

supervisory system The complete set of supervisory programs used on a given system.

support programs A set of programs needed to install a system, including diagnostics, testing aids, data generator programs, terminal simulators, etc.

suppressed carrier transmission That method of communication in which the carrier frequency is suppressed either partially or the maximum degree possible. One or both of the sidebands may be transmitted.

SURFNET Dutch University NETwork.

surge A short duration increase of current or voltage in a circuit.

Survivable Adaptable Fiber Embedded Network (SAFENET) A U.S. Navy project which uses fiber optic technology aboard ships to obtain continued operations in the event a ship is hit during battle.

SVC Switched Virtual Circuit.

SVD Simultaneous Voice/Data.

SWIFT Society for Worldwide Interbank Financial Telecommunications

switch 1. Informal for data PABX. 2. In packet-switched networks, the device used to direct packets, usually located at one of the nodes on the network's backbone.

switch hook The switch on a telephone that requests service from the Central Office.

switchboard A device designed to provide an interconnection capability between telephones. Each telephone station is connected to a common switchboard and a switchboard operator either uses switches or cords with plugs for insertion into jacks connected to the ends of the lines from the switchboard to telephones to connect them to one another.

switched carrier A modem-to-modem handshaking technique, in which RTS generates CD at the other end. Can be used full- or half-duplex.

switched line A communications link for which the physical path may vary with each usage, such as the public, switched telephone network.

switched major node Same as switched SNA major node (IBM's VTAM).

switched network A network which is shared among many users, any one of whom can potentially establish communications with any other when required.

switched network (public) It includes all of the integrated network components required to provide a telecommunications service.

Switched Network Backup (SNBU) In IBM's SNA, an option where a switched, or dial-up line is used as an alternate path if the primary, typically a leased line, is unavailable.

switched SNA major node In IBM's ACF/VTAM, a major node whose minor nodes are physical units and logical units attached by switched SDLC links.

switched virtual call In packet switching, a temporary association between end users in which logical channel numbers (LCNs) at each X.25 interface are dynamically assigned.

Switched Virtual Circuit (SVC) In a packet switching environment, a type of connection established between two network devices via a CALL request command. The circuit is temporary for the duration of the call.

switching The process of transferring a connection from one device to another by connecting the two circuits.

switching center An installation in a communications system in which switching equipment is used to provide exchange telephone service for a given geographical area. A switching center is also called a switching facility, switching exchange or Central Office.

switching matrix In LAN technology, the electronic equivalent of a cross-bar switch.

Switching Module (SM) An AT&T 5ESS switch unit composed of a module controller/time slot interface unit along with a number of peripheral units. It performs 95 percent of all switching done in the 5ESS switch.

switching multiplexer A multiplexer able to switch channels from one high-speed, multiplexed circuit to another, or among any of several low-speed circuits.

switching node 1. In circuit switched systems, a location that terminates multiple circuits and is capable of physically interconnecting circuits for the transfer of traffic. 2. In message or packet switched systems, a location which terminates multiple circuits and which is capable of storing and forwarding messages, or packets, that are in transit.

switching office Same as Central Office.

switching processor Communications processor specialized for handling the function of switching messages/packets in a network.

switchover When a failure occurs in the equipment, a switch may occur to an alternative component. This may be an alternative communication line or an alternative computer. This switchover process may be automatic or manual.

symbol In data transmission, a symbol is a discrete waveform, usually representing binary digits, modulated as appropriate to be understood by the receiver.

symbolic name A means used to identify a collection of stations (as in an access group) or computer ports (as in a resource class).

symmetric cipher A cipher function that requires

the same key to decrypt as was used to encrypt.

symmetric flow A flow pattern with a symmetric traffic matrix and, on each line of the network, an (s,t) flow component equal and opposite to the (t,s) flow component for each pair s,t terminal points. Thus the overall flow is symmetric and so also is the routing of each flow component.

symptom string In IBM's SNA, a structured character string written to a file when VTAM detects certain error conditions.

SYNAD exit routine In IBM's SNA, a synchronous EXLST exit routine that is entered when a physical error is detected.

sync Short for synchronous or for synchronous transmission.

synchronization, synchronizing The process of making the receiver be "in step" with the transmitter. Usually achieved by having a constant time interval between successive bits, by having a predefined sequence of overhead bits and information bits, and by having a clock.

synchronizing character (SYNC) A character used to synchronize the receiver so it can accept the transmitted data coherently.

synchronous Having a constant time interval between successive bits or characters. The term implies that all equipment in the system is in step. Also called bit synchronous.

synchronous data channel A communications channel capable of transmitting timing information in addition to data. Sometimes called an isochronous data channel.

Synchronous Data Link Control (SDLC) A communications line discipline, associated with the IBM system network architecture SNA: initiates, controls, checks, and terminates information exchanges or communications lines. Designed for full-duplex operations simultaneously sending and receiving data. It can also be used for half-duplex transmission. SDLC is a bit-oriented protocol. Instead of using a control character set as does bisynchronous, SDLC uses a variety of bit patterns to flag the beginning and end of a frame. Other bit patterns are used for the address, control and packet header fields which route the frame through a network to its destination. Typical SDLC transmission frames are shown in the illustration.

synchronous data network A data network in which the timing of all components of the network is controlled by a single timing source.

synchronous idle In synchronous transmission, a control character used to maintain synchronization and as a time fill in the absence of data. The sequence of two SYN characters in succession is used to maintain synchronization following each line turnaround.

synchronous modem A modem which can transmit timing information in addition to data. It must be synchronized with its associated terminal equipment by the exchange of timing signals. Sometimes called an isochronous modem.

synchronous network A network in which all communications links are synchronized to a common clock.

synchronous operation In IBM's ACF/VTAM, a communication, or other operation in which ACF/VTAM, after receiving the request for the operation, does not return control to the program until the operation is completed.

Synchronous Optical Network (SONET) A Bellcore proposed synchronous optical transmission protocol standard that accommodates the capability to add and drop lower bit rate signals from the higher bit rate signal without requiring electrical demultiplexing. The standard defines a set of transmission rates, signals and interfaces for fiber-

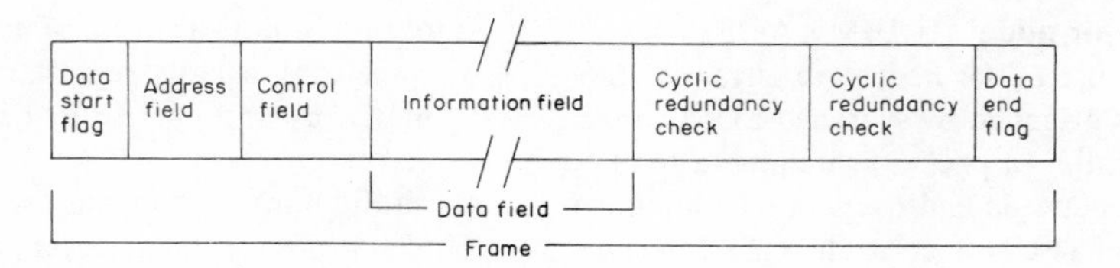

Synchronous Data Link Control (SDLC) frame

optic transmission facilities. The base transmission rate of SONET, Optical Carrier 1 (OC-1) is approximately 51 times the bandwidth of a standard 1.544 Mbps T1 lines.

synchronous request In IBM's ACF/VTAM, a request for a synchronous operation.

synchronous routine A set of computer instructions that is called into service at periodic intervals.

synchronous system A system in which sending and receiving equipment is operating continuously at the same frequency.

synchronous terminal A data terminal that operates at a fixed rate with transmitter and receiver in synchronization.

synchronous transmission Transmission in which the data characters and bits are transmitted at a fixed rate with the transmitter and receiver synchronized. This eliminates the need for individual start bits and stop bits surrounding each byte, thus providing greater efficiency.

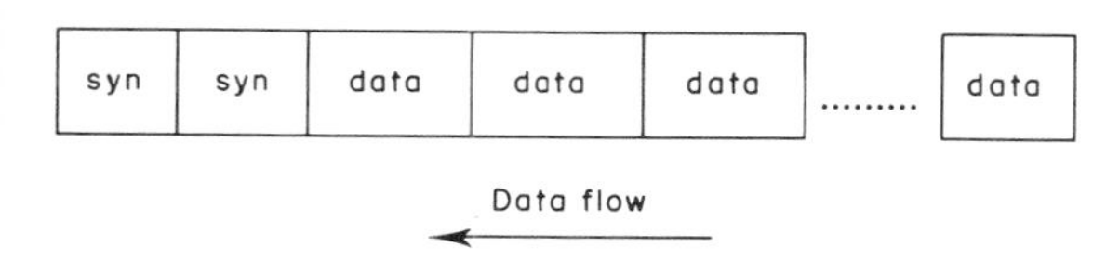

synchronous transmission

Sync-up A trademark of Universal Data Systems of Huntsville, AL, as well as a series of adapter cards designed for use in an IBM PC or compatible personal computer which contain a popular modem and emulation hardware.

syntax The rules of grammar in any language, including computer language.

Syntran A restructured electrical DS3 (45-Mb/s) signal format for efficient synchronous transmission designed to add/drop DS1 (1.544-Mbps) signals without demultiplexing the DS3 signal.

SYSGEN SYStem GENeration.

SYSOP SYStem OPerator.

system An implementation of a computer that supports the Network layer, the Transport layer, and the Session Control layer. Each system has a unique Network layer address.

SYSTEM 7 CCITT Common Channel Signaling System No. 7. The current international standard for out-of-band call control signaling on digital networks.

SYSTEM 12 A telephone switch marketed by Alcatel.

System Application Architecture (SAA) A set of specifications developed by IBM which describes how application programs, communications programs, and users interface with one another. SAA represents an attempt by IBM to standardize the look and feel of applications.

system clock The source designated as the reference for all clocking in a network of electronic devices, such as a multiplexer or transmission facilities management system.

system control programming IBM-supplied programming that is fundamental to the operation and maintenance of the system. It serves as an interface with program products and user programs and is available without additional charge.

System Generation (SYSGEN) The process of loading an operating system in a computer.

System Management Facilities (SMF) An optional control program feature of OS/360 and OS/VS that provides the means for gathering and recording information that can be used to evaluate system usage.

System Message Number (SMN) A unique number assigned by Telemail to each message.

System Modification Program (SMP) In IBM's SNA, a program that facilitates the process of installing and servicing an MVS system. For NPDA, either SMP or SMP/E is required for installation in an MVS system.

System Modification Program Extended (SMP/E) In IBM's SNA, a program that facilitates the process of installing and servicing an MVS system. For NPDA, either SMP or SMP/E is required for installation in an MVS system.

system operator The person responsible for operating a computer system.

system restart A restart that allows reuse of previously initialized input and output work queues. Synonymous with warm start.

System Service Control Point (SSCP) In IBM's SNA, a host-based network entity that manages the network configuration.

System Support Program (SSP) An IBM product program, which is made up of a collection of utilities and small programs, that supports and is required for the operation of the NCP.

Systematic Intrusion Detection A Nynex software program designed to help telephone companies track down computer users who program their PCs to make a series of calls using random number sequences until they identify a valid long-distance calling card number.

Système Interbancaire de Télécompensation (SIT) A network in France designed for interbank transactions that uses Transpac, the nationwide X.25 packet switching network operated by France Telecom, the state-controlled network operator.

Systems Network Architecture (SNA) IBM's total description of the logical structure, formats, protocols, and operational sequences for transmitting information units between IBM software and hardware devices. Data communications system functions are separated into three discrete areas: the application layer, the function management layer, and the transmission subsystem layer. The structure of SNA allows the ultimate origins and destinations of information—that is, the end users—to be independent of, and unaffected by, the specific data communications system services and facilities used for information exchange.

Systems Network Architecture-TO-X.25 (STX) Software marketed by Telenet that can be installed in a host for 3270-to-asynchronous conversion.

systems software Programs or routines that belong to the system and that usually perform a support function.

systems technology management A joint venture between Electronic Data Systems Corp., Dallas, and the Korean Lucky-Goldstar Group.

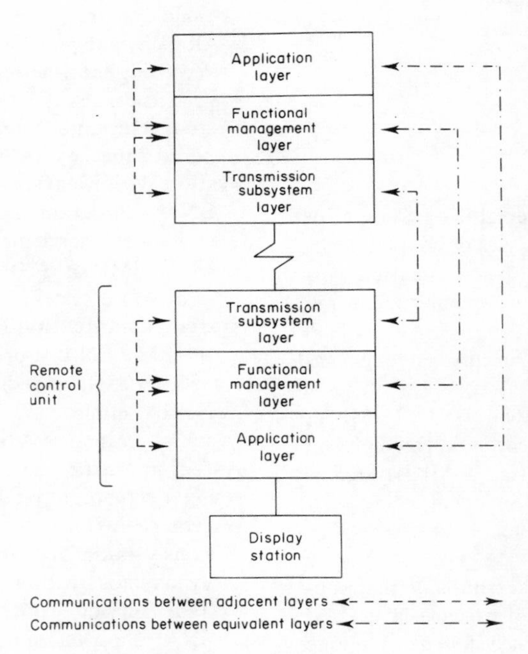

Structure of SNA

Systems Network Architecture (SNA)

T

T1 Time-division multiplexing, level 1. The original, and still more common, name for DS1 Service. T1 describes a digital, time-division multiplexed bit stream of 1.544 Mbps (in North America) or 2.048 Mbps (in Europe).

T1 Committee A committee of the American National Standards Institute (ANSI) organized to set U.S. standards for digital telephony, especially ISDN. *Note*: Despite its name, Committee T1 is not chartered to set standards for T1 circuits; this is a point of frequent confusion.

ISDN Activity Working Groups include:

T1S1.1 (formerly T1D1.1)—	ISDN architecture and services
T1S1.2 (formerly T1D1.2)—	Signaling and switching protocols
T1S1.3 (formerly T1X1.1)—	Common channel signalling
T1S1.4 (formerly T1X1.2)—	Individual channel signalling

T1 line A digital transmission line that carries data at a rate of 1.544 Mbps in North America (DS1 level). In a digital service it is used for short haul links (Intracity Megaroute).

T1 timer In packetswitched networks, used to measure timeout intervals in link initialization and data exchanges.

T2 An American Telephone & Telegraph Company digital facility using the DS2 format and operating at a speed of 6.312 Mbps.

TA Terminal Adapter.

table A body of information (data structure) used to describe the connectivity, equipment characteristics, or operating parameters of a device in a network.

table-driven Pertaining to a logical computer process frequently used in network routing, access security, and modem operation. In a table-driven process, a user-entered variable is matched against an array of predefined values. If a match occurs, another variable in a table associated with the user-entered variable is selected, resulting in the process also being called a table-lookup.

TABS Telemetry Asychronous Block Serial protocol.

TAC 1. Technical Assistance Center. 2. Telenet Access Controller.

TACL Tandem Advanced Command Language.

TACT Terminal-Activated Channel Test.

TAF Terminal Access Facility (IBM's SNA).

tail circuit A feeder circuit or extension of an existing communication link to a network node, normally a leased line. In the example illustrated, there is a point-to-point line from a corporate headquarters in Chicago to a regional office in Boston with three terminals. There are two modems with built-in mutiplexers to accomplish data transfer. We now need a line that connects the computer in Chicago to a newly opened office in Providence, RI, having one terminal. It is possible to connect the modem in Boston to another modem in Providence, via a tail circuit. Doing so saves the cost of another point-to-point line directly connecting Chicago to Providence. The data between Chicago and Providence will now go through the modem in Boston and be transmitted on to Providence.

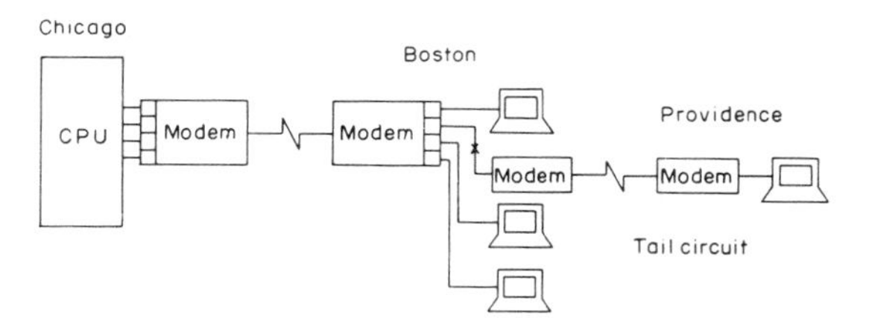

tail circuit

TAMS Telenet Access Management System.

Tandem Advanced Command Language (TACL) A more powerful command interpreter for the Tandem non-stop systems.

tandem data circuit A channel connecting two data

circuit-terminating equipment (DCE) devices in series.

tandem office A telephone company office with switching equipment which is used to interconnect Central Offices over tandem trunks in a densely settled exchange area where it is uneconomical to provide direct interconnection between all Central Offices. The tandem office completes all calls between the Central Offices but is not directly connected to subscribers.

Tandem Operating System Known as Guardian, the operating system of the Tandem non-stop computer used by the Telemail system.

tandem switch A special class of telephone company trunk to trunk switching system typically used in large metropolitan areas to interconnect Central Offices. Tandems are often classed as toll switching systems although a large portion of the connection may be within the exchange area of a Central Office.

tap 1. In cable-based LANs, a connection to the main transmission medium.

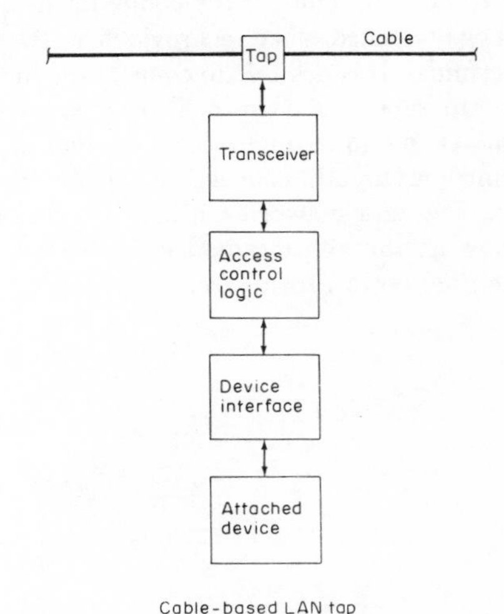

Cable-based LAN tap

tap

2. Trace Analysis Program (IBM's SNA).

TARA Threshold Analysis and Remote Access (IBM's SNA).

tariff The published schedule of rates for specific equipments, facilities, or services offered by a common carrier; also, the vehicle by which regulatory agencies approve the rates. Thus, a contract between the customer and the common carrier.

Tariff 9 A tariff which enables AT&T to offer C and D conditioning on leased lines.

Tariff 12 A tariff which enables AT&T to offer custom designed networks that incorporate voice and/or data services at a fixed price for large users.

Tariff 15 A tariff which provided AT&T with the ability to lower its switched network rates selectively. This tariff enables AT&T to negotiate rates with large customer accounts.

TASI Time Assignment Speech Interpolation.

task A sequence of instructions that has a starting point, an ending point, and performs some identifiable function.

task group A named set of one or more tasks that has a common set of resources.

TAT-8 The eighth transatlantic telephone cable and the first to use single-mode fiber optic technology.

TAT-9 A transatlantic fiber cable scheduled to begin operations in 1991 that will link the United States and Canada to the United Kingdom, France and Spain.

T-BERd Trademark of Telecommunications Techniques Corporation. A hand-held T1 BERT tester.

T30bis A new specification drafted by the CCITT to add error-correction techniques to FAX transmission. Under the proposed standard images to be faxed are broken down into data blocks of 64 or 256 bytes. Each block is further divided into smaller frames with a cyclic redundancy check (CRC) character added to each frame.

TC 1. Terminal Controller. 2. Transmission Control.

T-1C An American Telephone & Telegraph Company digital facility using the DS-1C format and operating at a speed of 3.152 Mbps.

TCAM Telecommunications Access Method.

T-carrier A time division multiplexed digital transmission facility operating at an aggregate rate of 1.544 Mbps.

TCAS Terminal Control Address Space (IBM's SNA).

TCC Technical Control Center.

TCM Time-Compression Multiplexing.

TCO Telenet Central Office.

TCP/IP Transmission Control Protocol/Internet Protocol.

TCU Transmission Control Unit.

TD 1. Transmitted Data. 2. Transmitter-Distributor.

TDB Terminal Descriptor Block.

TDM Time-Division Multiplexer or Time Division Multiplexing.

T1DM T1 Data Multiplexer.

TDMA Time Division Multiple Access.

TDR Time Domain Reflectometer.

TDT2 Telenet Diagnostic Tool 2.

TE Terminal Equipment.

TE-820 Trademark of Tekelec, Inc. A T1 framing tester, designed to test Central Office trunks.

Technical and Office Protocols (TOP) A Boeing Corporation version of the Manufacturing Automation Protocol (MAP) aimed at office and engineering applications.

Technical Control Center (TCC) A centralized electronic system from which trained personnel monitor, test, and control the operation of a large data communication network.

Tekelec, Inc. A Calabasas, CA, manufacturer of data communications test equipment. Manufacturer of the Chameleon 32 and the TE-820. A partly owned subsidiary of France's Tekelec-Airtronic.

Ttel Plus A series of key telephone systems manufactured by Siemens Information Systems which permits modem pooling, continuation of service during power failures, and the use of cartridge-type program modules for software enhancements.

tel set Telephone Set.

Telaction Corporation A subsidiary of J. C. Penny which markets a videotex service that was co-developed with Cableshare of London, ON, Canada. Telaction combines telephone and cable in an interactive service that users control with ordinary tone-dial telephones.

telco A general term for telephone common carrier or for a telephone company Central Office.

Telebit Corporation A company located in Cupertino, CA, which specializes in the manufacture of high-speed modems under the Trailblazer trademark that are designed for operation on the switched telephone network.

Telebox The electronic mail service offered by Deutsche Bundespost in West Germany.

telecommunication line Any physical medium such as a wire or microwave beam, that is used to transmit data. Synonymous with transmission line.

telecommunications A term encompassing the transmission or reception of signals, images, sounds, or information by wire, radio, optic, or infrared media.

Telecommunications Access Method (TCAM) IBM teleprocessing access methods that controls the transfer of messages between the application program and the remote terminals and provides the high-level message control language. TCAM macro instructions can be used to construct a message control program that controls messages between remote stations and application programs.

telecommunity Society in which the technologies of the information age allow anyone, anywhere, at any time, to send or receive any kind of information without technical barriers.

Telecoms Integrated Management System (TIMS) A Swedish Telecom-designed system which provides computerized assistance for handling subscriber requests for service, trouble handling, resource planning, line and telephone number assignment, and directory service.

teleconferencing The process of conferring between persons in separate geographic areas by using telephonic means. Also refers to a type of communications (e.g., electronic mail) conducted by computers.

telecopier Facsimile machine.

Telefon Treff A pilot program between Neumann Electronik in Mulheim and the German Bundespost that offers to set up conference calls involving as many as eight participants.

Telefonica Spain's national communications carrier.

telegraph A system employing the interruption of, or change in, the polarity of dc current signaling to convey coded informaton.

telegraph distortion Distortion which alters the duration of signal elements.

telegraphy Data transmission technique characterized by data rate of 75 bps where the direction, or polarity, of dc current flow is reversed to indicate bit states.

Telemail The Telenet computer-based electronic messaging service.

Telemail Local Community (TM-LC) A group of subsystems responsible for accessing Telemail and delivering messages to local users.

Telemail Message Transfer Agent (TM-MTA) A

subsystem that handles connections with other Telemail systems. Also referred to as the Interconnect MTA.

telemarketing A marketing system which combines telecommunications technology with management information systems for planned, controlled sales and service programs. Used effectively by small and large businesses to accomplish specific marketing goals.

telemessage The British Telecom International modern "telegram" service that operates in the U.K. and the U.S.

telemetry Transmission of coded analog data, often real-time parameters, from a remote site.

Telemetry Asynchronous Block Serial Protocol (TABS) AT&T's proprietary protocol for the Link Data Channel (LDC) on Extended Superframe Format (ESF) circuits. Used to convey statistical information on signal quality and hardware failure rates (retrieves ESF data from customer service units (CSUs)).

Telenet Value added network service proved by GTE Telenet Corporation.

Telenet Access Management System (TAMS) A network-level security system that screens virtual calls being placed in the network.

Telenet Central Office (TCO) A location where Telenet network equipment is installed. There are four "classes" or levels of TCOs. Class I TCOs are hubs that are part of the network backbone. Class II TCOs are those TCOs that must, according to the rules of network architecture, be connected to Class I TCOs. Class III TCOs are large enough to contain several pieces of equipment; however, these TCOs in most cases must be connected to a Class II TCO to be connected to the network. Class IV TCOs are small, asynchronous-only locations having only a single equipment cabinet.

Telenet Central Office, Class I (Class I TCO) The backbone hub site that connects to other Class I sites. This site also contains Class II and Class III types of equipment and services.

Telenet Central Office, Class II (Class II TCO) An asynchronous/synchronous office. It connects to Class I and other Class II TCOs. It services Class III TCOs by providing access to the backbone.

Telenet Central Office, Class III (Class III TCO) An asynchronous site only. It connects to a Class II TCO for access to the backbone.

Telenet Central Office, Class IV (Class IV TCO) An asynchronous site only. This is a single-cabinet site that replaces the normal Foreign Exchange (FX) service when FX costs or services prohibit the use of the FX services.

Telenet Diagnostic Tool 2 (TDT2) A program used in the Network Control Center (NCC) to remotely diagnose and correct problems in Telenet Processor 3000s (TP3s) and Telenet Processor 4000s (TP4s).

Telenet Internal Network Protocol (TINP) A virtual circuit-based, proprietary, backbone, network protocol. TINP is a superset of the CCITT X.75 gateway protocol.

Telenet Processor (TP) A data communications processor that interfaces terminals and host computers to the Telenet network. Generally, no changes to the customer's software or hardware are needed. The TP is available in different models designed to support various user requirements. The TP may be used as a terminal concentrator. Ports are provided by the TP to allow customer terminals access to the network. These access ports may be directly cable-connected to nearby customer or authorized user terminals, or they may be connected by leased or dial-in communications channels to distant terminals. TP4s are used as packet switches within the network to set up and manage virtual calls.

Telenet Processor, 3000-Series (TP3) A series of microprocessor-based concentrators in a Telenet Public Data Network (PDN). The TP3 is used for data transmission among hosts, terminals, and other TP devices in the network.

Telenet Processor, 4000-Series (TP4) A series of multi microprocessor-based packet switches and concentrators used for data transmission among hosts, terminals, and other TP devices in the network.

Telenet Processor, 5000-Series (TP5) A prime minicomputer-based Network Management System (NMS) located in the Network Control Center (NCC) and connected to a Telenet network as the X.25-interface host processor. The TP5 is used to monitor and control network facilities and usage.

Telenet Processor Operating System (TPOS) TPOS controls process scheduling, buffer allocation, and intercard communications. The software also contains a set of subroutines for queue management and provides debug port and buffer management. Each card on a Telenet Processor 4000 (TP4) has its own TPOS.

Telenet Processor Reporting Facility (TRPF) A software program in the Network Control Center (NCC) that receives messages regarding alarms or events sent by TPs. If the messages indicate problems, the Telenet Diagnostic Tool 2 (TDT2) program is used to rectify them.

telephone A component of a voice transmission system which converts voice power into electrical power by the use of a diaphram which makes the resistance of carbon granules vary at the same frequency as the sound wave. The change in resistance is used to vary the current flow on a transmission line.

telephone transmitter The portion of a telephone which converts sound waves into electric current which varies by waveform and frequency to changes in the sound waves.

telephony A general term for voice telecommunications.

Telephony User Part (TUP) The higher layer protocol in Common Channel Signalling System No. 7 that deals with end user signalling for voice telephony.

Teleplan An AT&T sponsored marketing agreement with hotels located outside of the U.S. which reduces the surcharge on calls to the US.

teleport A telecommunications service wholesaler, usually involved in local loop bypass, who serves a particular geographical region with specific classes of service, e.g., satellite access.

teleprinter A terminal without a CRT that consists of a keyboard and a printer.

Teleprinter Exchange Service (TELEX) A network of teleprinters connected over an international public switched network. Uses Baudot code.

teleprocessing A form of information handling in which a data processing system utilizes communication facilities. (Originally, but no longer, an IBM trademark.) Synonymous with data communications.

Teleprocessing Access Method (TPAM) Access method using a monitor specially made to interface teleprocessing devices. Examples include BTAM, TCAM, and VTAM.

teleset A trademark of Aspect Telecommunications of San Jose, CA, as well as a proprietary digital telephone designed by that company.

Teletel The French videotext service.

Teletex The new CCITT standard for text and message communications which is intended to replace ASCII Telex. Teletex operates at a high speed (2400 bps), can accommodate upper-case and lowercase characters, and has a well-defined format for transmission and presentation of text.

teletext A system for the one-way transmission of graphics and text for display on subscriber television sets. More limited than two-way videotex.

teletraining Using the interactive audio and/or graphics and video capability of the telephone network to provide remote interactive training. Businesses use teletraining instead of sending employees to a training site. Similarly, schools and universities use teletraining to allow handicapped or seriously ill students to participate in classes.

Teletype Trademark of Teletype Corporation. Commonly used to refer to one of their series of teleprinters or to compatible devices manufactured by other vendors.

teletype grade The lowest type of communications circuit, in terms of speed, cost, and accuracy. The term is used to establish a distinction between this type of service and voice-grade service. Teletype grade is also called teleprinter grade.

teletypewriter A generic term for a start–stop signalling device that consists of a keyboard transmitter and printing receiver.

Teletypewriter Exchange Service (TWX) An AT&T public-switched teletypewriter service in which suitably arranged teletypewriter stations are provided with lines to a Central Office for access to other such stations throughout the United States and Canada. Both Baudot and ASCII coded machines are used. Business machines may also be used, with certain restrictions.

Televerket Sweden's national communications carrier (PT&T).

telex The public-switched low-speed (telegraph) data network which is used worldwide for the transmission of administrative messages. Uses Baudot code, however, numerous code conversion facilities are available for sending data on the Telex network.

Telex network Same as Telex.

Telex Plus A telex service offered by British Telecom which allows subscribers to send a message to up to 100 addresses from a single call from their machine.

Telpak The name given to a now obsolete pricing arrangement by AT&T in which many voice-grade telephone lines were leased as a group between two points.

template In Digital Equipment Corporation Network Architecture (DECnet), a named collection of module-specific parameters which can be referenced by a client of the module without knowing their individual significance.

temporary error In IBM's NPDA, the resource failure that can be resolved by error recovery programs. Synonymous with performance error.

Temporary Text Delay (TTD) The TTD control sequence (STX ENQ) transmitted by a sending station when it wants to retain the line but is not ready to transmit.

ter Appended to a CCITT standard, it identifies a third version of the standard.

terminal A device for sending and/or receiving data on a communication channel. A wide variety of terminal devices have been built, including teleprinters, special keyboards, light displays, cathode tubes, personal computers, telephones, etc.

terminal access facility In IBM's NCCF, a facility that allows network operators to control a number of subsystems. In a full-screen or operator control session, operators can control any combination of such subsystems simultaneously.

terminal adapter A key element in Integrated Services Digital Network (ISDN) which permits non-ISDN terminals to be connected to an ISDN network.

terminal cluster A group of terminals, usually geographically co-located and controlled by a single unit (a cluster controller).

terminal component A separately addressable part of a terminal that performs an input or output function, such as the display component of a keyboard–display device or a printer component of a keyboard–printer device.

Terminal Control Address Space (TCAS) In IBM's SNA, the part of TSO/VTAM that provides logon services for TSO/VTAM users.

terminal control unit Same as cluster control unit.

Terminal Descriptor Block (TDB) A list of the port parameters residing in software that describes the terminal with which the computer communicates.

terminal emulation In protocol testing, a technique in which the protocol analyzer terminates a circuit and performs the role of that circuit's normal terminal device.

terminal emulator A program enabling a personal computer to imitate another computer system and execute the programs written for that other system as though they were written for the personal computer.

terminal handler A part of a data communication network which serves simple, character stream terminals. It has other names such as terminal processor, terminal interface processor (TIP) in the ARPA network and packet assembler and disassembler (PAD) in public packet networks.

terminal node In IBM's SNA, a peripheral node that is not user-programmable, having less intelligence and processing capability than a cluster controller node.

terminal polling Same as polling.

terminal processor In a packet switching network it is convenient to treat terminals like other processors needing communication. To this end, each terminal has a process looking after it. This can be regarded as residing in a terminal processor (a separate processor is not essential—it could be part of an interface computer, for example).

terminal server In LAN technology, a device that allows one or more terminals or other devices to connect to an Ethernet system.

terminate In IBM's SNA, a request unit that is sent by an LU to its SSCP to cause the SSCP to start a procedure to end one or more designated LU–LU sessions.

terminated line A circuit with resistance at the far end equal to the characteristic impedance of the line so no reflections or standing waves are present when a signal is entered at the near end.

termination Placement of a connector on a cable.

ternary Having three possible values. There are ternary number representations using, for example, the digits 0, 1, 2 and there are ternary signals which nominally take three possible values, for example +1 volt, 0, −1 volt.

terrestrial Pertaining to long-distance transmission that uses earth-bound transmission facilities as opposed to satellite transmission.

test center A facility to detect and diagnose faults and problems with communications lines and equipment. Also called network control center.

Test Interface Module (TIM) The component of Atlantic Research Corporation's Interview 7000 Series protocol analyzers that provides the physical interface to the circuit under test. Test interface modules are available for RS-232-C, V.24, and T1. The T1 TIM provides physical access to ISDN primary rate circuits.

test mode A condition of a modem or DSU in which its transmitter and receiver are inoperative because of a test in progress on the line.

text The part of a message between a start of text sequence and an end of text sequence and includes the data to be processed.

Text Direct A service offered by British Telecom which enables customers to send and receive Telexes without a dedicated Telex terminal.

text editor Program to facilitate user preparation of text at a terminal.

Text Search Service A service of Dow Jones News/Retrieval Service which provides on-line full-text coverage of the *Wall Street Journal*, the *Washington Post*, and several other publications.

TG Transmission Group (IBM's SNA).

TGID Transmission Group IDentifier (IBM's SNA).

TH Transmission Header.

thermal noise A type of electromagnetic noise in conductors or in electronic circuitry which is proportional to temperature. Also called Gaussian noise.

thermal printer A printer that uses heated wires to melt ink on a ribbon and deposit it on paper to form characters and graphics.

Thick Ethernet Standard Ethernet cabling 0.5 inches in diameter that is yellow coated and considerably heavier than Thin Ethernet.

Thin Ethernet A lighter, black-coated variation of Ethernet cable that is 0.2 inches in diameter. This cable is specified under the IEE 802.3 standard and is used to save cable and installation costs but is restricted in effective distance. Also known informally as "Cheapernet."

thin-film waveguide A film of transparent material that provides a path about one ten-thousandth of an inch thick for light waves. Some of the uses for integrated optical circuits include transmission, switching, and filtering.

thrashing Unfortunate state in which overhead is consuming so much of a processor's attention that no useful work is being accomplished.

threshold In IBM's NPDA, refers to a percentage value set for a resource and compared to a calculated error-to-traffic ratio.

Threshold Analysis and Remote Access (TARA) In IBM's SNA, the NPDA feature that can notify a central operator about network problems and errors. It provides remote control of IBM 3600 and 4700 controllers and can record, analyze, and display performance and status data on IBM 3600 and 4700 Finance Communications Systems.

through channels The term given to channels that are routed through a multiplexer, i.e., from an aggregate to another aggregate without being demultiplexed.

throughput A measurement of processing or handling ability which measures the amount of data accepted as input and processed as output by a device, link, network, or a system.

throughput delay The length of time required to accept input and transmit it as output.

TICC Terminal-Initiated asynchronous Channel Configuration.

TIE Time Internal Error.

tie line A private-line communications channel of the type provided by communications common carriers for linking two or more points together.

tie line/tie trunk A private line communications channel provided to link two or more switchboards or PBX systems. Interconnection may be on a manual or dial access basis. A configuration whereby two or more tie trunks are connected together is known as a tandem tie trunk network.

tightly coupled Pertaining to the interrelationship of processing units that share real storage and are controlled by the same control program.

TIM Test Interface Module.

timbre of sound A term used to express the quality of a particular sound which helps to identify the object, instrument, or person that is its source.

Time Assignment Speech Interpolation (TASI) A technique for making trunk circuits more efficient by combining portions of conversations on the same circuit. This technique makes use of the fact that typical conversations have quiet periods during which the circuit can be used for other conversations.

time call A call between two subscribers, where the called subscriber is within the caller's area code but outside the caller's free calling area.

Time Division Multiplexers (TDMs) 100% digital, dividing high speed digital channels, such as a modem RS-232C output, into digital subchannels. TDMs either bit interleave or character interleave data.

Bit interleaving is used primarily for synchronous multiplexing of protocols (Bisync, HDLC, X.25, SDLC, etc.). Bit interleaving maintains the order and number of bits from input at one end of the channel to output at the other end. Synchronous protocols require maintaining the number and order of bits in each clock to ensure correct calculation of the block check characters at the receiving ends, and to ensure that the receiving end divides the receiving data into correct 8-bit bytes.

Character interleaving is used primarily for asynchronous data. Asynchronous data is packaged in start bits (0s) and stop bits (1s). This makes it easy to distinguish each character and place it in a slot. Character interleaved multiplexers strip start and stop bits, achieving about 20 percent increased efficiency. The extra space is used for more channels or to pass RS-232C control signals. For instance, the RS-232C Request to Send control may be passed to the other end as Data Carrier Detect, this simulates the switched carrier modem function for one or more low speed channels.

Time Domain Reflectometer (TDR) A test device used to place a pulse of energy on a cable and connect the cable to a display where an electron beam moves from left to right as a function of time. By measuring the position of the beam deflection, one can determine the location of impedance discontinuity which indicates a cable fault.

Time Interval Error (TIE) A measure of the phase variation of a given signal with respect to an ideal timing source over a defined observation interval. The TIE provides an indication of the magnitude and direction of a signal's drift over a defined time interval.

time of day reconfiguration The ability of an unattended multiplexer to automatically install another configuration in place of an existing configuration at a predetermined time, either daily or on a specific date. This feature is often employed to optimize the multiplexer bandwidth use. For example, a daytime configuration that consists primarily of voice communication being replaced by a configuration that allows a high-speed CPU to CPU transfer at night.

time sharing The sharing of an equipment between several processes by giving the processes access to the equipment in turn, i.e., sharing out its time.

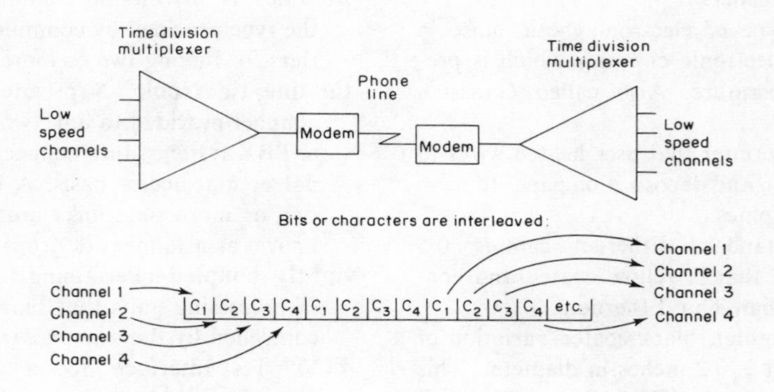

Time Division Multiplexers (TDMs)

Usually applied to processor time, but TDM is also a form of time sharing.

Time Sharing Option (TSO) An IBM Host application.

Time Sharing Option for VTAM (TSO/VTAM) An optional configuration of the operating system that provides conversational time sharing from remote stations in a network using VTAM.

time slot 1. In LAN technology, an assigned period of time or an assigned position in a sequence. 2. In multiplexer technology, the reserved time in the data stream for a specific device connected to a time division multiplexer.

Time-Compression Multiplexing (TCM) A digital transmission technique that permits full-duplex data transmission by sending compressed bursts of data in an alternating or "ping-pong" fashion.

time-derived channel Any of the channels obtained from multiplexing a channel by time division.

Time-Division Multiple Access (TDMA) In LAN technology, a high-speed, burst mode of operation that can be used to interconnect LANs. First used as a multiplexing technique on shared communications satellites where several earth stations have use of the total transponder bandwidth, with each station in sequence transmitting in short bursts.

Time-Division Multiplexer (TDM) A multiplexer which apportions the time available on its composite link between its channels, usually interleaving bits or bytes of data from successive channels.

Time-Division Multiplexing (TDM) A multiplexing method in which the time on the multiplexed channel is allocated at different times to different constituent channels. The allocation may be repeated regularly (fixed cycle) or may be made according to demand (dynamic).

time-division switching Switching method for a TDM channel requiring the shifting of data from one slot to another in the TDM frame. The slot in question may carry a bit or byte (or, in principle, any other unit of data).

timeout 1. The expiration of a predefined interval which then triggers some action, such as a disconnection that occurs following 30 seconds without any data activity (in a 30-second, no-activity timeout). 2. The length or existence of such an interval.

timesharing A method of computer operation that allows several interactive terminals to use a computer and its facilities; although the terminals are actually served in sequence, the high speed of the computer makes it appear as if all terminals were being served simultaneously.

TIMS 1. Telecoms Integrated Management System. 2. Transmission Impairment Measuring Set.

TINP Telenet Internal Network Protocol.

T-Interface In ISDN, the four-wire physical interface at the T-reference point, between NT1 and NT2. In configurations where NT1 and NT2 are parts of the same physical device, the T-reference point resides inside the hardware, and there is no T-interface. According to current ISDN standards, the T- and S-interfaces are physically and logically identical. This interface can only be about 1 kilometer long.

TIP Terminal Interface Processor.

tip and ring Names for the two conductors in a conventional two-wire local loop.

TLP Transmission Level Point.

TM-LC TeleMail Local Community.

TMS Transmission Message Unit.

TNC A threaded connector for miniature coax. TNC is said to be short for threaded-Neill-Concelman.

TNS Transaction Network Service.

toggle Activation or deactivation of a function or mode.

token In LAN technology, a packet (or part of a packet) used in explicit access LANs. The station that "owns" the token is the station that controls the transmission medium.

token bus, token-passing bus In LAN technology, a bus topology LAN that uses a token for explicit access.

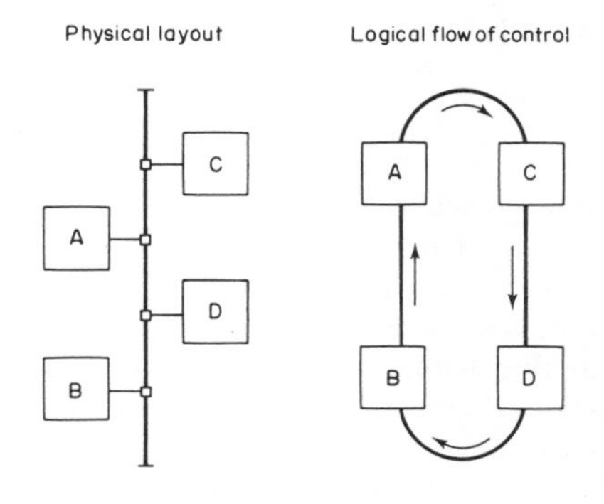

token bus, token-passing bus

token passing A method of controlling traffic on local area networks. A special message called a "token," or control packet is transmitted from node to node. When a node receives the token, it can transmit messages before transmitting the token on the next node.

token ring A local area network topology in which a control packet or token is passed from station to station in sequential order. Stations wishing access must wait for the token before transmitting data. In a token ring, the next logical station receiving the token is also the next physical station, as opposed to token bus.

token-ring repeater A device that extends the allowable distance between IBM Multistation Access Units on an IBM Token-Ring Network wired with the IBM Cabling System data grade media.

toll cable A cable whose outermost covering is made of lead. In general, a toll cable is used in permanent installations for long-distance transmission and may be either strung overhead on poles or installed underground.

toll center A Class 4 Central Office where channels and toll message circuits terminate. While this is usually one particular Central Office in a city, larger cities may have several offices where toll message circuits terminate. Also called "toll office" and "toll point."

toll circuit 1. (*American*) Same as trunk circuit. 2. (*British*) A circuit connecting two exchanges in different localities.

toll office A switching office where trunks are interconnected to serve toll calls. Toll offices are in a hierarchical structure as follows:

Regional center	Class 1
Sectional center	Class 2
Primary center	Class 3
Toll center	Class 4

toll service A telephone service across Central Office boundaries, or otherwise as provided in a relevant tariff, such that call duration and distance determine the cost billed to the subscriber.

toll switching trunk (*American*) A line connecting a trunk exchange to a local exchange and permitting a trunk operator to call a subscriber to establish a trunk call. Same as trunk junction (*British*).

toll-free Pertaining to any of a variety of services which use 800 in place of an area code. Calls made to 800 numbers are toll-free to the caller and are paid for by the receiver, usually at a lesser cost than normal long distance rates or collect calls. Most companies who wish to encourage customers to call them to place orders, to inquire for further information, or other customer services, use 800 services.

toll-free directory assistance In the U.S. you can dial 1-800-555-1212 to obtain the telephone number of organizations providing toll-free telephone service.

tone dialing Pushbutton dialing.

toner A black, powdered ink used in laser printers and cetain types of duplication equipment.

top 1. A network design tool of DMW Commercial Systems of Ann Arbor, MI. 2. Technical and Office Protocols.

topology The shape of the arrangement of network components and of the interconnections between them. Common network topologies include linear bus, multiple bus, a circular ring, and a star.

topology of networks The layout of the nodes (switches, concentrators) and lines of a network. The strict meaning refers to their pattern of connection but the word is used to include distance and geography.

torn-tape switching center A location where operators tear off incoming punched paper tape and transfer it manually to a tape reader connected to the proper outgoing circuit.

touch tone A trademark of AT&T, referring to tone or pushbutton, rather than pulse or rotary, dialing.

touch-call Proprietary term of GT&E used to refer to pushbutton dialing.

tower A protocol sequence along with associated address and protocol-specific information.

TP Telenet Processor.

TPAM TeleProcessing Access Method.

T-Pause The data flow control mechanism used by Tandem Non-Stop Computer Systems designed to eliminate the chance of data loss due to buffer overflow.

TPDU Transport layer Protocol Data Unit.

TPOS Telenet Processor Operating System.

TPRF Telenet Processor Reporting Facility.

Trace Analysis Program (TAP) In IBM's SNA, an SSP program service aid that assists in analyzing trace data produced by VTAM, TCAM, and NCP and provides network data traffic and network error reports.

trace packet A special kind of packet in the ARPA network which functions as a normal packet but because its 'trace' bit is set causes a report of each stage of its progress to be sent to the network control center.

traffic The volume and intensity of transmitted and received messages over a communications facility.

traffic control A means to stop and restart data transmission without losing data.

traffic engineering The science of designing communications facilities to meet user requirements.

traffic matrix A matrix of which the (i,j) element contains the amount of traffic originated at node i and destined for node j. The unit of measurement could be calls, or packets per second, for example, depending on the kind of network.

Trailblazer A modem manufactured by Telebit Corporation that can operate at data rates up to 19.2 Kbps over the switched telephone network.

trailer/trace block Control information which is transmitted after the text of a message. It is used for tracing error events, timing, and recovering blocks after system failures.

train The adjusting to the conditions of a particular line segment by a receiving modem. These conditions include amplitude response, delay distortion, and timing recovery.

training The process in which a receiving modem achieves equalization with a transmitting modem.

training pattern The sequence of signals used in training.

training time The time that the modem with automatic equalizer uses to adjust its equalization parameters. Also called learning time.

transaction A message directed to an application program. In batch or remote job entry (RJE), a job or job step.

Transaction Network Service (TNS) A common-user switched network service.

transaction processing A real-time of data processing in which individual tasks or items of data (transactions) are processed as they occur—with no primary editing or sorting.

transaction set A standardized electronic message format used in Electronic Data Interchange (EDI) communications. Transaction sets specify formats for business documents to include purchase orders, invoices and bills of lading.

transceiver A generic term for a single device that combines the function of a TRANSmitter and a reCEIVER.

Transcode An early 6-bit transmission code used with an airline reservation system version of IBM's Binary Synchronous Communications protocol.

Transcom The French switched 64 Kbps service.

transducer A device that converts signals from one form to another.

transients Intermittent, short-duration signal impairments.

Transistor–Transistor Logic (TTL) A common set of electrical characteristics used between various levels of integrated circuits, and, occasionally, as the interface between terminals and transmission equipment (e.g., modems).

transit exchange European version of tandem exchange.

transit network The highest level of a switched network. The transit network is well provided with links between its switching centers so that it rarely needs intermediate switching.

Transit Network Identifier Code (TNIC) The code that identifies a transmitted country network in international calls. This code is used for accounting purposes.

transit switch A Telenet Processor 4000 (TP4) used for routing and switching functions over trunk lines using the Telenet Internal Network Protocol (TINP). A TP4 transit switch cannot be equipped with Packet Assembler/Disassembler (PAD) software and does not communicate with X.25/X.75 Data Terminal Equipment (DTEs).

translator A device that converts information from one system of representation into equivalent information in another system. In telephone equipment, it is the device that converts dialed digits into call routing information.

transmission The passage of information through a communications medium.

transmission block A sequence of continuous data characters or bytes transmitted as a unit, over which a coding procedure is usually applied for synchronization or error control purposes.

Transmission Control (TC) Category of control characters intended to control or facilitate transmission of information over telecommunication networks. Samples of TC characters are: acknowledgment (ACK), data link escape (DLE), enquiry (ENQ), end of transmission block (ETB), negative acknowledgement (NAK), start of header (SOH), start of text (STX), and synchronization (SYN).

transmission control character Any control character used to control or facilitate transmission of data between data terminal equipment. Synonymous with communication control character.

Transmission Control (TC) layer In IBM's SNA, the layer within a half-session that synchronizes and paces session-level data traffic, checks session sequence numbers of requests, and enciphers and deciphers end user data. Transmission control has two components: the connection point manager and session control.

Transmission Control Protocol/Internetwork Protocol (TCP/IP) A set of *de facto* networking standards commonly used over Ethernet wiring and X.25 networks. TCP/IP was originally developed by the U.S. government. TCP/IP functions at the 3rd and 4th layers of the OSI Model.

Transmission Control Unit (TCU) A control unit (such as an IBM 2703) whose operations are controlled solely by programmed instructions from the computing system to which the unit is attached. No program is stored or executed in the unit.

transmission facilities The equipment that a communications common carrier uses to provide a stated type of service. Some examples of the equipment are links, switching centers, and other devices.

Transmission Group (TG) In IBM's SNA, a group of links between adjacent subarea nodes, appearing as a single logical link for routing of messages. *Note*: A transmission group may consist of one or more SDLC links (parallel links) or of a single System/370 channel.

Transmission Group Identifier (TGID) In IBM's SNA, a set of three values, unique for each transmission group, consisting of the subarea addresses of the two adjacent nodes connected by the transmission group, and the transmission group number (1–255).

Transmission Header (TH) In IBM's SNA, control information, optionally followed by a basic information unit (BIU) or a BIU segment, that is created and used by path control to route message units and to control their flow within the network.

Transmission Level Point (TLP) The transmission level of any point in a transmission system is the ratio (in dB) of the power of a signal at that point to the power of the same signal at the reference point. In the direction of transmission, each end of the channel is said to be the 0 dB Transmission Level Point (0TLP). The 0TLP gives the maximum power applicable at this point. All other level points on the overall circuit are referenced to the 0TLP, they are identical in numbers throughout the system. For example, −13 dBm0 means that the test tone is 13 dB below the 0 reference when measured anywhere in the circuit. The 0TLP was at one time a point accessible to probes and measuring instruments, but is seldom so today. As a consequence of improving transmission, it is now normal to consider the outgoing side of the toll transmitting switch as −3 TLP. Signal magnitudes measured at this point are 3 dB lower than would be measured at the reference level point if such a measurement were possible.

transmission line Synonym for telecommunication line (IBM's SNA).

transmission loss Total loss encountered in transmission through a system.

transmission media Twisted-pair wire, fiber-optic cable, microwaves, radio, and broadband or baseband coaxial cable.

transmission priority In IBM's SNA, a rank assigned to a path information unit (PIU) that determines its precedence for being selected by the transmission group control component of path control for forwarding to the next subarea node of the route used by the PIU.

transmission protocol A set of rules for the exchange of data over a communications network.

Transmission Services (TS) profile In IBM's SNA, a specification in a session activation request (and optionally, in the responses) of transmission control (TC) protocols (such as session-level pacing and the usage of session-level requests) to be supported by a particular session. Each defined transmission services profile is identified by a number.

transmission speed The number of information elements sent per unit time, usually expressed as bits, characters, or words. Preferred expression is bits per second (bps).

Transmission Subsystem Component (TSC) In IBM's SNA, the component of VTAM that comprises the transmission control, path control, and data link control layers of SNA.

transmission window The wavelength at which a fiber lightguide is most transparent.

transmissive star A fiber-optic transmission system that allows a single input light signal to be transmitted on multiple output fibers. Used primarily in fiber-optic local networks.

transmit flow control A transmission procedure which controls the rate at which data may be transmitted from one terminal point so that it is equal to the rate at which it can be received by the remote terminal point.

Transmitted Data (TD) An RS-232 data signal (sent from DTE to DCE on pin 2).

transmitter A device which inserts data into a communications channel.

Transmitter-Distributor (TD) The device in a teletypewriter terminal which makes and breaks the line in timed sequence. Usage of the term can refer to a paper tape transmitter.

Transpac The French packet switched data network.

transparency The ability of a communications system to pass control signals or codes as data to the receiving unit, without affecting the communications system.

transparent A mode of transmission in which the transmission medium will not recognize control characters or initiate any control function.

transparent mode A transmission technique that places no restrictions on the format of user data.

transponder In satellite communications, a circuit that receives an up-link signal, translates it to another, higher, frequency, amplifies it, and then retransmits it as the down-link signal.

Transport Connection A virtual connection at the Transport layer between two Transport service users.

transport delay The amount of time it takes to carry information through a network. Its value depends on the propagation delay and the switching method used in the network.

Transport layer The fourth layer in the OSI model. Ensures error-free, end-to-end delivery.

transport protocol The basic level of protocol which is concerned with the transport of messages. The software which carried out this protocol was called a transport station in the Cyclades network and the term has been widely adopted.

transport service user A user of the service provided by the Transport layer. For example, Session Control.

transversal filter In this filter the input is passed through a delay network and delayed versions of the signal, through suitable attenuators, are added to generate the output. The attenuators must be able to invert the signal.

transverse parity check A type of parity error checking performed on a group of bits in a transverse direction for each frame.

tree A network topology, with only one route between any two network nodes. A network that resembles a branching tree, such as CATV networks.

tree network A complex form of bus network in which there are branches in the cable but only a single transmission path between any two stations.

tree topology A topology in which there is only one route between any two network nodes.

trellis coding A method of forward error correction used in some high-speed modems whereby each signal element (baud) is assigned a coded binary value to represent the element's phase and amplitude. Due to a convoluting coding scheme, the receiving modem can determine if the signal element was received in error and, if so, correct the error.

trellis encoding An advanced modulation technique which provides greater throughput and reliable transmission rates for speeds above 9600 bps. With trellis encoding, coding information is added to the traditional modulation scheme to provide a record of successive dependencies between transmitted signal points. By continuously looking backwards, comparing received data with newly presented information, trellis encoding provides a greater tolerance to noise for a given block error rate.

tributary A circuit path connecting one or more stations, terminals, or devices to a network backbone, or to a centralized switching system.

tributary station A non-control station in a multipoint configuration.

Trintex A joint venture of Sears, Roebuck and Co. and IBM for a videotex service. The name was changed to Prodigy on 1 June, 1988.

TR-TSY-000385 The Bellcore standard definition

for billing transmission interface.

true power Power in a circuit when the load is purely resistive. This condition occurs when the phase angle between current and voltage is zero, resulting in the product of voltage and current (power) becoming zero.

trunk A telephone circuit connecting two or more telephone company (or PTT) central offices. A trunk may be either a high-speed digital circuit or a wideband analog circuit, however, all trunks now being installed are digital, using DS1 or higher rates.

trunk circuit 1. (*British*) Same as toll circuit. 2. (*American*) A circuit connecting two exchanges in different localities. *Note*: In Britain, a trunk circuit is approximately 15 miles long or more. A circuit connecting two exchanges less than 15 miles apart is called a junction circuit.

trunk exchange A telephone office primarily for switching trunks.

trunk group A group of telephone circuits treated as a unit and connecting PBXs, Central Offices, or other switching devices.

trunk junction (*British*) A line connecting a trunk exchange to a local exchange and permitting a trunk operator to call a subscriber to establish a trunk call. Same as toll switching trunk (*American*).

trunking protocol The protocol, or rules of operation, that apply to the transmission, or trunking, of data across a digital facility.

TSC Transmission Subsystem Component (IBM's SNA).

T-span A telephone channel through which a T-carrier operates.

TSO Time Sharing Option.

TSO/VTAM Time Sharing Option for VTAM.

T-tap A passive line interface used for monitoring data flowing in a circuit.

TTD Temporary Text Delay.

TTL 1. Transistor–Transistor Logic. 2. Transmission Test Line.

TTY TeleTYpewriter.

TTY transmission Teletypewriter communications. Usually asynchronous ASCII data communications.

TUCC flat cable AT&T trademark for Telephone Under Carpet Cable.

tunable laser A laser that can be made to vary the frequency of its light.

TUP Telephony User Part.

Turbocom A data compression program marketed by Datran Corporation of La Crescenta, CA.

turn key A system sold by a vendor (often an OEM) which is self-contained and used intact by the customer. A computer system that can be used by an untrained person.

turnaround time The actual time required to reverse the direction of transmission from sender to receiver or vice versa when using a half-duplex circuit. Time is required for line propagation effects, modem timing and computer reaction.

twinaxial cable A shielded coaxial cable with two center conducting leads.

twisted pair Two insulated copper wires twisted together. The twists, or "lays," are varied in length to reduce the potential for signal interference between pairs. In cable greater than 25 pairs, the twisted pairs are grouped and bound together in a common cable sheath. Twisted pair cable is the most common type of transmission media.

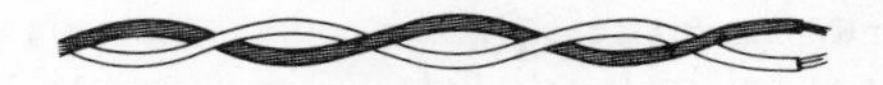

twisted pair

two-way alternate Synonym for half-duplex operation.

two-way loss The capability of measuring the loss of a circuit in both the transmit and receive directions.

two-way simultaneous Synonym for full-duplex operation.

two-wire circuit A circuit formed of two conductors, which are insulated from each other, that provide a "go" and "return" channel in the same frequency.

TWX TeletypeWriter eXchange service.

TxC Transmit clock, an interface timing signal that synchronizes the transfer of Transmit Data (TxD), provided by DCE.

TXD Transmitted Data.

Tymnet Value added network service provided by Tymnet Corporation (McAUTO).

Tymusa An enhanced gateway access service from

the Tymnet public data network which permits fast and easy access of U.S.-based host computers from foreign locations, without the administrative burden of obtaining an account from the local PTTs.

Tymview A software interface marketed by the Tymnet public data network which allows Tymnet network equipment to be managed by IBM's host based Netview network management system.

Tymvisa An enhanced gateway access service marketed by the Tymnet public data network which provides easy access to host computers in foreign countries without requiring the use of long number sequences.

TYM-X.25 Access A Tymnet dial-up service that provides users with the benefits of end-to-end synchronous communication, without the cost of expensive leased lines. Error detection is performed throughout the entire session using the X.25 protocol, and 2400 bps data transmission speeds cut connect-time charges. Personal computers, minicomputers, and Packet Assemblers/Disassemblers (PADs) are provided switched access to applications based on host computers connected to the TYMNET network.

Tymnet Asynchronous Outdial Service A service of Tymnet which provides the means to communicate with asynchronous devices not directly connected to the Tymnet network. Outdial can be used to send information from a central computer to remote asynchronous devices, from a PC to another PC, or from a terminal or PC to an asynchronous central computer. No customer-dedicated facilities are required. The Outdial service is provided through public ports in selected Outdial nodes throughout the domestic U.S.

Type A Coax In IBM 3270 systems, a serial transmission protocol operating at 2.35 Mbps which provides for the transfer of data between a 3274 control unit and attached display stations or printers.

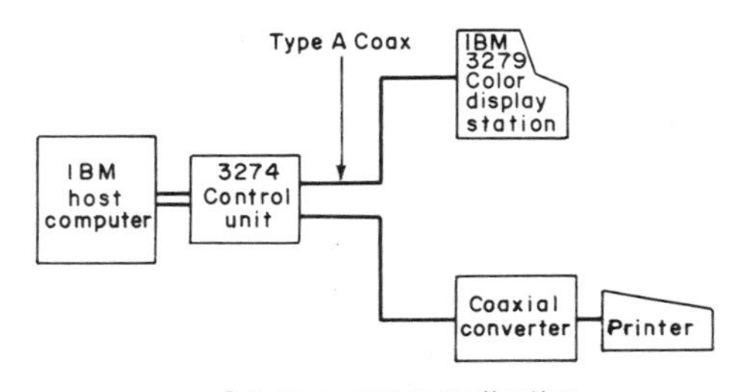

Type A Coax

Type 105 test line A test line in a telephone Central Office with the ability to allow access to a responding type unit.

T1DM (T1 Data Multiplexer) A multiplexer used for time division multiplexing for up to twenty-four 64 Kbps channels with synchronizing information into a DS1 line.

U

U-interface (ISDN) In the U.S. the U-interface is the two-wire physical interface between the NT1 and the provider's local loop. The U-interface exists only to satisfy legal requirements that the provider's network be managed separately from customer premises equipment.

UA 1. Unnumbered Acknowledgment (response). 2. User Agent.

UART Universal Asynchronous Receiver/Transmitter.

UDI Unrestricted Digital Information.

UDLC Univac Data Link Control.

UDR User Destination Routing.

UFD User File Directory.

UHF Ultra High Frequency. Ranges from 300 MHz to about 3 GHz. Includes television channels 14 through 83, and cellular radio frequencies.

UI frame An unnumbered information frame in HDLC used to carry data which is not subject to flow control or error recovery.

UIS Universal Information Services.

UL Underwriters Laboratory.

ULSI Ultra Large-Scale Integration.

Ultra Large-Scale Integration (ULSI) A term used to describe a multifunction semiconductor device with an ultra-high density (over 10 000 circuits) of electronic circuitry contained on a single silicon chip.

ultrasonics Frequencies above the audible range, normally 20 000 Hz or above.

UNA Universitats-Netz Austria.

unattended file transfer A feature of a communications program which permits the transmission and reception of messages on an unattended basis.

unattended messaging The ability to preprogram transmissions of electronic messages, voice mail, or file transfers.

unattended mode A term that describes the operation of a device, such as an auto-answer modem, designed to operate without the manual intervention of an operator.

unattended operations The automatic features of a station's operation permit the transmission and reception of messages on an unattended basis.

unbalanced-to-ground (TWO-WIRE) The impedance-to-ground on one wire is measurably different from that of the other.

unbalanced line A transmission line in which the magnitudes of the voltages on the two conductors are not equal with respect to ground; for example, a coaxial line.

unbind In IBM's SNA, a request to deactivate a session between two logical units (LUs).

unbundling Separation of vendor provided services.

uncontrolled terminal A user terminal that is on line all the time and does not contain control logic for polling or calling.

undel message A message sent to the originator of an electronic message indicating that the message could not be delivered.

Underwriters Laboratories (UL) A private testing laboratory concerned with electrical and fire hazards of equipment.

unformatted In IBM's VTAM, pertaining to commands (such as LOGON or LOGOFF) entered by an end user and sent by a logical unit in character form. The character-coded command must be in the syntax defined in the user's unformatted system services definition table. Synonymous with character-coded.

Unformatted System Services (USS) In IBM's SNA products, a system services control point (SSCP) facility that translates a character-coded request, such as a logon or logoff request into a field-formatted request for processing by formatted system services and translates field-formatted replies and responses into character-coded requests for processing by a logical unit.

unframed BERT A reference pattern for Bit Error Rate Testing that uses the full bandwidth of the circuit under test without following framing conventions.

Unified Network Management Architecture (UNMA) An AT&T integrated network management system based upon the International Standards Organization's (ISO) evolving Open Systems Interconnection network management standards.

Uninet A common carrier offering a X.25 PDN that merged with Telenet.

Uninett Nordic University Network.

uninterpreted name In IBM's SNA, a character string that an SSCP is able to convert into the network name of an LU. *Note*: Typically, an uninterpreted name is used in a logon or initiate request from an SLU to identify the PLU with which the session is requested. The SSCP interprets the name into the network name of the PLU in order to set the session. When the PLU eventually sends a BIND to the SLU, the BIND contains the original uninterpreted name.

Uninterruptible Power System (UPS) Consists of a battery package and the necessary power conversion and inversion equipment, switch, gear, etc. The system carries the design load at all times. This system will switch to battery power upon commercial power loss. System will run for 10 to 15 minutes on battery power before system shut-down. If an Emergency Power System (EPS) is present, it will take over before UPS shuts down.

unipolar non-return to zero signaling A simple type of signaling used for early key telegraphy. Currently used with private line teletypewriter systems as well as representing the signal pattern used by RS-232 and V.24 interfaces. In this signal scheme, a dc current or voltage represents a mark, while the absence of current or voltage represents a space. When used with transmission systems, line sampling determines the presence or absence of current, which is then translated into an equivalent mark or space.

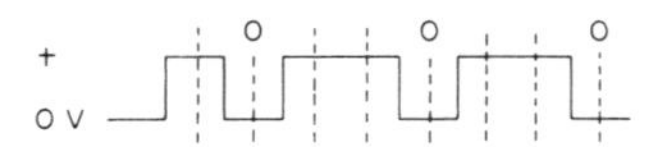

unipolar non-return to zero signaling

unipolar return to zero signaling A variation of unipolar non-return to zero signaling is unipolar return to zero. Here the signal always returns to zero after every "1" bit. While this signal is easier to sample, it requires more circuitry to implement and is not commonly used.

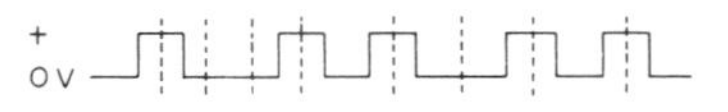

unipolar return to zero signaling

unipolar/bipolar signals A digital signal technique that uses a positive or negative excursion and ground as the two binary signal states for unipolar signals where bipolar has amplitude in both directions from the reference axis.

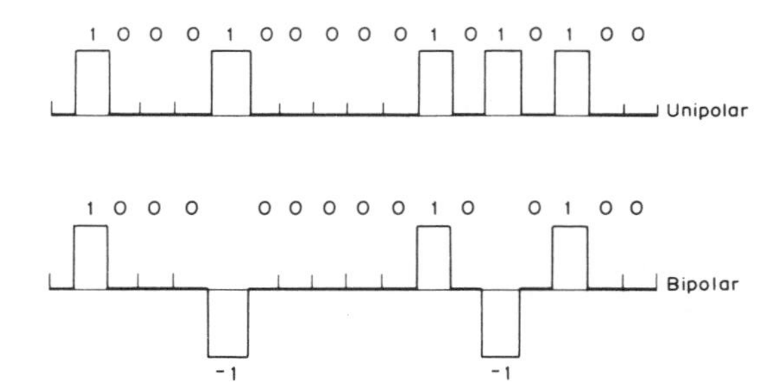

unipolar/bipolar signals

unit element A signal element of one unit element duration.

unit interval A unit interval is the duration of the shortest nominal signal element. The duration of the unit interval in seconds is the reciprocal of the telegraph speed expressed in baud.

United States Independent Telephone Association The association of telephone companies excluding AT&T and its former operating company subsidiaries.

Universal Asynchronous Receiver/Transmitter (USART) An integrated circuit fabricated as a chip which is designed to interface with asynchronous circuits on behalf of a central processing unit, such as a microprocessor. The USART performs basic functions such as character assembly/disassembly operations and the conversion of data in parallel form from the CPU into several forms for transmission. The USART combines the functions of a UART and a USRT.

Universal Character Set (UCS) A printer feature that permits the use of a variety of character arrays.

Universal Information Services (UIS) AT&T vision for the future that will permit users of telecommunications networks to receive any kind of voice, data, or image service in any combination, with a maximum of convenience and economy.

Universal Product Code (UPC) Bar code used to identify products and their manufacturers in inventory systems.

Universal Protocol Platform (UPP) Software from Excelan, Inc., of San Jose, CA, which separates hardware specific components from the underlying transport and network functions, permitting multiple protocol suites to simultaneously operate on the same hardware. This eases the migration from one network protocl set to another.

universal service The policy of providing basic telephone service at a cost which is affordable by the entire U.S. population.

Universal Synchronous Receiver/Transmitter (USRT) An integrated circuit designed to interface with synchronous circuits on behalf of a central processing unit, such as a microprocessor, and to perform basic functions such as character assembly/disassembly, and the conversion of parallel data into serial form for transmission over a synchronous data channel.

universally administrated address A permanently encoded address in a token-ring adapter card which is unique. This address is administered by the IEEE.

UNIX An operating system developed by AT&T Bell Laboratories. Originally UNIX was designed for communicating, multi-user, 32-bit minicomputers but now operates on many microcomputers and mainframe computers.

unloaded line A pair of wires running from one building to another, using no loading coils or switching equipment. A dc continuity line.

UNMA Unified Network Management Architecture.

unsolicited message In IBM's SNA, a message, from VTAM to a program operator, that is unrelated to any command entered by the program operator.

UPC Universal Product Code.

uplink The earth-to-satellite transmission channel used to transmit information to a geosynchronous satellite. Complement of downlink.

up-loop A fixed pattern (repeating 10000) which forces a receiving CSU to disconnect a signal loopback.

UPP Universal Protocol Platform.

UPS Uninterruptible Power Supply.

upstream 1. In the direction of data flow from the end user to the host. 2. The direction opposite to message flow in a token-ring network.

upstream device For the IBM 3710 Network Controller, a device located in a network such that the device is positioned between the 3710 and a host. A communication controller upstream from the 3710 is an example of an upstream device.

upstream line A telecommunication line attaching an IBM 3710 Network Controller to an upstream device.

uptime Uninterrupted period of time that networks or computer resources are accessible and available to a user.

upward compatible Application that can be configured to function in some vendor or protocol enhanced mode. As opposed to downward compatible.

US Access A subsidiary of Telephone Electronics Corporation which operates its own fiber optic network between Denver and Colorado Springs and is constructing digital microwave links in Mississippi and Tennessee.

USA Direct A service mark of AT&T, this service permits persons outside the United States to rapidly and economically make calls to the U.S. from certain foreign locations. Callers can user their AT&T card or call collect. USA Direct can be used from any telephone in certain foreign locations or from designated USA Direct phones which places you in contact with an AT&T operator in the U.S. who places your call. Dial access countries and their telephone numbers are listed in the following table.

Australia	0014-881-011
Belgium	11-0010
Br. Virgin Is	1 800 872-2881
Denmark	0430-0010
France	19-0011
Germany (FRG)	0130-0010

Hong Kong	008-1111
Japan	0039-111
Netherlands	06-022-9111
Sweden	020-795-611
United Kingdom	0800-89-0011

USART Universal Synchronous/Asynchronous Receiver/Transmitter.

USASCII (USA Standard Code for Information Interexchange) Same as ASCII.

USENET USEr's NETwork.

user Anyone who requires the services of a computing system.

User Agent (UA) A set of processes that are used to do message-related functions, such as composing, editing, and displaying messages. It provides the commands to submit messages to the Message Transfer System (MTS) and retrieves messages that have been delivered to the user's mailbox.

user correlator In IBM's SNA, a 4-byte value supplied to VTAM by an application program when certain macro instructions such as REQSESS are issued. It is returned to the application program when subsequent events occur (such as entry to a SCIP exit routine upon receipt of BIND) that result from the procedure started by the original macro instruction.

User Destination Routing (UDR) A term originally used by Codex Corporation for the addition of asynchronous port switching with contention and queueing to some of its statistical multiplexers.

user dictionary A directory that contains unique information pertaining to each registered entity (user, node, bulletin board, etc.) within the Telemail system.

user exit A point in an IBM-supplied program at which a user exit routine may be given control.

user exit queue In IBM's SNA, a structure built by VTAM that is used to serialize the execution of application program exit routines. Only one exit routine on each user exit queue can run at a time.

User File Directory (UFD) A specific work area on the Telenet Processor 5000 (TP5) system that contains files and from which software programs are operated.

user proprietary channel Allocated to the user for input of information for use in maintenance activities and remote alarms external to the span equipment.

user-application network In IBM's SNA, a configuration of data processing products, such as processors, controllers, and terminals, established and operated by users for the purpose of data processing or information exchange, which may use services offered by communication common carriers or telecommunication administrations.

USITA United States Independent Telephone Association.

USOC Universal Service Ordering Code.

USRT Universal Synchronous Receiver/Transmitter.

USS Unformatted System Services (IBM's SNA).

utilization The fraction of the available capacity that is being used—reference processing, memory, and communications facilities.

UTL Universal Test Line.

UTS Universal Telephone Service.

UUCP Unix to Unix CoPy.

V

V CCITT code designation for standards dealing mainly with modem operations.

VAC Value Added Carrier.

validation Checking data for correctness, or compliance with applicable standards, rules, and conventions.

validity check The method used to check the accuracy of data after it has been received.

validity of data The status of information being processed by hardware component or communications equipment.

value added carrier A carrier that supplies specialized services, such as computer-oriented services or facsimile transmission.

Value Added Network (VAN) A network source that "adds value" to the common carrier's network services by adding computer control of communications.

Value Added Network Service (VANS) A data transmission network which routes messages according to available paths, assures that the message will be received as it was sent, provides for user security, high speed transmission and conferencing among terminals.

VAN Value Added Network.

VANS Value Added Network Services.

VAR Value Added Reseller.

variable In IBM's NCCF, a character string beginning with "&" that is coded in a command list and is assigned a value during execution of the command list.

Variable Quantizing Level (VQL) A method of speech encoding that quantizes and encodes an analog voice conversation for transmission on a digital network. VQL usually encodes a voice conversation at 32 Kbps.

varistor A circuit component designed to suppress voltage surges. A varistor is a type of diode that conducts current only when voltage above a certain threshold value is present.

VAX The trademarked name for a family of computers manufactured by Digital Equipment Corporation.

VC Virtual Call.

VDM Voice Data Multiplexer.

VDT Video Display Terminal.

VDU Video Display Unit. Same as VDT.

vendor code A term used to indicate that software was written by the same company that manufactured the computer system.

vent safe Unique design which continues to provide surge protection even in the remote circumstance of gas venting from the device to the atmosphere.

verify 1. To determine whether a transcription of data or other operation has been accurately accomplished. 2. To check the results of keypunching.

vertical parity Vertical parity checks character integrity. It is decided that the one bits in each character will add up to an odd or even number. As the character is transmitted, the one bits are counted

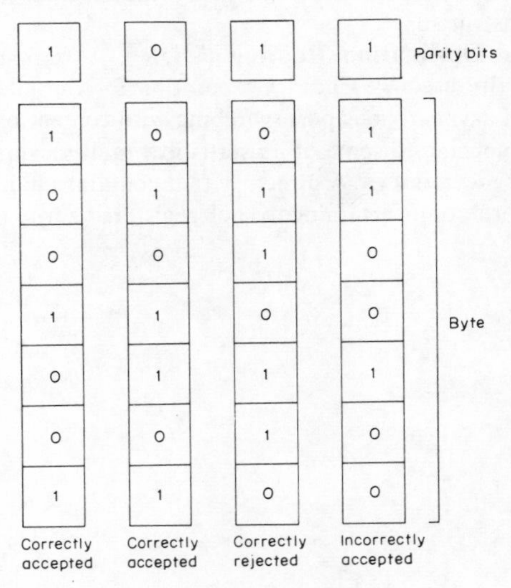

vertical parity

and if they do add up to the predetermined odd or even number, the data is considered to be error-free. This is not a foolproof system, however. If an even number of bits are destroyed, the character will be seen as correct even though it is not. The illustration below is an example of even vertical parity.

vertical parity (redundancy) check See parity check, vertical.

Vertical Redundancy Check (VRC) In ASCII-coded blocks, the parity test (even or odd) performed on each character in the block.

Vertical Tabulation (VT) A format effector which advances the active position to the same character position on the next predetermined line.

Very High Frequency (VHF) The portion of the electromagnetic spectrum with frequencies between approximately 30 and 300 MHz. This is the operating band for TV channels 2 through 13 and most FM radio.

Very Large-Scale Integration (VLSI) A term used to describe a multifunction semiconductor device with a very high density (up to 10 000 circuits) of electronic circuitry contained on a single silicon chip.

Very Low Frequency (VLF) That portion of the electromagnetic spectrum having continuous frequencies from approximately 3 to 30 kHz.

Very Severe Burst (VSB) An error condition on high-speed digital circuits. Defined as a period longer than 2.5 continuous seconds during which the Bit Error Rate exceeds 1 percent. A Very Severe Burst causes the receiving hardware to enter a Yellow Alarm Condition.

Very Small Aperture Terminal (VSAT) In satellite communications, a small diameter receiver station that normally operates in the KU frequency band.

VF Voice Frequency.

VFCT Voice Frequency Carrier Telegraph.

VHF Very High Frequency.

video 1. The portion of the frequency spectrum used for TV signals. 2. A signal with a bandwidth of approximately 5 MHz generated from TV scanning.

video attribute A description of the way video appears on a cathode ray tube display, such as reverse, blink, underline, low intensity, or a combination of the preceding.

video conferencing Two-way transmission of video between two or more locations. Requires a wide-band transmission facility, usually a satellite. Bandwidth ranges from 56 Kbps (freeze frame) to T1 (1.544 Mbps).

Video Display Terminal (VDT) Terminal providing user temporary display (not hard copy) of input and output.

video resolution The number of dots (pixels) that can be displayed on a video screen at one time. It is presented as the number of horizontal dots times the vertical dots (H * V).

Video Stream British Telecom's videoconferencing service. Both full-motion monochrome and color videoconferencing between customer's premises are offered.

Videotel Inc. A Houston-based company that has exclusive rights to the French Minitel technology in the United States. Videotel markets a modified version of the Minitel terminal manufactured by Alcatel.

Videotex An interactive communications system designed to allow users to access data bases using television sets or low-cost terminals.

viewdata The use of equipment based on teletext techniques to access data bases, through the telephone network, has been given the generic name of viewdata.

Viewtron An experimental videotex service marketed by Knight-Ridder.

V-interface (*ISDN*) The two-wire physical interface used for single-customer termination from a remote terminal.

VIP Visual Information Projection.

virtual A term generally used by vendors, implying infinite capacity, but permitting effective use.

virtual call Same as a virtual circuit.

virtual circuit In packet switching network facilities that give the appearance to the user of an actual end-to-end circuit established for the duration of a call, in contrast to a physical circuit. Also a variable network connection which enables transmission facilities to be shared by many virtual users. Also called a logical circuit.

virtual connection An apparent communications channel between two stations, whereby information or data transmitted by one station is automatically routed through the network via the most expeditious path to the other station. No long-haul circuit capacity is preassigned to a virtual connection; rather, capacity is made available only as data is transmitted by the stations.

virtual disk server In a local area network, a type of disk server which gives each user access to a section of the disk, called a volume. Sometimes referred to as disk server.

virtual file server In a local area network, a type of disk server which allows users to share files. Also referred to as file server.

Virtual Machine (VM) An IBM program product whose full name is the Virtual Machine/System Product. It is a software operating system controlling the execution of programs.

virtual machine facility An IBM system control program, essentially an operating system that controls the concurrent execution of multiple virtual machines on a single mainframe.

virtual memory The use of disk memory to hold portions of a computer's internal memory, permitting large programs to be segmented to operate in internal memory.

virtual private network A communications carrier provided service in which the public switched telephone network provides capabilities similar to those of private lines to include conditioning, error testing, and full-duplex, four-wire transmission.

Virtual Route (VR) In IBM's SNA, a logical connection (1) between two subarea nodes that is physically realized as a particular explicit route, or (2) that is contained wholly within a subarea node for intra-node sessions. A virtual route between distinct subarea nodes imposes a transmission priority on the underlying explicit route, provides flow control through virtual-route pacing, and provides data integrity through sequence numbering of path information units (PIUs).

Virtual Route Identifier (VRID) In IBM's SNA, a virtual route number and a transmission priority number that, when combined with the subarea addresses for the subareas at each end of a route, identify the virtual route.

virtual route pacing In IBM's SNA, a flow control technique used by the virtual route control component of path control at each end of a virtual route to control the rate at which path information units (PIUs) flow over the virtual route. VR pacing can be adjusted according to traffic congestion in any of the nodes along the route.

virtual route selection exit routine In IBM's VTAM, an optional installation exit routine that modifies the list of virtual routes associated with a particular class of service before a route is selected for a requested LU–LU session.

virtual route sequence number In IBM's SNA, a sequential identifier assigned by the virtual route control component of path control to each path information unit (PIU) that flows over a virtual route. It is stored in the transmission header of the PIU.

virtual storage An internal memory system which allows the computer to make efficient use of the available memory by drawing into the main memory only that portion of a program from a disk which is needed at a specific point of time. Also called virtual memory.

Virtual Storage Access Method (VSAM) In IBM's SNA, an access method for direct or sequential processing of fixed and variable-length records on direct access devices. The records in a VSAM data set or file can be organized in logical sequence by a key field (key sequence), in the physical sequence in which they are written on the data set or file (entry sequence), or by relative-record number.

Virtual Storage Extended (VSE) An IBM program product whose full name is the Virtual Storage Extended/Advanced Function. It is a software operating system controlling the execution of programs.

Virtual Telecommunications Access Method (VTAM) IBM teleprocessing access method that gives users at remote terminals access to applications programs. It also provides resource sharing, a technique for efficiently using a network to reduce transmission costs.

Virtual Telecommunications Access Method Entry (VTAME) An IBM program product that provides single-domain and multiple-domain network capability for 4300 systems using VSE. The set of programs control communication between terminals and application programs running under DOS/VS, OS/VS1 and OS/VS2.

virtual terminal A technique that allows a variety of terminals with different characteristics to be accommodated by the same network by converting to a standard network format.

virtual terminal protocol In the ISO Model, an Application layer protocol intended to facilitate the interconnection of specific terminal types with the more general constructs of the Model.

VIT VTAM Internal Trace.

VLF Very Low Frequency.

VLSI Very Large Scale Integration.

VM 1. Virtual Machine. 2. Virtual Machine Operating System (IBM's SNA).

VM/SP Virtual Machine/System Product. An IBM Operating System. Synonym for VM.

VOGAD Voice-Operated Gain-Adjusting Device.

voice annotated text A system in which a message can be delivered to a user's terminal and displayed on a screen so that certain portions of the message can be selected which then result in voice commentary being delivered through a speaker of telephone handset connected to the terminal.

voice compression A technique for reducing the bandwidth occupied by a voice telephone call. ADPCM and CVSD are voice-compression techniques.

voice/data PABX A device which combines the functions of a voice PABX and a data PABX, often with emphasis on the voice facilities.

voice digitization The conversion of analog voice signals into a digital form.

voice encoding The process of encoding an analog conversation into a digital data stream.

voice grade An access line suitable for voice, low-speed data, facsimile, or telegraph service. Generally, it has a frequency range of about 300–3000 Hz.

voice jack Most data modems in today's environment are installed using a voice jack. The common nomenclature for such a jack is RJ-11. The determination for the type of jack to be used rests with the manufacturer of the modem. Specifications for the jack will be determined by how the modem is registered with the Federal Communications Commission (FCC). Information concerning the installation and jack requirements will be found in the manufacturer's brochure which came with the modem. With a voice jack requirement, it is "required" that the modem transmits at a fixed level of −9.0 dBm.

voice mail A system generally associated with large PBX capabilities that allows callers to leave a voice message for the person called.

voice message service A service that permits a caller to send a one-way spoken message to a service user. The message is digitally stored in a message box.

voice PABX, voice-only PABX A PABX for voice circuits. A telephone exchange.

voice recognition The process by which a machine, such as a computer, accepts, decodes, and acts on human speech.

voice response The conversion of computer output into spoken words and phrases.

voice sounds Complex sounds which contain different sets of harmonics.

voice store-and-forward systems A system that enables a computer to accept a message and store it until a transmission path or receiver is available.

voice synthesis The process by which a machine creates human-sounding speech.

voiceband A bandwidth for audio transmission, generally 300 to 3300 Hz. Also, voice grade (VG), voice frequency (VF).

voice-frequency Any frequency within that part of the audio-frequency range essential for the transmission of speech of commercial quality, i.e., 300–3400 Hz.

voice-frequency carrier telegraphy That form of carrier telegraphy in which the carrier currents have frequencies such that the modulated currents may be transmitted over a voice-frequency telephone channel.

voice-frequency multichannel telegraphy Telegraphy using two or more carrier currents the frequencies of which are within the voice-frequency range. Voice-frequency telegraph systems permit the transmission of up to 24 channels over a single circuit by use of frequency-division multiplexing.

voice-grade channel, voice-grade line A channel or line that offers the minimum bandwidth suitable for voice frequencies, usually 300 to 3400 bps.

voice-operated device A device used on a telephone circuit to permit the presence of telephone currents to effect a desired control. Such a device is used in most echo suppressors.

Voice-Operated Gain-Adjusting Device (VOGAD) A device somewhat similar to a compandor and used on some radio systems. A voice-operated device which removes fluctuation from input speech and sends it out at a constant level. No restoring device is needed at the receiving end.

volatile Pertaining to a data storage device (memory)

that loses its contents when power is lost. Contrast with non-volatile.

volatile memory A storage medium that loses all data when power is removed.

volatile storage A storage device whose contents are lost when power fails.

voltage A measure of electrical potential expressed in units of volts and named after Count Alessandro Volta.

Volt-Ohm-Meter (VOM) A test device that is used to measure and display voltage, resistance and current.

volume A logical portion of a disk due to its partition. Each volume is treated like a separate hard disk by anyone using it.

VOM Volt-Ohm-Meter.

VQL Variable Quantizing Level.

VR Virtual Route.

VRC Vertical Redundancy Check.

VRID Virtual Route IDentifier.

VS Virtual Storage.

VSAM Virtual Storage Access Method.

VSAT Very Small Aperture Terminal.

VSB Very Severe Burst.

VSE Virtual Storage Extended operating system.

VSE/AF Virtual Storage Extended/Advanced Function operating system. Synonym for VSE.

V-Series A set of CCITT recommendations for the transmission of data over the public switched telephone network.

General

V.1 Equivalence between binary notation symbols and the significant conditions of a two-condition code.

V.2 Power levels for data transmission over telephone lines.

V.3 International Alphabet No. 5.

V.4 General structure of signals of International Alphabet No. 5 code for data transmission over public telephone networks.

V.5 Standardization of data signalling rates for synchronous data transmission in the general switched telephone network.

V.6 Standardization of data signalling rates for synchronous data transmission on leased telephone-type circuits.

V.7 Definitions of terms concerning data communication over the telephone network.

Interface and voice-band modems

V.10 Electrical characteristics for unbalanced double-current interchange circuits for general use with integrated circuit equipment in the field of data communications. Electrically similar to RS-423.

V.11 Electrical characteristics of balanced double-current interchange circuits for general use with integrated circuit equipment in the field of data communications. Electrically similar to RS-422.

V.15 Use of acoustic coupling for data transmission.

V.16 Medical analogue data transmission modems.

V.19 Modems for parallel data transmission using telephone signaling frequencies.

V.20 Parallel data transmission modems standardized for universal use in the general switched telephone network.

V.21 300 bps duplex modem standardized for use in the general switched telephone network. Similar to the Bell 103.

V.22 1200 bps duplex modem standardized for use on the general switched telephone network and on leased circuits. Similar to the Bell 212.

V.22 bis 2400 bps full-duplex two-wire.

V.23 600/1200 baud modem standardized for use in the general switched telephone network. Similar to the Bell 202.

V.24 List of definitions for interchange circuits between data terminal equipment and data circuit-terminating equipment. Similar to and operationally compatible with RS-232.

V.25 Automatic calling and/or answering equipment on the general switched telephone network, including disabling of echo suppressors on manually established calls. RS 366 parallel interface.

V.25 bis Serial RS-232 interface.

V.26 2400 bps modem standardized for use on four-wire leased telephone-type circuits. Similar to the Bell 201 B.

V.26 bis 2400/1200 bps modem standardized for use in the general switched telephone network. Similar to the Bell 201 C.

V.26 ter 2400 bps modem that uses echo cancellation techniques suitable for application in the general switched telephone network.

V.27 4800 bps modem with manual equalizer standardized for use on leased telephone-type circuits. Similar to the Bell 208 A.

V.27 bis 4800/2400 bps modem with automatic equalizer standardized for use on leased telephone-type circuits.

V.27 ter 4800/2400 bps modem standardized for use in general switched telephone telephone network.

Similar to the Bell 208 B.

V.28 Electrical characteristics for unbalanced double-current interchange circuits (defined by V.24; similar to and operational with RS-232).

V.29 9600 bps modem standardized for use on point-to-point four-wire leased telephone-type circuits. Similar to the Bell 209.

V.31 Electrical characteristics for single-current interchange circuits controlled by contact closure.

V.32 Family of 4800/9600 bps modems operating full-duplex over two-wire facilities.

V.33 14.4 Kbps modem standardized for use on point-to-point four-wire leased telephone-type circuits.

V.35 Data transmission at 48 Kbps using 60–108 kHz group band circuits. CCITT balanced interface specification for data transmission at 48 Kbps, using 60–108 kHz group band circuits. Usually implemented on a 34 pin M block type connector (M 34) used to interface to a high speed digital carrier such as DDS.

V.36 Modems for synchronous data transmission using 60–108 kHz group band circuits.

V.37 Synchronous data transmission at a data signaling rate higher than 72 Kbps using 60–108 kHz group band circuits.

Error control

V.40 Error indication with electromechanical equipment.

V.41 Code-independent error control system.

Transmission quality and maintenance

V.50 Standard limits for transmission quality of data transmission.

V.51 Organization of the maintenance of international telephone-type circuits used for data transmission.

V.52 Characteristics of distortion and error-rate measuring apparatus for data transmission.

V.53 Limits for the maintenance of telephone-type circuits used for data transmission.

V.54 Loop test devices for modems.

V.55 Specification for an impulse noise measuring instrument for telephone-type circuits.

V.56 Comparative tests of modems for use over telephone-type circuits.

VSPC Visual Storage Personal Computing.

VT Vertical Tabulation.

VT102 A Digital Equipment Corporation video display terminal which is similar to the firm's VT100 product but also supports a printer.

VT240 A Digital Equipment Corporation 132-column terminal. Many third party vendors market emulation software to enable personal computers to operate as this type of device.

VTAM Virtual Telecommunications Access Method.

VTAM Application Program In IBM's SNA, a program that has opened an ACB to identify itself to VTAM and can now issue VTAM macro instructions.

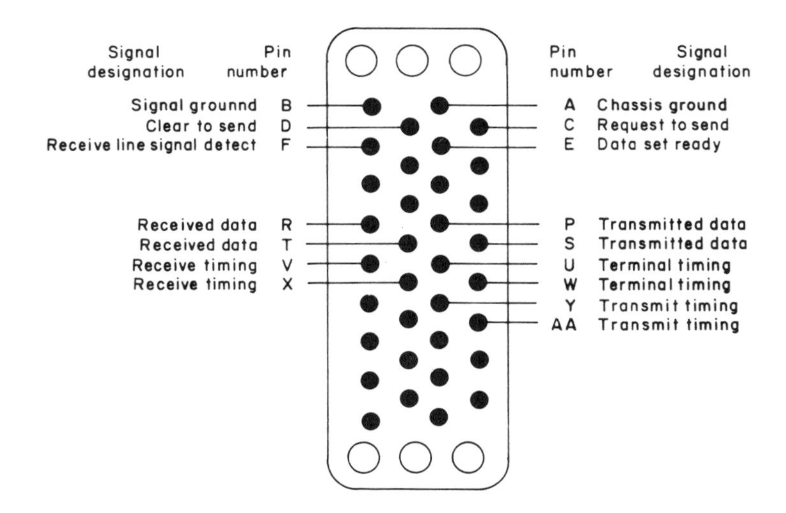

V.35 interface

VTAM definition In IBM's SNA, the process of defining the user application network to VTAM and modifying IBM-defined characteristics to suit the needs of the user.

VTAM definition library In IBM's SNA, the operating system files or data sets that contain the definition statements and start options filed during VTAM definition.

VTAM Internal Trace (VIT) A trace used in VTAM to collect data on channel I/O, use of locks, and storage management services.

VTAM operator In IBM's SNA, a person or program authorized to issue VTAM operator commands.

VTAM operator command In IBM's SNA, a command used to monitor or control a VTAM domain.

VTAM Terminal I/O Coordinator (VTIOC) In IBM's SNA, the part of TSO/VTAM that converts TSO TGET, TPUT, TPG, and terminal control macro instructions into SNA request units.

VTAME Advanced Communications Function for the Virtual Telecommunications Access Method Entry (IBM's SNA).

VTIOC VTAM Terminal I/O Coordinator (IBM's SNA).

VTP Virtual Terminal Protocol.

V&H Vertical and Horizontal grid coordinates are used by AT&T and other communications carriers in mathematical formulas to determine the airline distance between any two rate centers.

V+TU Voice Plus Teleprinter Unit.

W

WACK (wait before transmit) The WACK character sequence allows a receiving station to indicate a "temporarily not ready to receive" condition to the transmitting station.

wafer A thin disk of a purified crystalline semiconductor, typically silicon, that is divided into chips after processing. Typically, a wafer is about one fiftieth of an inch thick and four or five inches in diameter.

WAN Wide Area Network.

wander Jitter which occurs at a rate less than 10 Hz. The primary causes of wander are clock drift and propagation delay changes.

warm boot The complete reload of a "warm" computer's operating system. The computer is "warm" at the start because the electricity is on.

warm start In IBM's SNA, a synonym for system restart.

watchdog timer A timer set by a program to generate an interrupt to the processor after a given period of time, ensuring that a system does not lose track of buffers and communications lines because of a hardware error.

WATS Wide Area Telephone Service.

waveform The pattern resulting from plotting the voltage of a signal as a function of time.

waveguide A hollow conductor used to efficiently transmit high-energy, high-frequency waves in the centimeter to micrometer range. Fiber-optic cables are one form of solid conductors for micrometric and smaller frequencies.

wavelength The distance between successive peaks of a sinusoidal wave.

Wavelength Division Multiplexing (WDM) The simultaneous transmission of more than one information-carrying channel on a single fiber lightguide using two or more light sources of different wavelengths.

WDM Wavelength Division Multiplexing.

WECO 310 A large tip-and-ring jack connector used in digital telephony.

Weibull plot A statistical graph that is used to show data plotted as cumulative failure probability versus a load, stress, or time at failure. Used with cable testing.

Westar Communications satellites operated by Western Union.

wet T1 A T1 circuit with a Bell Operating Company (BOC) powered interface.

white line skipping A facsimile compression scheme in which scanned lines containing no information are encoded, rather than transmitted as a large number of blank bits.

white noise Background noise caused by thermal agitation of electrons. Also referred to as Gaussian noise.

Wide Area Network (WAN) A network that is spread over a larger geographic area than a local area network and where telecommunications links are typically implemented via common carrier. Examples of Wide Area Networks include packet switching, public data, and value-added networks.

Wide Area Telecommunications Service (WATS) Same as Wide Area Telephone Service (WATS).

Wide Area Telephone Service (WATS) A service provided by telephone companies in the U.S. which permits a customer by use of an access line to make calls to telephones in a specific zone for a flat charge per hour of call duration. Under the WATS arrangement, the U.S. is divided into six zones for interstate usage as well as 50 zones for intrastate usage.

wideband In general, having a large bandwidth. While "wideband" and "broadband" mean basically the same thing, "wideband" is the preferred term in telephony. "Broadband" is usually applied to communication at radio frequencies such as broadband local area networks. Generally, wideband circuits operate at data rates in excess of 9.6 Kbps and often at rates of 56 Kbps to 1.544 Mbps.

wideband channel A telecommunications channel

with a bandwidth greater than the bandwidth of a voice-grade channel.

Winchester disk A high density magnetic storage system originally developed by IBM in which the media is non-removable. Also called a fixed disk and hard disk.

window 1. A flow control mechanism whose size governs the number of units of data that may be transmitted before an acknowledgment is required. 2. In IBM's SNA: (a) The path information units (PIUs) that can be transmitted on a virtual route before a virtual route pacing response is received, indicating that the virtual route receiver is ready to accept more PIUs on the route. (b) The requests that can be transmitted on the normal flow in one direction in a session before a session-level pacing response is received, indicating that the receiver is ready to accept the next group of requests. A synonym for pacing group.

window size 1. In a packet switching environment, refers to a traffic flow mechanism which limits the number of elements which can be sent through the network before an acknowledgment is received. Window size is defined on both packet level and link level. 2. In IBM's SNA: (a) The number of path information units (PIUs) in a virtual route window. The window size varies according to traffic congestion along the virtual route. (b) The number of requests in a session-level window. The window size is set at session activation. A synonym for pacing group size.

windows A method of displaying information on a screen in which viewers see what appear to be several sheets of paper similar in appearance to a desktop. The viewer can shift the positions of the windows on the screen, resize different windows, and select a window to work with the data it contains.

wink The momentary interruption in a single frequency tone which indicates that a distant Central Office is ready to receive the digits you just dialed.

wire The twisted wire pair which provides telephone service. By extension, any telephone circuit which uses twisted wire pairs or coaxial cable for transmission.

wire center A building in which one or more local Central Office switching systems are installed and where the outside cable plant is connected to Central Office equipment.

wire center serving area The area of an exchange served by a single wire center.

wire fault An error condition caused by a break or short between the wires in a segment of a cable.

wire pair Two conductors, isolated from each other, that associate to form a communications channel.

wire pairs Transmission medium in which a pair of conducting wires form a circuit.

wiring blocks Molded plastic blocks designed in various pair configuration which terminate cable pairs and establish pair location.

wiring closet A closet within which a number of wires or cables connected to individual telephones or stations terminate. A wiring closet is normally used to facilitate the rearrangement of telephone sets to telephone lines.

Wizard Mail A trademark of H&W Computer Systems of Boise, ID, for an electronic mail program that operates on IBM DOS and OS computer systems.

word In telegraphy, six operations or characters (five characters plus one space). ("Group" is also used in place of "word.") In computing, a sequence of bits or characters treated as a unit and capable of being stored in one computer location.

word length The number of bits in a word, determined as an optimal size for processing, or transmission. Often based on internal operation of a computer (8-bit, 16-bit, 32-bit).

word processor Computer software used to create, edit, and format text. The availability of word processing software and personal computers has made a significant impact on the way businesses handle administrative functions.

Words Per Minute (WPM) A common measure of speed in telegraph systems.

work location wiring subsystem That part of a premises distribution system that includes the equipment and extension cords from the information outlet up to the terminal device connection.

working draft In ISO, a working draft is the initial stage of a standards document which describes the standard as envisioned by a working group of standards committee or subcommittees.

work station Input/output equipment at which an operator works. A station at which a user can send data to or receive from a computer for the purpose of performing a job. For example, a terminal or microcomputer connected to a mainframe or to a local or wide area network.

working diskette A computer diskette onto which programs and files are copied from an original diskette for use in everyday operation.

World Communications, Inc. (WORLDCOM) A former part of Western Union Corporation involved in international private-line business that was sold to Tele-Columbus AG of Baden, Switzerland.

WORLDCOM WORLD COMmunications, Inc.

WPM Words Per Minute.

wraparound The movement of the cursor as it reaches the right edge of screen, disappears, and "wraps around" to the beginning of the next line.

WRU The who-are-you character.

X

X.21 communication adapter An IBM 3710 Network Controller Communication adapter that can combine and send information on one line at speeds up to 64 Kbps, and conforms to CCITT X.21 standards.

X.25 NCP Packet Switching Interface (NPSI) The X.25 Network Control Program Packet Switching Interface, which is an IBM program product that allows SNA users to communicate over packet switched data networks that have interfaces complying with Recommendation X.25 (Geneva 1980) of the International Telegraph and Telephone Consultative Committee (CCITT). It allows SNA programs to communicate with SNA equipment or with non-SNA equipment over such networks. In addition, this product may be used to attach native X.25 equipment to SNA host systems without a packet network.

X.25 PAD A device that permits communication between non-X.25 devices and the devices in an X.25 network.

XC Cross-connect.

Xedit An IBM full-screen editor that can be used to create or change programs or files.

Xenix Microsoft Corporation trade name for a 16-bit microcomputer operating system that was derived from AT&T's UNIX operating system.

Xerox The company that developed Ethernet.

Xerox CIN Corporate INternet.

Xerox Network Services (XNS) Xerox Corporation's layered data communications protocols.

Xerox Network Systems' Internet Transport Protocol (XNS/ITP) In LAN technology, a special communications protocol used between networks. XNS/ITP functions at the 3rd and 4th layer of the OSI model. Similar to TCP/IP.

Xerox RIN Research INternet.

XID 1. eXchange station IDentification. 2. In IBM's SNA, a data link control command and response passed between adjacent nodes that allows the two nodes to exchange identification and other information necessary for operation over the data link.

XID frame A frame in HDLC used to exchange operational parameters between the participating stations.

XMIT Transmit.

XMODEM A popular public-domain personal computer link control protocol. This protocol blocks groups of asynchronous characters together for transmission and computes a checksum which is appended to the end of the block. The checksum is obtained by first summing the ASCII value of each data character in the block and dividing that sum by 255. Then, the quotient is discarded and the remainder is appended to the block as the checksum. The following figure illustrates the XMODEM protocol block format. The Start of Header is the ASCII SOH character whose bit composition is 00000001, while the one's complement of the block number is obtained by subtracting the block number from 255. The block number and its complement are contained at the beginning of each block to reduce the possibility of a line hit at the beginning of the transmission of a block causing the block to be retransmitted.

Start of Header	Block number	One's complement block number	128 data characters	Checksum

XMODEM protocol block format

XMODEM

XMODEM 1K A modified version of the XMODEM protocol which transmits 1K blocks to obtain faster data transfers.

XNS Xerox Network Services.

XNS/ITP Xerox Network Systems' Internet Transport Protocol.

XOFF Transmitter off. The communication control character that instructs a terminal to suspend transmission.

XON Transmitter on. The communication control character that instructs a terminal to start or to resume transmission.

X-ON/X-OFF A handshaking protocol. When the terminal's buffer is nearly full, it transmits a X-OFF to the computer to stop transmission when the buffer is almost empty, a X-ON is transmitted to the host to resume transmission.

X/Open Co. A consortium of computer companies promoting open systems.

X.PC An error connection protocol designed and sponsored by Tymnet which provides the functions typical of the OSI Network Layer 3.

X-Press Information Services A videotext service of McGraw-Hill and Telecommunications, Inc., that offers a news service called X-CHANGE which allows subscribers to access a variety of international newswire services.

X-Series A set of CCITT recommendations for data transmission in public data networks. The CCITT standards are tabulated below.

SERVICES AND FACILITIES

X.1 International user classes of service in public data networks.

X.2 International user services and facilities in public data networks.

X.3 Packet assembly/disassembly facility (PAD) in a public data network.

X.3 Parameters/X.25 Level 1

Parameter number	Parameter function	Values
1	PAD Recall Defined Character	0 or 1 2–127
2	Local Echo	0 or 1
3	Data Forwarding Characters	0,2,4,8,16, 32,64,126, 127,128+n, 254 or 255
4	Data Forwarding Timeout	0–255
5	PAD to Terminal Flow Control	0–4,10,11
6	Control of PAD Service Signals	0,1,4,5
7	PAD Action on Receipt of Break from Term	0,1,2,8,21
8	Discard Output	0 or 1
9	Padding After Carriage Return	0–7
10	Line Folding	0–255
11	Async. Speed (Read Only parameter)	0–14
12	Terminal to PAD Flow Control	0,1,4
13	Line Feed Instruction	0–15
14	Padding After Line Feed	0–7
15	Editing	0 or 1
16	Character Delete Defined Character	1–127
17	Buffer Delete Defined Character	1–127
18	Buffer Display Defined Character	1–127
19	Editing Service Signals	0,2,8,32-126
20	Echo Mask	0–128
21	Parity Treatment	0,1,2,3
22	Page Wait	0–255

X.4 General structure of signals of International Alphabet No. 5 code for data transmission over public data networks.

X.10 Categories of access for data terminal equipment to public data transmission services provided by PDNs and/or ISDN through terminal adapters.

X.15 Definitions of terms concerning public data networks.

Interfaces

X.20 Interface between data terminal equipment (DTE) and data circuit-terminating equipment (DCE) for start–stop transmission services on public data networks.

X.20 bis Use on public data networks of data terminal equipment (DTE) which is designed for interfacing to asynchronous duplex V-Series modems.

X.21 Interface between data terminal equipment (DTE) and data circuit-terminating equipment

(DCE) for synchronous operation on public data networks. This recommendation defines the physical characteristics and control procedures to set up calls, transfer data and terminate calls through the network. Switched or leased circuit communications is supported at data rates up to and beyond 64 Kbps. X.21 is a mix of three protocols: physical, link and network. The physical link is a 15-pin connector. The electrical specification (V.11) is capable of data rates of 64 Kbps and higher.

Interchange circuit	Name	Direction	
		To DCE	From DCE
G	Signal ground or common return		
Ga	DTE common return		
XT	Transmit	×	
R	Receive		×
C	Control	×	
I	Indication		×
S	Signal element timing		×
B	Byte timing		×

X.21 bis Use on public data networks of data terminal equipment (DTE) which is designed for interfacing to synchronous V-Series modems.

X.22 Multiplex DTE/DCE interface for user classes 3–6.

X.24 List of definitions for interchange circuits between data terminal equipment (DTE) and data circuit-terminating equipment (DCE) on public data networks.

X.25 Interface between data terminal equipment (DTE) and data circuit-terminating equipment (DCE) for terminals operating in the packet mode on public data networks.

X.26 Electrical characteristics for unbalanced double-current interchange circuits for general use with integrated circuit equipment in the field of data communications.

X.27 Electrical characteristics for balanced double-current interchange circuits for general use with integrated circuit equipment in the field of data communications.

X.28 DTE/DCE interface for a start–stop mode data terminal equipment accessing the packet assembly/disassembly facility (PAD) in a public data network situated in the same country.

X.29 Procedures for the exchange of control information and user data between a packet assembly/disassembly (PAD) and a packet mode DTE or another PAD.

X.30 Support of X.21 and X.21 bis based DTEs by an ISDN.

X.31 Support of packet mode terminal equipment by an ISDN.

X.32 Interface between DTE and DCE for terminals operating in the packet mode and accessing a PSPDN through a PSTN or a Circuit Switched Public Data Network (CSPDN). Dial X.25 connection.

TRANSMISSION, SIGNALLING AND SWITCHING

X.40 Standardization of frequency-shift modulated transmission systems for the provision of telegraph and data channels by frequency division of a group.

X.50 Fundamental parameters of a multiplexing scheme for the international interface between synchronous data networks.

X.50 bis Fundamental parameters of a 46 Kbps user data signaling rate transmission scheme for the international interface between synchronous data networks.

X.51 Fundamental parameters of a multiplexing scheme for the international interface between synchronous data networks using 10-bit envelope structure.

X.51 bis Fundamental parameters of a 48 Kbps user data signaling rate transmission scheme for the international interface between synchronous data networks using 10-bit envelope structure.

X.52 Method of encoding anisochronous signals into a synchronous user bearer.

X.53 Numbering of channels on international multiplex links at 64 Kbps.

X.54 Allocation of channels on international multiplex links at 64 Kbps.

X.55 Interface between synchronous data networks using 6+2 envelope structure and SCPC-satellite channels.

X.56 Interface between synchronous data networks using an 8+2 envelope structure and SCPC-satellite channels.

X.57 Method of transmitting a single lower speed data channel on a 64 Kbps data stream.

X.60 Common channel signaling for circuit switched data applications.
X.61 Signaling System No. 7—Data user part.
X.70 Terminal and transit control signaling system for start–stop services on international circuits between anisochronous data networks.
X.71 Decentralized terminal and transit control signaling system on international circuits between synchronous data networks.
X.75 Terminal and transit call control procedures and data transfer system on international circuits between packet switched data networks.
X.80 Internetworking of interexchange signaling systems for circuit switched data services.
X.87 Principles and procedures for realization of international user facilities and network utilities in public data networks.

NETWORK ASPECTS

X.92 Hypothetical reference connections for public synchronous data networks.
X.96 Call progress signals in public data networks.
X.110 Routing principles for international public data services through switched public data networks of the same type.
X.121 International numbering plan for public data networks.
X.130 Provisional objectives for call set-up and clear-down times in public synchronous data networks (circuit switching).
X.131 Provisional objectives for grade of service in international data communications over circuit switched public data networks.
X.132 Provisional objectives for grade of service in international data communications over circuit switched public data networks.
X.135 Delay aspects of grade of service for public data networks when providing international packet switched data services.
X.136 Blocking aspects of grade of service for public data networks when providing international packet switched services.
X.140 General quality of service parameters for communications via public data networks.
X.141 General principles for the detection and correction of errors in public data networks.

MAINTENANCE

X.150 DTE and DCE test loops for public data networks.

ADMINISTRATIVE ARRANGEMENTS

X.180 Administrative arrangements for international closed user groups (CUGs).
X.181 Administrative arrangements for the provision of international permanent virtual circuits (PVCs).
X.200 Series of standards defining and structuring ISO seven layer interconnect architecture.
X.210 OSI layer service definition conventions.
X.213 Network service definition for Open Systems Interconnection for CCITT applications.
X.214 Transport service definition for Open Systems Interconnection for CCITT applications.
X.215 Session service definition for Open Systems Interconnection for CCITT applications.
X.224 Transport protocol specification for Open Systems Interconnection for CCITT applications.
X.225 Session protocol specification for Open Systems Interconnection for CCITT applications.
X.244 Procedure for the exchange of protocol identification during virtual call establishment on packet switched public data networks.

SYSTEM DESCRIPTION TECHNIQUES

X.250 Formal description techniques for data communications protocols and services.

INTERNETWORKING BETWEEN NETWORKS

X.300 General principles and arrangements for internetworking between public data networks, and between public data networks and other public data networks.
X.310 Procedures and arrangements for DTEs accessing circuit switched digital data services through analog telephone networks.
X.350 General requirements to be met for data transmission in the maritime satellite service.
X.351 Special requirements to be met for packet assembly/disassembly facilities (PADs) located at or in association with coast earth stations in the maritime satellite service.
X.352 Internetworking between public packet switched data networks and the maritime satellite data transmission system.
X.353 Routing principles for interconnecting the maritime satellite data transmission system with public data networks.

MESSAGE HANDLING SYSTEMS

X.400 High level protocol recommendation for message handling providing protocol and technical specifications for interconnection of separate computer-based message handling system.

X.401 Basic service elements and options user facilities.

X.408 Encoded information type conversion rules.

X.409 Presentation transfer syntax and notation.

X.410 Remote operations and reliable transfer server.

X.411 Message transfer layer.

X.420 Interpersonal messaging user agent layer.

X.430 Access protocol for teletex terminals.

X.500 The electronic mail directory handling standard.

X3T5 Information Processing Technical Committee. Working Groups of the X3T5 committee are organized as follows:

X3T5.1— OSI Architecture and the Reference Model.

X3T5.4— OSI Management Protocols and the Directory.

X3T5.5— OSI Session, Presentation, and Application Layers.

XTC eXternal Transmit Clock.

Y

Yellow Alarm An alarm condition on a T1 circuit. The hardware at one end of a T1 circuit enters the Yellow Alarm condition when it receives either a Red Alarm signal or a Very Severe Burst over the circuit. In ESF, a device signals a Yellow alarm by transmitting a continuous pattern of Hex FF Hex 00 over the Facility Data Link.

YMODEM A version of the XMODEM protocol that sends data in 1024-byte blocks with a 2-byte CRC. Also known as XMODEM 1K.

YMODEM G A streaming protocol that sends an entire file prior to waiting for an acknowledgment. This protocol is designed for use with modems that have built-in error-correction capability, such as Hayes Microcomputer products V-Series.

Z

zap To eradicate all or part of a program or data base.

ZBTSI Zero Bit Time-Slot Insertion.

zero bit insertion A technique in bit-oriented protocols such as HDLC/SDLC to achieve transparency. A zero is inserted into sequences of one bits that would cause false flag detection. Then, the inserted zero is removed at the receiving end of the data link. Also called bit stuffing.

Zero Bit Time-Slot Insertion (ZBTSI) The most complex technique for maintaining ones density on framed T1 circuits in which an area in the ESF frame carries information about the location of all-zero bytes consisting of eight consecutive zeros in the data stream. This technique requires intelligent rearrangement of the bit stream at both ends of the circuit and is rarely used.

zero code suppression The insertion of a "one" bit to prevent the transmission of eight or more consecutive "zero" bits. Used with digital T1 and related facilities.

zero insertion In SDLC, the process of including a binary 0 in a transmitted data stream to avoid confusing data and SYN characters; the inserted 0 is removed at the receiving end.

zero slot LAN A local area network that uses the built-in serial port of a personal computer instead of a separate network card. This reduces the cost of the LAN, although it operates slower then conventional LANs.

Zero Transmission Level Point (0TPL) A reference point for measuring the signal power gain and loss of a circuit.

0–9

029 An IBM card punch.

059 An IBM card verifier.

083 An IBM punch card sorter.

1A2 multi-line phone A type of telephone with lighted buttons to select up to 5, 9 or 19 phone lines, plus a hold button. It's easily identified by its thick 25-pair cable that ends in a $3\frac{1}{2}''$ long connector to attach the phone to the line.

1-persistent In LAN technology, same as persistent.

2B1Q Two binary, one quaternary. This means that two binary bits of data are compressed and transmitted in one time state as a four-level code. It is the T-1 Committee's recommended draft standard for the line code for echo cancellation devices used in North America. This ISDN standard for digital line code sends a 160-Kbps duplex signal over a copper pair using a quaternary pulse and echo cancellation.

4B3T A line code for echo cancelers advocated by the West German standards group, in which four binary data bits are compressed to transmit in three time states.

10BASE5 An IEEE 802.3 media standard abbreviation for 10 Mbps, baseband, 500 meters.

10NET A registered trademark of Digital Communications Associates as well as a local area network marketed by that vendor's 10NET Communications Division in Dayton, OH.

31-bit storage addressing In IBM's SNA, the storage address structure available in an MVS/XA operating system.

42A This is the most common connector used with four-wire private line circuits. This is the type of connection that was found on all telephones prior to the advent of the RJ11C modular connector. The four wires found under the cover of the 42A block are terminated with screw down terminals. Red and green wires are the transmit pair and black and yellow wires are the receive pair.

357 An IBM Badge and/or Serial Card Reader Input Unit for a 357 Data Collection System.

1287 An IBM optical reader.

1288 An IBM optical page reader.

2501 An IBM card reader.

2701 IBM Data Adapter Unit.

2702 IBM Transmission Control Unit.

3088 An IBM Multisystem Channel Communications Unit used for interprocessor communications over block multiplexer channels.

3101 An IBM standalone CRT display terminal that uses ASCII code and has an asynchronous communications interface.

3104 An IBM CRT display terminal that can be attached to that vendor's 8100 Information System or 4331 Processor.

3151 An IBM ASCII display station.

3174 The most recent addition to IBM's series of cluster controllers. Members of the 3174 control unit series can support 8, 16 or 32 display stations and have either a token-ring network adapter or an asynchronous emulation adapter installed in the device.

3178 An IBM compact, lightweight, monochrome display station.

3179 An IBM compact, lightweight, color display station.

3180 An IBM low priced, multiple screen format, monochrome display station.

3191 An IBM monochrome display station which has a built-in printer port on some models, is manufactured with 12, 14, or 15 inch screens, supports 102-, 104-, and 122-key keyboards, and contains a record/play/pause function which permits up to 1500 characters to be stored for later recall.

3192 An IBM color display station which has a built-in parallel printer port on some models, is manufactured with 14- or 15-inch screens, supports 102-, 104-, and 122-key keyboards and contains a record/play/pause function which permits data to be stored in RAM for later recall.

3193 An IBM high-resolution, portrait-like, monochrome display station. The Model 1 has a 122-key typewriter keyboard while the Model 2 has a 102-key IBM enhanced keyboard.

3194 An IBM series of color and monochrome display stations. The members of the 3194 series support four host windows, two notepads and screen management functions. The 3194 display stations have a built-in parallel printer port, support the use of 102-, 104-, and 122-key keyboards, and have a record/play/pause feature which permits up to 30 000 characters to be stored in RAM for recall.

3210 An IBM selectric console typewriter that is used as an I/O unit for certain S/370 computers.

3251 An IBM interactive computer graphics display station that can be used with a S370, 30XX or 43XX processor.

3258 An IBM channel attached control unit which supports up to 16 3251 display stations.

3262 A series of IBM line printers, used with S/34 and S/38 minicomputers that can be attached to 3274 and 3276 control units and other IBM products.

3270 data stream A coded character data stream.

3270, 3270 information display system A very popular IBM data entry and display system which consists of control units, display stations, printers, and other equipment.

3274 An IBM standalone control unit. Locally attached models 21A, 21B, 21D, 31A 31D, 41A, and 41D can control clusters of up to 32 display stations and printers. Remotely attached models 21C, 31C, and 41C can control clusters of up to 32 display stations and printers. Remotely attached models 51C and 61C control mid-sized clusters, with the model 51C supporting up to 12 display stations and printers while the model 61C can control up to 16.

3276 An IBM tabletop control unit which is integrated into a display station and controls up to eight display stations and printers. Models 1, 2, 3, 4 operate at 1200, 2400, 4800, 7200 bps using the BSC protocol. Models 11, 12, 13, 14 operate at 1200, 2400, 4800, 7200, 9600 bps using the SNA/SDLC protocol.

3278 An older monochrome display station no longer manufactured by IBM.

3279 An older color display station no longer manufactured by IBM.

3290 An IBM display station that features a large, flat plasma panel as its visual display medium.

3299 An IBM terminal multiplexer which permits up to eight terminals to be connected to an IBM 3174 or 3274 control unit via a single coaxial cable.

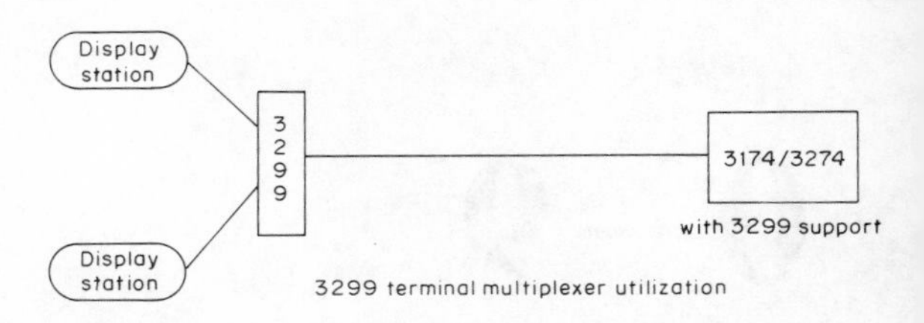

3299

3601 An IBM programmable communication controller which is used to attach 3600 Finance Communication System terminals to several types of IBM processors.

3663 An IBM supermarket terminal.

3667 An IBM checkout scanner.

3683 An IBM point of sale terminal.

3684 An IBM point of sale control unit.

3704 A low-cost programmable communications processor announced by IBM in 1974.

3705 IBM's first programmable communications processor that was released in 1972.

3710 An IBM network controller capable of handling up to 31 lines using mixed protocols to include X.25.

3725 An IBM programmable communications processor announced in March 1983, which can also be used as a node processor or remote concentrator.

3726 The expansion chassis for the IBM 3725.

3737 IBM's Remote Channel-to-Channel Unit. This device is a standalone control unit that provides host-to-host connectivity through a high-capacity communications facility at data rates up to 1.544 Mbps.

3745 An IBM programmable communications processor which supports up to eight T1 circuits.

3747 An IBM data converter used to convert batched data from diskette to one-half inch magnetic tape.

3780 An IBM data transmission terminal that uses the bisysnchronous transmission protocol.

3800 compatibility mode Operating the IBM 3800 model 3 as a 3800 model 1.

3800 model 3 startup That process part of system initialization when the IBM 3800 model 3 is initializing.

3812 An IBM multifunction, non-impact page printer of tabletop design that can be connected to an IBM 3270 Information Display System.

3814 An IBM Switching Management System that is used to switch processor channels among I/O control units in a data processing center.

3833 An IBM 2.4 Kbps modem designed for use on leased lines.

3834 An IBM 4.8 Kbps modem designed for use on leased lines.

3845 An IBM data encryption device which supports asynchronous, bisynchronous and SDLC communications at data rates from 110 to 19 200 bps.

3846 An IBM rack-mounted data encryption similar to the IBM 3845.

3863 model 1 An IBM 2400 bps modem designed for use on leased lines.

3864 model 1 An IBM 4800 bps modem designed for use on leased lines.

3865 An IBM 9600 bps modem designed for use on leased lines.

3866 An IBM multimodem enclosure that provides housing, ventilation and power for certain IBM rack-mounted modems.

3868 model 1 An IBM 2400 bps modem designed for use on leased lines.

3868 model 2 An IBM 4800 bps modem designed for use on leased line.

3868 model 3 An IBM 9600 bps modem designed for use on leased line.

3868 MODEL 4 An IBM 9600 bps modem designed for use on leased line.

4224 A series of IBM serial dot-matrix impact printers that attaches to an IBM 3270 Information Display System through a 3174, 3274 or 3276 control unit.

4234 A heavy duty IBM intermediate-speed impact-matrix printer.

4245 An IBM high-speed printer which includes an optical character recognition (OCR) feature.

4250 A high-resolution IBM printer which produces camera-ready print masters with text and line-out graphics intermixed.

4745 An Amdahl Corporation front-end processor designed to compete against IBM's 3745 communications controller.

4829 An IBM 2.4 Kbps modem developed on a half size adapter card that is designed for use inside an IBM PC or compatible computer.

4860 The IBM PC jr home computer.

5150 The original IBM personal computer.

5151 An IBM monochrome display designed for use with the IBM PC series of personal computers.

5152 An IBM graphics printer marketed for use with the IBM PC series of personal computers.

5155 The IBM portable personal computer.

5160 The IBM PC XT.

5161 An IBM expansion unit that can be connected to the IBM PC or IBM PC XT.

5210 A desktop IBM impact printer that uses a bidirectional printwheel to produce letter-quality output.

5540 An IBM multiworkstation that can function as a Japanese-language personal computer for business applications, a Japanese-language word processor, and a Japanese-language on-line communications terminal.

5550 An IBM multiworkstation similar to the 5540.

5560 An IBM multiworkstation similar to the 5540.

5811 1. An IBM baseband modem capable of operating at 2400, 4800, 9600 or 19 200 bps. 2. An IBM limited distance modem.

5812 An IBM limited distance modem.

5852 An IBM standalone 2.4 Kbps modem designed for use with personal computers.

5865 An IBM 9600 bps modem which operates with the vendor's Communications Network Management Facility, which is part of Netview.

5866 An IBM 14.4 Kbps modem which operates with the vendor's Communications Network Management Facility, which is part of Netview.

5868 An IBM 9600/14 400 bps modem.

5979 model L41 An IBM baseband modem capable of operating at 2400, 4800, 9600 or 19 200 bps.

6150 The IBM RT personal computer.

6151 The IBM RT personal computer which features a 32-bit reduced instruction set microprocessor.

7170 An IBM protocol converter.

7172 An IBM protocol converter.

7820 An IBM terminal adapter which attaches synchronous host computer channels, controllers or terminals to networks with ISDN Basic Rate services.

7860 A series of IBM modems that operate at data rates from 4.8 to 19.2 Kbps and which are designed to work with Netview.

8218 An IBM repeater designed for use with copper wire type 3 media on a Token-Ring network.

8219 An IBM optical fiber repeater that can be used with type 5 cable on a Token-Ring network.

8228 An IBM multistation access unit which permits up to eight token-ring devices to be connected to a token-ring network.

8525 The IBM Personal System/2 Model 25 with a 3278/3279 Emulation Adapter.

8530 The IBM Personal System/2 Model 30 with a 3278/3279 Emulation Adapter.

8550 The IBM Personal System/2 Model 50 with a 3270 connection.

8560 The IBM Personal System/2 Model 60 with a 3270 connection.

8570 The IBM Personal System/2 Model 70 with a 3270 connection.

8580 The IBM Personal System/2 Model 80 with a 3270 connection.

9370 An IBM series of computer systems designed for departmental usage.